Fundamentals of Radio Astronomy

Astrophysics

Series in Astronomy and Astrophysics

The *Series in Astronomy and Astrophysics* includes books on all aspects of theoretical and experimental astronomy and astrophysics. Books in the series range in level from textbooks and handbooks to more advanced expositions of current research.

Recent books in the series

Dark Sky, Dark Matter
J M Overduin and P S Wesson

Dust in the Galactic Environment, 2nd Edition
D C B Whittet

The Physics of Interstellar Dust
E Krügel

Very High Energy Gamma-Ray Astronomy
T C Weekes

Numerical Methods in Astrophysics: An Introduction
P Bodenheimer, G P Laughlin, M Rózyczka, H W Yorke

An Introduction to the Physics of Interstellar Dust
Endrik Krugel

Astrobiology: An Introduction
Alan Longstaff

Fundamentals of Radio Astronomy: Observational Methods
Jonathan M Marr, Ronald L Snell, and Stanley E Kurtz

Stellar Explosions: Hydrodynamics and Nucleosynthesis
Jordi José

Cosmology for Physicists
David Lyth

Cosmology
Nicola Vittorio

Cosmology and the Early Universe
Pasquale Di Bari

Fundamentals of Radio Astronomy: Astrophysics
Ronald L. Snell, Stanley E. Kurtz, and Jonathan M. Marr

Fundamentals of Radio Astronomy

Astrophysics

Ronald L. Snell
University of Massachusetts, Amherst, Massachusetts, USA

Stanley E. Kurtz
National Autonomous University of Mexico, Morelia, Michoacan, Mexico

Jonathan M. Marr
Union College, Schenectady, New York, USA

CRC Press
Taylor & Francis Group
Boca Raton London New York

CRC Press is an imprint of the
Taylor & Francis Group, an **informa** business

CRC Press
Taylor & Francis Group
6000 Broken Sound Parkway NW, Suite 300
Boca Raton, FL 33487-2742

Printed on acid-free paper

International Standard Book Number-13: 978-1-4987-2577-4 (Hardback)

Library of Congress Cataloging-in-Publication Data

Names: Snell, Ronald Lee, author. | Kurtz, Stanley, author. | Marr, Jonathan M., author.
Title: Fundamentals of radio astronomy : astrophysics / Ronald L. Snell, Stanley E. Kurtz, Jonathan M. Marr.
Other titles: Series in astronomy and astrophysics.
Description: Boca Raton, FL : CRC Press, Taylor & Francis Group, [2019] |
Series: Series in astronomy and astrophysics | Includes bibliographical references and index.
Identifiers: LCCN 2018051118| ISBN 9781498725774 (hbk ; alk. paper) |
ISBN 1498725775 (hbk ; alk. paper)
Subjects: LCSH: Radio astronomy. | Astrophysics.
Classification: LCC QB476.5 .S64 2019 | DDC 522/.682--dc23
LC record available at https://lccn.loc.gov/2018051118

Visit the Taylor & Francis Web site at
http://www.taylorandfrancis.com

and the CRC Press Web site at
http://www.crcpress.com

Contents

Preface

THIS is the second volume of a two-volume textbook on radio astronomy. The authors have all taught courses in radio astronomy at the undergraduate level, and although there are a number of reference books on radio astronomy available, none were found ideal for use as an undergraduate textbook. Thus we found the need to develop our own teaching material and this textbook is based on these notes and our teaching experiences. Unlike many reference books on radio astronomy, our goal is to develop an understanding in the students of the underlying principles. Toward this end we have provided the background and derivations for most of the equations used in the textbook. Although we wrote this textbook with the undergraduate student in mind, we feel these volumes will also serve graduate students who are just getting started in radio astronomy.

The first volume, *Fundamentals of Radio Astronomy: Observational Methods*, discusses radio astronomy instrumentation and the techniques that a radio observer should know in order to make successful observations. In this, the second volume, *Fundamentals of Radio Astronomy: Astrophysics*, we discuss the physical processes that give rise to radio emission, present examples of astronomical objects that emit by these mechanisms, and illustrate how the relevant physical parameters of astronomical sources can be obtained from the radio observations.

We recognize that these two volumes cover more material than can be reasonably covered in a one semester course on radio astronomy. We provide the material in two volumes to serve courses that either emphasize radio wavelength instrumentation and observational techniques or that emphasize the astrophysics of radio sources. The two volumes are largely independent of each other, so instructors can pick and choose topics from the two volumes that best fit their courses. Despite devoting an entire volume to the astrophysics at radio wavelengths, we still could not cover all possible topics. We have, for instance, excluded planetary radio astronomy. Radio emission from planets, asteroids and comets is a rich and exciting area of study, but one that requires a volume in itself to be properly presented. Furthermore, discussion of the topics that we do present is necessarily limited in scope in order to provide a textbook of reasonable length and to avoid immersing the undergraduate student in overly complex details.

In this volume we assume students have completed at least a beginning sequence in physics that includes a good background in electricity and magnetism and some knowledge of modern physics. We also assume that students have had both differential and integral calculus. Some of the more complex and lengthy derivations have been placed in appendices, so as not to distract from the presentation of the astrophysics, while still being available for readers who want to work through them. We also include in the appendices several important mathematical concepts. Additionally, in our experiences we have found that students benefit from examples; hence we have included an ample number of examples dispersed throughout the chapters. Each chapter also concludes with a list of questions and problems.

In the presentation of the material, we devote the first four chapters to covering the relevant physics, with an extensive discussion of radiation processes. The physics concepts and equations can then be applied as needed in the presentation of the astrophysics. We believe that this approach yields a deeper understanding and teaches the student how to use the tools of the trade more generally. Additionally, we have written the astrophysics chapters (Chapters 5-10) to be independent of each other. Instructors should not feel the need to cover Ch. 5 in order to discuss Ch. 6, for example, nor do these chapters need to be presented in sequential order.

When we wrote the first volume, we chose primarily to use SI units, as this system is used in most physics textbooks and historically was used by early radio astronomers. We often provided the conversions needed to change to cgs units. However, results presented in the astronomical literature have generally used cgs units. As such, when we considered introducing quantities in this volume such as column density, we felt that the cgs unit of cm^{-2}, and not the SI unit of m^{-2}, was more appropriate for teaching students about the field of astrophysics. For this reason all equations and examples in this volume are in cgs units. The first appendix lists constants in both cgs and SI units, and Section 1.1.1 provides a discussion of the unique challenge of converting the equations of electricity and magnetism between cgs and SI units. We hope this approach for units in the two volumes is not too confusing for students. If nothing else it might help prepare them for a professional life that almost certainly includes mutiple systems of units.

Over its 80 year history, radio astronomy has contributed greatly to our understanding of the Universe, including four Nobel Prize winning discoveries, and we believe courses covering this subject are important in the undergraduate curriculum. We hope that our efforts in making this material available to others will enhance both the teaching and the learning of radio astronomy, and that you will find the study of this topic equally as rewarding as we have.

Ronald L. Snell
University of Massachusetts, Amherst, Massachusetts

Stanley E. Kurtz
National Autonomous University of Mexico, Morelia, Mexico

Jonathan M. Marr
Union College, Schenectady, New York

Acknowledgments

We have received generous and insightful help from a number of people whose selfless contributions of time deserve to be acknowledged. We are very grateful for the advice, feedback, and other assistance provided by Esteban Araya, Chris Carilli, Ed Churchwell, Mark Claussen, Paul Goldsmith, David Hough, Jay Lockman, Amy Lovell, Jeff Mangum, Chris O'Dea, Bruce Partridge, Luis Rodriguez, Leise van Zee, Joel Weisberg, Francis Wilkin, Grant Wilson, and Min Yun. We also wish to thank the following who have allowed us to use their previously published material or have provided new materials for this volume: Esteban Araya, Dan Clemens, Scott Croom, Tom Dame, John Dickey, Rob Gutermuth, Mark Heyer, Adam Leroy, Jay Lockman, Michael McCrackan, Maura McLaughlin, Rick Perley, Peter Schloerb, Gregory Taylor, Joe Taylor, Steven Tremblay, Fabian Walter, Grant Wilson, and Min Yun. We acknowledge the use of the Legacy Archive for Microwave Background Data Analysis (LAMBDA), a service of the NASA Goddard Space Flight Center, the ESA/Planck Collaboration and the NRAO/AUI image gallery websites. We also made use of the NASA/IPAC Extragalactic Database (NED), which is operated by the Jet Propulsion Laboratory, California Institute of Technology, under contract with the National Aeronautics and Space Administration. Finally we wish to acknowledge the feedback we have received from all of the students who used our lecture notes as their reading material through the years.

CHAPTER 1

Introductory Material

IN 1932 Karl Jansky detected radio signals serendipitously from an extraterrestrial source, marking the beginning of the field of radio astronomy. Although Jansky deduced that the radio signals originated from beyond the solar system, more than two decades passed before we understood that the origin of this emission was due largely to synchrotron emission produced by high energy cosmic ray electrons interacting with the magnetic field of the Milky Way. The field of radio astronomy has advanced significantly since those early days, and observations at radio wavelengths today play a vital role in the study of astronomical objects ranging from solar system bodies to distant quasars. In the first volume of the Fundamentals of Radio Astronomy, we addressed how radio telescopes function and the methods used to make observations. In this second volume, we discuss what we can learn about the astrophysics of the Universe from radio wavelength observations.

The full range of electromagnetic radiation is enormous, spanning wavelengths from the very shortest gamma-ray radiation, through the x-ray, ultraviolet, optical and infrared, and ending with the very long radio wavelengths. Where the infrared (or far-infrared) wavelengths end and radio wavelengths start is ill-defined. However for the discussion in this book, we place the division at a wavelength of about 0.3 mm. Even the radio portion of the electromagnetic spectrum is quite broad, spanning over 4 orders of magnitude in wavelength, from wavelengths as short as 0.3 mm to wavelengths longer than 20 m, corresponding to frequencies as high as 1×10^{12} Hz (or 1,000 GHz) to lower than 1.5×10^{7} Hz (or 15 MHz). Remember that the basic unit of frequency is hertz, where 1 Hz = 1 cycle per second, therefore 1 MHz corresponds to 10^6 Hz and 1 GHz corresponds to 10^9 Hz. Jansky's original observations were made at the relatively low frequency of 20.5 MHz (or a wavelength of about 15 m); however astronomers today with modern telescopes and detectors routinely obtain observations at any frequency in the full radio window.

At radio wavelengths, just as at other wavelengths, we can make both broadband and spectroscopic measurements. Broadband measurements are designed to detect the emission produced by continuum emission processes as they measure the average emission over a broad range of wavelengths or frequencies. At optical wavelengths this type of measurement is often called photometry and is accomplished by passing the light through filters that transmit a well-defined but broad band of wavelengths for detection. Spectroscopic measurement requires much higher wavelength resolution, sufficient to detect individual spectral lines. The techniques used for filtering, dispersing and detecting the light at optical and radio wavelengths can be quite different. The techniques used at radio wavelengths to make both photometric and spectroscopic observations are described in detail in the first volume of this book. The present volume concentrates on the study of astronomical sources at radio wavelengths using both photometric and spectroscopic observations.

Before we discuss radio observations of different types of astronomical sources, we review some basic physics and astronomy. In the following sections we cover units used by radio astronomers, measures of the radiation from astronomical sources, sky coordinates, the Doppler effect and the cosmological redshift. In Chapter 2, we discuss the propagation of radiation through a medium and in Chapters 3 and 4 we present the physical processes responsible for producing both broadband radio continuum emission and radio spectral line emission. In subsequent chapters we apply these physical principles to discuss how radio observations can provide important insights into the properties of astronomical sources. The scope of radio studies in astronomy is very broad, and it is impossible to cover all aspects; therefore we have selected a subset of topics for inclusion in this volume.

1.1 UNITS AND NOMENCLATURE

Lengths and masses involved in astronomical research are often orders of magnitude larger than those used in Earth-based lab measurements. Hence, neither the SI system (meters, kilograms, and seconds — also called mks), nor the cgs (centimeters, grams, and seconds) system offers any particular advantage or convenience. Although most undergraduate physics texts use the SI system, astronomers have historically used the cgs system and for that reason we will generally use cgs units here. We provide in Appendix A both the cgs and SI values of important physical constants that will be used throughout the book. Although it is easy to convert between SI and cgs units in Newtonian mechanics, the fundamental equations in electricity and magnetism differ in these two systems. It can be tricky to convert units properly when electromagnetism is involved and this is a common source of error in calculations. We provide further discussion about this issue in the next section. In addition, there are several units that are defined strictly for astronomical purposes, which make the mathematics less cumbersome; these are introduced in Section 1.1.2.

1.1.1 Issues with Units of Electricity and Magnetism

We begin with Coulomb's law, the equation for the electric force between two charges q_1 and q_2, separated by a distance r_{12}. In a vacuum this is given in the SI system as

$$F_{\text{elec}} = \frac{1}{4\pi\epsilon_o}\frac{q_1\, q_2}{r_{12}^2},$$

where the unit of charge is the coulomb (C) and ϵ_o is called the electric permittivity of free space and has a value of $\epsilon_o = 8.854\times10^{-12}$ C^2 N^{-1} m^{-2}. In the cgs system (using the Gaussian electromagnetic units) the same force equation is written as

$$F_{\text{elec}} = \frac{q_1 q_2}{r_{12}^2}, \tag{1.1}$$

where the unit of charge is called the electrostatic unit (esu) or the statcoulomb. The absence of a coefficient in Equation 1.1 is the obvious difference between these two formulations. In the cgs system, the coupling constant that relates the force to the product of the charges divided by distance squared is unity and dimensionless, while in the SI system the coupling constant is given by $1/(4\pi\epsilon_o)$ which does have dimensions. Therefore, the fundamental unit of charge in the two systems must be dimensionally different. The cgs unit of charge, the esu, has physical dimensions such that 1 esu = 1 g$^{1/2}$ cm$^{3/2}$ s^{-1}. The SI unit of charge, the coulomb (C), is defined by the SI base units the ampere and the second, and the C^2 of the product $q_1 q_2$ is cancelled by the C^2 in ϵ_o. Because of the different coupling constant, one has to be very careful when converting from cgs units of esu to the SI unit of coulomb and vice versa. The correct conversion of charge from one system to the other is given by

$$q \text{ (in esu)} = \left(\frac{10^5 \text{ dyne}}{1 \text{ N}}\right)^{1/2} \left(\frac{100 \text{ cm}}{1 \text{ m}}\right) \frac{1}{\sqrt{4\pi\epsilon_o}} q \text{ (in C)}.$$

Note there are two parts to this unit conversion: the first involves changing from SI to cgs units (force and distance) while the second involves changing the dimensions of charge between the two systems. The net result is that 1 C is equivalent to 3.00×10^9 esu.

Now consider the Lorentz force equation which combines the electric and magnetic forces on a charge moving with velocity $\vec{v}$ in an electric field ($\vec{E}$) and a magnetic field ($\vec{B}$). In the SI system the force is given by

$$\vec{F} = q(\vec{E} + \vec{v} \times \vec{B}),$$

while in the cgs system it is given by

$$\vec{F} = q(\vec{E} + \frac{\vec{v}}{c} \times \vec{B}). \quad (1.2)$$

Inspection of Equation 1.2 shows that the dimensions of the electric and magnetic fields in the cgs system are the same, *i.e.* force per charge, or dyne esu^{-1}. Substituting in the dimensions of dyne and esu, from above, we see that the electric and magnetic fields have dimensions of $\text{g}^{1/2}\ \text{cm}^{-1/2}\ \text{s}^{-1}$. In the SI system, the electric field has dimensions of force per charge, or N C^{-1}. Although the dimensions of the electric field in the two systems have the same form (force per unit charge), remember that the dimensions of charge are different and therefore the dimensions of the electric field must be different as well. The conversion of units of electric field between the two systems is given by

$$E \text{ (in dyne esu}^{-1}) = \left(\frac{10^5 \text{ dyne}}{1 \text{ N}}\right) \left(\frac{1 \text{ C}}{3 \times 10^9 \text{ esu}}\right) E \text{ (in N C}^{-1}).$$

Therefore an electric field of 1 N C^{-1} is equivalent to 3.33×10^{-5} dyne esu^{-1}. As was the case for charge, the unit conversion consists of two parts.

The unit of magnetic field in the cgs system is the gauss (1 gauss = 1 esu cm^{-2} = 1 $\text{g}^{1/2}$ $\text{cm}^{-1/2}$ s^{-1}), while for the SI system it is the tesla (1 T = 1 kg C^{-1} s^{-1}). The conversion of magnetic field between the cgs and SI units is given by

$$B \text{ (in gauss)} = \left(\frac{1000 \text{ gm}}{1 \text{ kg}}\right)^{1/2} \left(\frac{1 \text{ m}}{100 \text{ cm}}\right)^{1/2} \left(\frac{4\pi}{\mu_o}\right)^{1/2} B \text{ (in T)},$$

where μ_o is called the magnetic permeability of free space and is $4\pi \times 10^{-7}$ kg m C^{-2} = 1.256×10^{-6} kg m C^{-2}. As was the case for charge and electric field, the unit conversion consists of two parts, first changing from SI to cgs and second changing between the different dimensions of the magnetic field in the two systems. The net result is that a 1 tesla magnetic field is equivalent to a 10^4 gauss magnetic field. In SI units the magnetic permeability is related to the electric permittivity and the speed of light by the relation

$$\epsilon_o\mu_o = \frac{1}{c^2}.$$

In cgs units, both ϵ_o and μ_o are dimensionless, and the above equation does not apply.

Again, remember that the dimensions of the charge, magnetic field and electric field are different for the two systems as the conversion requires two steps.

The salient point here is that when dealing with the physics of electricity and magnetism, the process of converting from one system of units to another can be tricky. We strongly suggest that one avoid such conversions as much as possible by using only the equations that correspond to the chosen system. Physics and engineering textbooks almost all use SI units. Since the roots of radio astronomy are in electrical engineering, it is often the case that when astronomers discuss radio telescopes and receivers they use SI units and for this reason Volume I of this series used SI units. However, when the discussion turns to astrophysics, the cgs system is more widely used, so we will primarily use cgs units in this volume.

1.1.2 Astronomy Units

Although cgs units are used by most astronomers, specialized units defined just for astronomy are often used as well. These units are defined primarily because results expressed in either cgs or SI units can be rather large. A good example is distance: Astronomers usually express distances not in centimeters or meters but in special astronomy units. For planetary systems, distances are often given in *Astronomical Units* (AU), which is defined as the average distance between the Earth and the Sun, and is about 1.50×10^{13} cm or 1.50×10^{11} m. For greater distances, astronomers use either the *light-year* or the *parsec*. A light-year (ly) is the distance that light travels in one year, which equals 9.46×10^{17} cm or 9.46×10^{15} m. The more commonly used unit of distance is the parsec (pc) and is defined as the distance at which a length of 1 AU has an angular size of 1 arcsecond. This may seem like an odd definition, but it is the natural unit when applying the most fundamental method for measuring distances called trigonometric parallax. This is a geometrically determined distance, and uses the Earth's orbit about the Sun (radius 1 AU) as the baseline for measuring the parallax of a star. The parallax is the apparent angular displacement of a foreground star relative to more distant stars as the the Earth orbits the Sun. The parsec refers to the distance of a star with a stellar parallax of 1 second of arc. A parsec is equal to

$$1 \text{ pc} = 3.25 \text{ ly} = 206,265 \text{ AU} = 3.09 \times 10^{18} \text{ cm} = 3.09 \times 10^{16} \text{ m}.$$

Because of how parallax is defined, the number 206,265 is also the number of arcseconds in a radian. The distances to the nearest stars are on the order of parsecs, distances to galaxies in the Local Group are of order hundreds of kpc (kiloparsecs, where 1 kpc = 1000 pc), distances to nearby galaxies beyond the Local Group are of order tens of Mpc (megaparsecs, where 1 Mpc = 10^6 pc) and the most distant objects known are of order several Gpc (gigaparsecs, where 1 Gpc = 10^9 pc).

Some units in astronomy are set by using the Sun as a standard. The Sun is a typical star and one whose parameters we know exceedingly well, so the units of mass and luminosity are often expressed in comparison to the Sun's mass and luminosity. We call these units *solar masses* and *solar luminosities* and they are symbolized by $M_\odot$ and $L_\odot$, respectively, where $\odot$ is the symbol for the Sun. The values of these units are the following

$$\text{Solar Mass } (M_\odot) = 1.99 \times 10^{33} \text{ g} = 1.99 \times 10^{30} \text{ kg, and}$$

$$\text{Solar Luminosity } (L_\odot) = 3.83 \times 10^{33} \text{ erg s}^{-1} = 3.83 \times 10^{26} \text{ watts.}$$

The luminosity of an object is a measure of the total energy radiated by the object per unit time and will be discussed further in Section 1.2. These astronomy units are also summarized in Appendix A.

1.1.3 Nomenclature for Atomic Ionization States

We will use the astronomical nomenclature for specifying the ionization state of an atom. This differs somewhat from the nomenclature used by chemists or physicists. We refer to a neutral hydrogen atom as HI. If the hydrogen atom is ionized, and thus its one electron has been stripped away, then it is referred to as HII (this is also sometimes indicated by H^+). For instance, the gas surrounding a luminous hot star can be ionized by the photons emitted by the star, thus producing a region of ionized hydrogen; this is commonly referred to as an HII region. We can use this same nomenclature for other atoms. For instance, CI refers to a neutral carbon atom in which all six electrons remain bound to the carbon nucleus, CII refers to a carbon ion in which one of its six electrons has been stripped away and CIV refers to a carbon ion in which three of its six electrons have been removed.

1.2 RADIATION MEASURES

The various measures of the radiation from astronomical sources are discussed in detail in Chapter 2 of Volume I; in this section we provide a brief refresher of these measures. What we measure when we observe an astronomical source is the rate of electromagnetic energy (or power) that our telescope collects and we need to discuss how this is quantified. However, what we really want to know about an astronomical source is the rate it emits energy (or luminosity). Therefore, we need to know how to relate the rate of energy collected to the total rate of energy emitted by the source. We start our discussion with a brief overview of the luminosity of an astronomical source.

1.2.1 Luminosity

Instead of referring to the total amount of energy an astronomical source emits over its lifetime, we usually refer to the rate at which a source emits energy, which we call the source's *luminosity* (L). The luminosity of a source may change with time and what we measure is its current value. In cgs units luminosity is measured in erg s^{-1}; in SI units in J s^{-1} (or equivalently, watts). Luminosity can also be measured in units of the Sun's luminosity or solar luminosities. Since it is often the case that the light emitted in different regimes of the electromagnetic spectrum arises from different physical processes, it sometimes makes sense to talk only about a source's optical luminosity or radio luminosity. Unfortunately, the luminosity of an astronomical source cannot be measured directly because we cannot collect all of the radiation that a source emits, since most of the energy is emitted in directions other than toward our telescope. Thus we need a new measure.

1.2.2 Flux

The rate at which we collect energy (or the power collected) from an astronomical source depends on its luminosity and distance, as well as the collecting area of our telescope. The collecting area of most radio telescopes is determined by the size of their primary reflector. If the telescope has a collecting area A, and we collect a total amount of energy E over time δt from an astronomical source, then we define the *flux*, F, of that source to be

$$F = \frac{E}{A\,\delta t}. \tag{1.3}$$

Flux has units of erg s^{-1} cm^{-2} or in SI units W m^{-2}. The amount of energy we collect is proportional to the collecting area of the telescope, but because we divide by the area, the flux is independent of the size of the telescope; however, it still depends on the distance and luminosity of the source.

To compute the luminosity of the source from the measured flux, we generally assume that the astronomical source emits light isotropically (equally in all directions); thus we as observers do not lie in any preferred direction. Imagine surrounding the astronomical source with a sphere of radius d (the distance to the source). The flux measured anywhere on that sphere would be the same (see Figure 1.1). Since the total surface area of the sphere is $4\,\pi\,d^2$, the total rate of energy crossing the sphere is the flux times the total area. Therefore, for an isotropically emitting source, there is a simple relation between flux and luminosity given by

$$F = \frac{L}{4\pi d^2}. \tag{1.4}$$

Another way of thinking about flux is that it is a measure of the rate that light energy crosses a unit cross-sectional area at a given distance from the source. It is clear from the above equation that the flux varies as the inverse square of the distance, the so-called inverse square law nature of light.

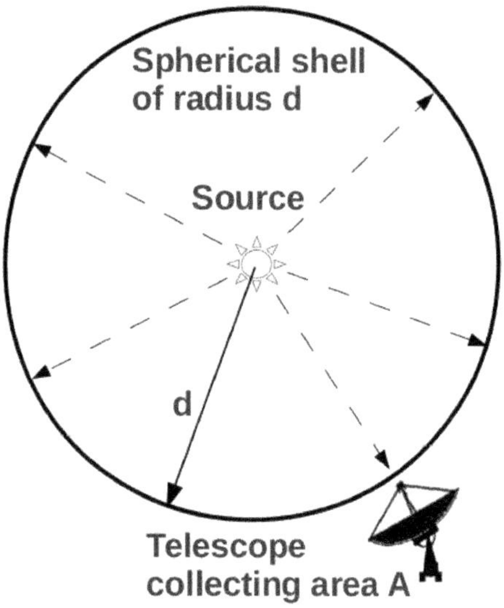

Figure 1.1: The rate energy is collected by a telescope is the fraction of the total luminosity emitted by the astronomical source that enters the small collecting area of the telescope relative to the entire area of a spherical shell of radius d surrounding the source.

In reality, flux is not a measurable quantity, because no telescope can measure the rate energy is received over the entire electromagnetic spectrum. Telescopes and their detectors are limited in what wavelengths they are designed to collect and detect; they only detect the radiation over a very small fraction of the electromagnetic spectrum. In fact, since the radio regime covers such a broad range of wavelengths, most radio telescopes can only detect at any one time a very small fraction of the radio wavelengths. Therefore we need a new measure, which we discuss next.

1.2.3 Flux Density

Flux density is the flux per unit frequency (F_ν) or per unit wavelength (F_λ) interval at a particular frequency or wavelength. We make measurements of the flux from an astronomical source over a frequency interval ($\Delta\nu$) or wavelength interval ($\Delta\lambda$), and therefore we can normalize the rate of energy detected, not only by the collecting area of the telescope, but also by the frequency or wavelength interval over which the measurement was made. We define the flux density as:

$$F_\nu = \frac{F}{\Delta\nu}, \quad \text{or} \quad F_\lambda = \frac{F}{\Delta\lambda}.$$

The cgs units of F_ν are ergs s^{-1} cm^{-2} Hz^{-1} and the SI units are watts m^{-2} Hz^{-1}. You might be tempted to cancel the s^{-1} with the Hz^{-1}; although dimensionally correct, it obscures the fact that flux density is power per unit frequency interval. Radio astronomers often call the frequency interval, $\Delta\nu$, the *bandwidth* of their observation. The cgs units of F_λ are ergs s^{-1} cm^{-2} cm^{-1} or ergs s^{-1} cm^{-3} and in SI units are watts m^{-3}. Although these two flux densities are the same conceptually, they are not the same measures. The relation between these two quantities is given by

$$F_\lambda = \frac{c}{\lambda^2} F_\nu = \frac{\nu^2}{c} F_\nu. \tag{1.5}$$

Note that this relation is not just a conversion of wavelength to frequency using $\lambda\nu = c$. The two flux densities have fundamentally different dimensions. The fact that there are two different ways of defining flux density can be confusing, and the subscripts ν and λ are important to remind one which definition is in use (the units will also help). Unfortunately, astronomers use both definitions. When working at optical wavelengths they tend to measure the flux density per unit wavelength interval (F_λ) and not frequency interval, while at radio wavelengths, they usually measure flux density per unit frequency interval (F_ν). It is also common for radio astronomers to use the symbol S_ν, rather than F_ν, to represent the flux density per unit frequency. Throughout this volume we will use F_ν for flux density defined per unit frequency interval.

The term flux density describes the density of flux in spectrum space. The flux density of a source varies with frequency, or wavelength, and so the total flux of the source is given by the integral of flux density over the entire electromagnetic spectrum, i.e.

$$F = \int F_\nu \, d\nu = \int F_\lambda \, d\lambda,$$

where the integral is over all possible frequencies or wavelengths.

Unlike flux, flux density is a quantity we can measure directly. You may be familiar with photometry at optical wavelengths, where astronomers use filters to restrict the wavelength or frequency range being detected, and these filters define the photometric bands. For example, the Johnson system is labeled by the letters U (for ultraviolet), B (for blue), V (for visible), R (for red) and I (for infrared). Optical astronomers often quote the flux density of an astronomical source in magnitudes, where a magnitude is a logarithmic measure of flux density of the source relative to some fiducial standard. Optical astronomers will often refer to the B-band magnitude or R-band magnitude of an astronomical source. These are logarithmic measures of the relative flux density of the source at wavelengths in the blue and red, respectively.

Radio astronomers never adopted this practice of using magnitudes for flux density, but instead use real physical units of flux density. For astronomical sources at radio wavelengths, the flux density, or the amount of energy we receive per unit time per unit area and per unit frequency interval is very small, so in either cgs or SI units the flux density is an awkwardly small number. For this reason a new unit of flux density was defined, and this unit was called the jansky (Jy), in honor of Karl Jansky. A jansky is defined as

$$1\ \text{Jy} = 1 \times 10^{-23}\ \text{erg s}^{-1}\ \text{cm}^{-2}\ \text{Hz}^{-1} = 1 \times 10^{-26}\ \text{W}^{-1}\ \text{m}^{-2}\ \text{Hz}^{-1}.$$

Note that a jansky is a measure of flux per unit frequency interval. The jansky is also commonly used at infrared wavelengths to measure flux density.

Example 1.1:

The bright radio source Cas A has a flux density at a frequency of 1 GHz of 3110 Jy. If we observe this source with a radio telescope of collecting area 100 m^2 or 10^6 cm^2 over a frequency interval (bandwidth) of 100 MHz centered on a frequency of 1 GHz, how much energy will we detect in 1 second?

Answer:
First, the flux we measure will be the product of the flux density and the bandwidth. Second, from Equation 1.3, the energy detected is the product of the flux, the collecting area of the telescope and the time interval. A flux density of 3310 Jy is equivalent to 3.31×10^{-20} erg s^{-1} cm^{-2} Hz^{-1}. Putting this all together we have

$$E = F\,A\,\delta t = F_\nu\,\Delta\nu\,A\,\delta t, \quad \text{therefore}$$

$$E = (3.31\times10^{-20}\ \text{erg s}^{-1}\ \text{cm}^{-2}\ \text{Hz}^{-1})\ (10^8\ \text{Hz})\ (10^6\ \text{cm}^2)\ (1\ \text{s})$$

$$E = 3.31\times10^{-6}\ \text{erg}.$$

1.2.4 Intensity

The most descriptive quantity regarding the amount of radiation received from an astronomical source is called *intensity* (also called specific intensity and sometimes surface brightness). The intensity of a source is its flux density per unit *solid angle*, and depending on whether flux density is defined per unit frequency or per unit wavelength interval, the intensity is denoted as either I_ν or I_λ. For those readers who are not familiar with solid angle, we give a brief explanation (more details are provided in Chapter 1 of Volume I).

A solid angle is a two-dimensional angle and is a measure of the opening angle of a cone. Before discussing solid angles, it is helpful to review simple angles and the concept of the angular size of an object. Imagine a circular astronomical source with radius r located at a distance d. For small angles (as is appropriate for most astronomical sources), we can approximate the angular diameter of the source in radians by $\Theta \approx 2r/d$, where the radian is the natural unit for angle (1 radian is approximately 57.3 degrees). If we draw a sphere of radius d centered on the observer it will pass through the astronomical source. Now imagine a cone originating at the center of the sphere that just encompasses the circular source (see Figure 1.2). This cone delineates an area, A, on the sphere. The solid angle of the cone is defined to be

$$\Omega = \frac{A}{d^2}, \tag{1.6}$$

and the units of solid angle are called *steradians* (sr). A solid angle therefore defines an angular area on the sky, and, like a radian, the units in the numerator and denominator cancel and so it is dimensionless. Since the area of a shell of radius d is $4\pi d^2$, the entire sky has 4π sr. For small solid angles there is little difference between the area of the cap (the part of the sphere intercepted by the cone) and a flat area represented by the area of the source πr^2. Therefore, in the small solid angle approximation we can write the solid angle of the source as

$$\Omega \approx \frac{\pi\,r^2}{d^2}. \tag{1.7}$$

Substituting for the angular diameter of the source Θ, we find that

$$\Omega \approx \frac{\pi}{4}\,\Theta^2. \qquad (1.8)$$

From this expression it is obvious that a solid angle has dimensions of a simple angle squared, i.e. steradians. Since solid angle defines an angular area on the sky, intensity is the flux density that we receive from each unit angular area of the source.

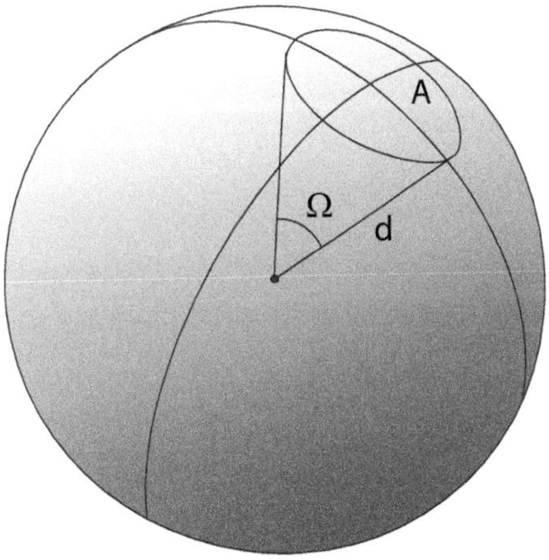

Figure 1.2: The solid angle (Ω) defines the opening angle of a cone. In the small angle approximation, a circular astronomical source of area A at a distance d will have, or subtend, a solid angle of $\Omega = A/d^2$.

The cgs units of I_ν are erg s^{-1} cm^{-2} Hz^{-1} sr^{-1} or in SI units they are watt m^{-2} Hz^{-1} sr^{-1}. Of course the units of I_ν could also be expressed as Jy sr^{-1}. The cgs units of I_λ are erg s^{-1} cm^{-2} cm^{-1} sr^{-1} or erg s^{-1} cm^{-3} sr^{-1} and in SI units they are watt m^{-2} m^{-1} sr^{-1} or watt m^{-3} sr^{-1}. The relationship between the flux density of a source and its intensity is

$$F_\nu = \int I_\nu \cos\theta \; \mathrm{d}\Omega,$$

where the integral is over the entire solid angle of the source and θ is the angle between the direction of the incident radiation and the direction normal to our detector surface. However, since the angular size of astronomical sources is relatively small, θ in the above equation is small and thus cos θ is approximately 1; therefore we can rewrite the above equation as

$$F_\nu \approx \int I_\nu \, d\Omega, \qquad (1.9)$$

Intensity provides more information about the astronomical source than does the flux density. Intensity is the amount of radiation emanating from each small piece of the object and provides information on how the radiation varies with position on the source. Intensity is directly related to the microscopic processes responsible for producing the radiation. We can not know from the measure of a large flux density alone, whether the source involves very strong emission, has a very large surface, or is very close. Clearly, we can only measure the intensity of a source if our telescope can resolve the source, meaning we can distinguish the emission from different parts of the source. For unresolved sources, such as most stars, the radiation we receive appears to come from a single point on the sky. For these unresolved sources (or point sources) we can only measure the source's flux density. There are many astronomical sources that are resolved at radio wavelengths, particularly using interferometric techniques (see Chapters 5 and 6 in Volume I), and therefore we can form an image of the source. The brightness, or the amount of flux density in each unit solid angle, of that image is its intensity.

An important fact about intensity is that it is distance independent. We discussed earlier that the flux density of a source depends on the inverse square of its distance. And, we saw in Equation 1.6 that the solid angle also depends on the inverse square of distance. The flux density per solid angle, then, must be independent of distance. This also means that we can view the distance dependence of flux density as coming from the distance dependence of solid angle.

Example 1.2:

The radio source Cas A from Example 1.1 is roughly a circular source with an angular diameter of about 3 arcminutes. What is the average intensity of the source at a frequency of 1 GHz?

Answer:
First we must convert the angular diameter from arcminutes to radians. Angular diameter in radians = 3 arcminutes (1 degree/60 arcminutes) (1 radian/57.3 degrees) = 8.7×10^{-4} radians. The source solid angle is given by Equation 1.8, thus

$$\Omega_{\text{source}} \approx \frac{\pi}{4}\,\Theta^2 = 6.0 \times 10^{-7} \text{ sr.}$$

Finally, the average intensity is just the flux density divided by the source solid angle (see Equation 1.9), therefore

$$I_\nu = (3.31 \times 10^{-20} \text{ ergs s}^{-1} \text{ cm}^{-2} \text{ Hz}^{-1})/(6.0 \times 10^{-7} \text{ sr})$$

$$I_\nu = 5.5 \times 10^{-14} \text{ ergs s}^{-1} \text{ cm}^{-2} \text{ Hz}^{-1} \text{ sr}^{-1}.$$

1.2.5 Polarization

Another characteristic of light is its polarization, which informs us of any net orientation of the electric field of the light. Remember that light is an electromagnetic wave consisting of oscillating electric and magnetic fields. As an electromagnetic wave travels through a vacuum, the direction of wave propagation, the direction of the electric field and the direction of the magnetic field are all mutually perpendicular. This is illustrated in Figure 1.3 which shows an electromagnetic wave propagating in the z-direction with an electric field oscillating in the y-direction and a magnetic field oscillating in the x-direction. Note that we have defined the x-, y-, and z-axes following the right-hand rule, so that the cross-product $\vec{E} \times \vec{B}$ points in the positive z-direction. If all of the electromagnetic waves detected from an astronomical source were oriented as shown in Figure 1.3, then the light would be perfectly polarized with the electric field from each wave oriented in the same direction. Since the net electric field oscillates along one axis, this radiation is what we call *linearly polarized.* As we discuss below, light can also be *circularly polarized* or *elliptically polarized.* We focus, first, on linear polarization.

Often, the electric fields of electromagnetic waves emitted from astronomical sources are randomly oriented, so there is no preferred orientation of the electric field, and therefore light is very often unpolarized. However, some radiation processes naturally produce light that is polarized (for instance, the synchrotron emission discussed in Section 3.3), and in some cases the propagation of the light through certain media can alter the polarization (as discussed in Section 2.2).

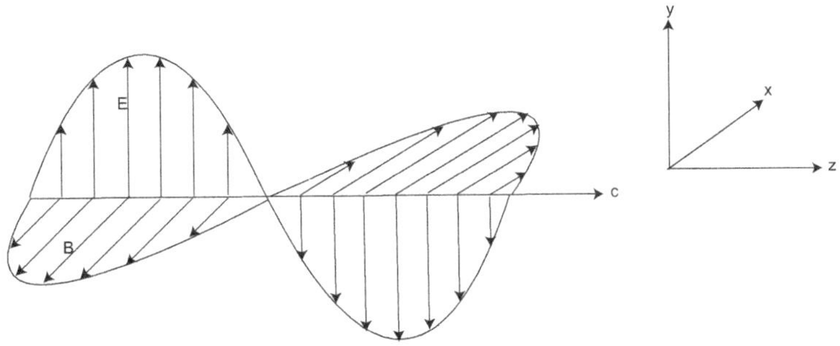

Figure 1.3: Propagation of an electromagnetic wave along the z-axis with the electric field oscillating along the y-axis and the magnetic field oscillating along the x-axis.

We can measure the degree of linear polarization by measuring the intensity of light through a polarizing element. The polarizing element permits light of only one orientation of the electric field or one polarization to pass, and this element can be rotated to measure the intensity of light at different angles or different directions of polarization. If, as the polarizing element is rotated, there is no change in the intensity of light, then the light source is unpolarized. However, if the intensity varies, then there is a preferred direction of the electric field, and the light is linearly polarized, As the polarizing element is rotated, one can measure the maximum intensity of light (I_{max}) and the minimum intensity of light (I_{min}), and the percent polarization, $P(\%)$, is then defined to be

$$P(\%) = 100 \frac{I_{max} - I_{min}}{I_{max} + I_{min}}.$$

Measurements of whether, and to what degree, light is linearly polarized provide useful clues to the processes responsible for producing the radiation.

The concept of circular polarization takes a little more explanation. Let's start with two waves, one whose electric field is oriented in the x-direction and a second whose electric field is oriented in the y-direction. The total electric field is then the vector sum of the electric fields of the two waves, see Figure 1.4. If the two waves are in phase, then the net electric field will oscillate in the plane indicated by $\vec{E}$ in Figure 1.4. However, there is no reason that these two waves need to be in phase. Imagine if the two waves are out of phase by one-quarter of a cycle or $\pm\pi/2$ radians. Then, the resultant electric field vector at various locations along the wave is depicted in Figure 1.5. At the right end of Figure 1.5, the electric field in the y-direction is at a maximum, while the electric field in the x-direction is zero. At positions to the left, the electric field in the y-direction decreases and the electric field in the x-direction increases. When the electric field in the x-direction reaches a maximum, the electric field in the y-direction is zero. The vector sum of these two electric fields therefore rotates around the z-axis, tracing out a circle as is illustrated in Figure 1.5, and so in this case the light is *circularly polarized.*

Depending on the sign of the phase difference the resultant waves can be either *right circularly polarized* (RCP) or *left circularly polarized* (LCP). Unfortunately, the definition of right and left circular polarization depends on your point of view and this choice varies with

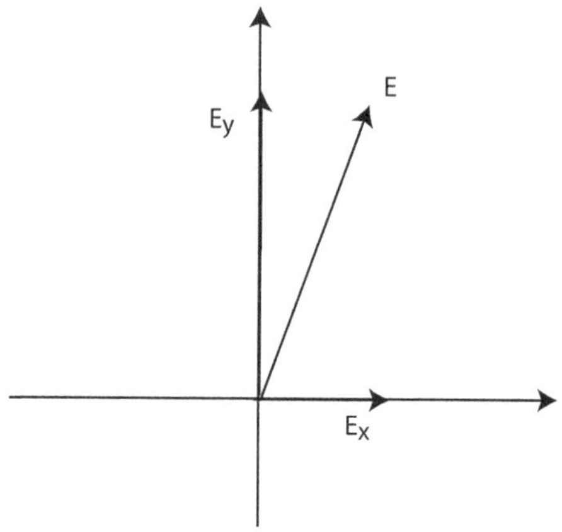

Figure 1.4: The total electric field vector of two linearly polarized waves can be described by their vector sum. If the two waves are in phase with each other, the orientation of the electric field vector remains constant.

discipline. Radio astronomers (and electrical engineers) define circular polarization from the point of view of the light emitter. In this perspective, if the electric field is rotating clockwise it is right circularly polarized and if rotating counter-clockwise it is left circularly polarized. Most physics textbooks adopt the perspective of the observer of the light. Right circular polarization still corresponds to clockwise rotation, but since this is seen from the opposite point-of-view, it corresponds to the radio astronomer's counter-clockwise rotation or left circular polarization, and vice versa. Be careful when reading about circularly polarized light to be certain you know which convention is being used.

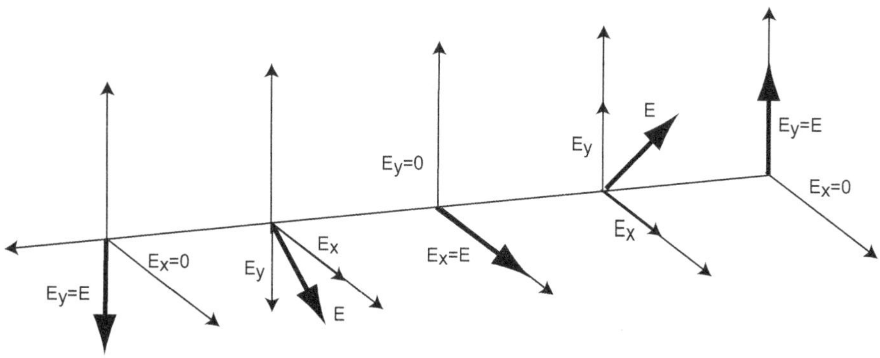

Figure 1.5: The sum of two linearly polarized waves, with electric fields oscillating in the x- and y-directions and out of phase which cause the total electric field vector to rotate around the direction of propagation.

The two waves illustrated in Figure 1.5 have equal electric field amplitudes. If their amplitudes were unequal, then the electric field would still rotate around the axis of propagation, but instead of tracing out a circle, it would trace an ellipse. In this case we say the light is *elliptically polarized.* Note that if the phase difference were zero or π radians, then the resultant electric field would oscillate in a fixed direction in the xy-plane, and we recover linear polarization. Therefore, you can consider elliptical polarization as the most general form of polarization, with circular and linear polarization as special cases.

An alternative approach that is equally valid is to describe any polarization state as the sum of two basis vectors, one with right circular polarization and a second with left circular polarization. We will make use of this approach in Section 2.2.

As was the case for linear polarization, measurements of the circular polarization of light provide insights into the emission mechanisms and the properties of the medium that the light is propagating through. The techniques for measuring circular polarization of light are somewhat more involved than for linear polarization, but can be readily made, thus producing a measure of the degree of right or left circular polarization of the light. Another method of quantifying the polarization of light is by the use of the Stokes parameters; these are explained in detail in Section 2.7.1 of Volume I.

1.3 SKY COORDINATES

Astronomers use several different coordinate systems to define the location of an astronomical object on the sky or on what we often call the celestial sphere. The two most commonly used systems are equatorial coordinates and Galactic coordinates. In later chapters, when we present radio astronomy results, some sky images are presented in equatorial coordinates, and some in Galactic coordinates; therefore, we need to be comfortable with both and so we describe each of these systems in this section.

1.3.1 Equatorial Coordinate System

On the Earth we locate a position by its longitude and latitude and these are defined relative to the rotation axis of the Earth. Lines of latitude encircle the Earth parallel to the equator and lines of longitude lie perpendicular to lines of latitude and connect the north pole to the south pole. The equatorial sky coordinate system is a projection of the lines of longitude and latitude onto the sky (see Figure 1.6), producing lines of *right ascension* (often denoted by RA or α) and lines of *declination* (often denoted by Dec or δ).

The declination coordinate is measured in degrees (°), minutes of arc (') and seconds of arc ("), where there are 60 minutes of arc in a degree and 60 seconds of arc in an arcminute. The projection of the equator (latitude 0°) onto the sky defines the *celestial equator* and following the convention for latitude, the celestial equator is defined to be the 0° declination line. The projections of the Earth's north and south poles on to the sky define the *north celestial pole* (at declination +90°) and the *south celestial pole* (at declination −90°), respectively. Following the definition of latitude, the value of the declination is the angle between the celestial equator and the line of declination as viewed from the center of the Earth. Angles north of the celestial equator have positive values of declination, while angles south are negative.

Lines of longitude projected onto the sky define the coordinate called right ascension. However, because the Earth rotates on its axis, the projected lines of longitude shift with time relative to the celestial sphere, so the definition of right ascension is a little more complicated than declination. The fiducial point of right ascension is defined as the position of the Sun when it crosses the celestial equator in the Spring, a point on the sky known as the *vernal equinox*. You may be familiar with the vernal equinox as the day in the year when the Sun crosses the equator, around March 21, but it is also defined as the position in the sky of the Sun at that time. A star's right ascension, then, is a measure of its angular distance eastward from the vernal equinox. One can compute, for any given moment in time, the angular offset (or time offset) between the fiducial line of right ascension and the fiducial line of longitude, and therefore we can locate any source on the sky. As a consequence of the

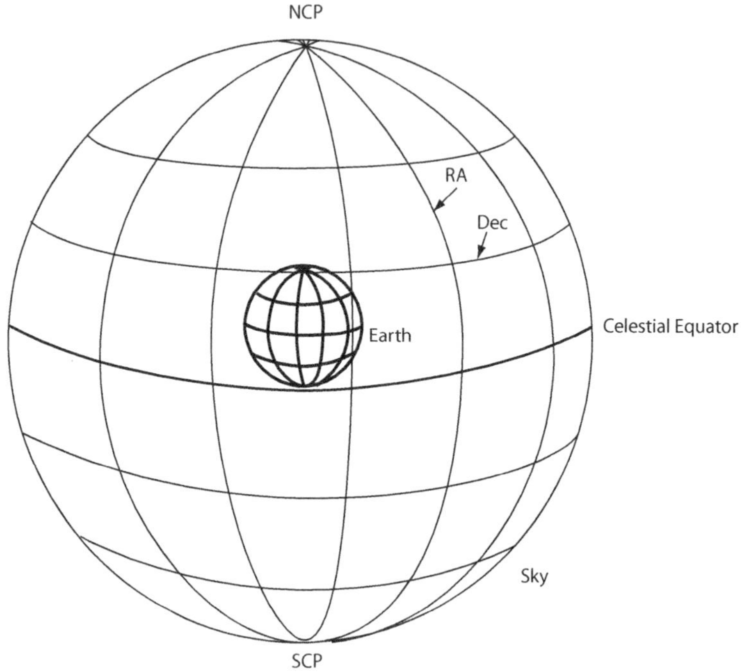

Figure 1.6: The celestial sphere showing the grid of right ascension and declination lines that make up the equatorial coordinate system. The celestial equator is a projection of the Earth's equator onto the sky, and the north celestial pole (NCP) and the south celestial pole (SCP) are the projections of the Earth's poles onto the sky. Lines of declination run parallel to the celestial equator, while lines of right ascension are perpendicular to the celestial equator and converge at the NCP and SCP.

rotation of the Earth relative to the sky, the values of right ascension are not measured in angle units, but are more conveniently given in units of time, with 24 hours representing a full circle at a fixed declination. Thus, 0 hours right ascension corresponds to 0°, 6 hours to 90°, 12 hours to 180° and 18 hours to 270°. Just as for time, each hour of right ascension is divided into 60 minutes and each minute is divided into 60 seconds. The use of time makes some sense, as the projected lines of longitude are constantly shifting with respect to the celestial sphere as the Earth rotates.

There is one more complication with equatorial coordinates and that is due to the precession of the Earth's rotation axis. The direction of the Earth's rotation axis changes, completing a circular path on the sky with an angular diameter of 47° over a time span of about 26,000 years. Although precession is slow, it changes the alignment of longitude and latitude with right ascension and declination and thus the equatorial coordinates of astronomical objects. Therefore when one gives a set of equatorial coordinates, we need to know the epoch in which they were computed and then correct for precession to point our telescope in the correct direction on the sky. The most current epoch in which equatorial coordinates are computed is called J2000, thus relative to Julian year 2000. So one will see coordinates today given as J2000 coordinates. In older books and papers you may find coordinates given in earlier epochs.

In summary, the position of an astronomical source in equatorial coordinates is given by its declination, right ascension and epoch. The declination is generally specified in degrees, minutes of arc, and seconds of arc, while right ascension is specified in hours, minutes and seconds of time. For example, the equatorial coordinates for one of the brightest radio sources in the sky, Cas A, are RA (J2000) = $23^h\ 23^m\ 24.0^s$ and Dec (J2000) = +58° 48' 54". Knowing these coordinates, along with the date and time, we can determine the source location on the sky.

1.3.2 Galactic Coordinate System

For observations of objects located within the Milky Way, and even for objects located well beyond our Galaxy, it is often more convenient to use a different coordinate system defined relative to the geometry of the Galaxy, called *Galactic coordinates.* Our Galaxy has the geometry of a flattened rotating disk in which the Sun is in the disk plane, but well-removed from the center, and therefore from our perspective we see the Milky Way disk encircle us. The Galactic coordinate system is centered on the Earth and forms a coordinate grid much like right ascension and declination, but the orientation of the coordinate lines is defined relative to the plane of the Galaxy, rather than related to the rotation axis of the Earth.

The Galactic coordinates of an object are defined by its *Galactic longitude* (denoted by ℓ) and *Galactic latitude* (denoted by b), which are analogous to right ascension and declination. Both Galactic longitude and latitude are measured in decimal degrees. The direction towards the center of our Galaxy is defined to be Galactic longitude 0° and the Galactic plane defines the 0° line of Galactic latitude. Although the Galactic plane is tilted relative to the celestial equator by about 60°, the north Galactic pole is defined to be in the same hemisphere as the north celestial pole. The northern Galactic hemisphere is often described as being "above the Galactic plane" and the southern hemisphere as "below the plane" (even though this has nothing to do with "up" or "down" as generally defined by gravity). The lines of Galactic longitude, then, are defined, as viewed from above the Earth looking down on the Galactic plane from Galactic north. Galactic longitude increases counter-clockwise as shown in the upper panel of Figure 1.7. Galactic latitude is defined by the angle above (positive) or below (negative) the Galactic plane as shown in the lower panel of Figure 1.7. The north Galactic pole has a Galactic latitude of +90° and the south Galactic pole has Galactic latitude of −90°. As in the equatorial coordinate system, any direction in the sky can be uniquely described by its Galactic longitude and latitude.

The Galactic coordinates of an object are not affected by the precession of the Earth and so remain constant, but the conversion from Galactic to equatorial coordinates does depend on the epoch of the equatorial coordinates. The location of the Galactic center in equatorial coordinates (J2000) is RA = $17^h\ 45^m\ 40.04^s$ and Dec = −29° 00' 28.1" and the coordinates of the north Galactic pole (J2000) are RA = $12^h\ 51^m\ 26.28^s$ and Dec = +27° 07' 42.0". We can also specify the location of Cas A, a supernova remnant within the Milky Way, by its Galactic coordinates, which are $\ell = 111.7347°$ and $b = -2.1296°$.

1.4 DOPPLER EFFECT

The Austrian physicist Christian Doppler, in 1842, explained why the pitch of a train whistle sounds higher when the train is approaching the observer than when the train is receding from the observer. The effect is a fairly simple concept whose explanation requires only an understanding of waves. Similarly, the apparent frequency (or wavelength) of light waves will also be shifted by the motion of the emitter of the light, relative to the observer. The precise

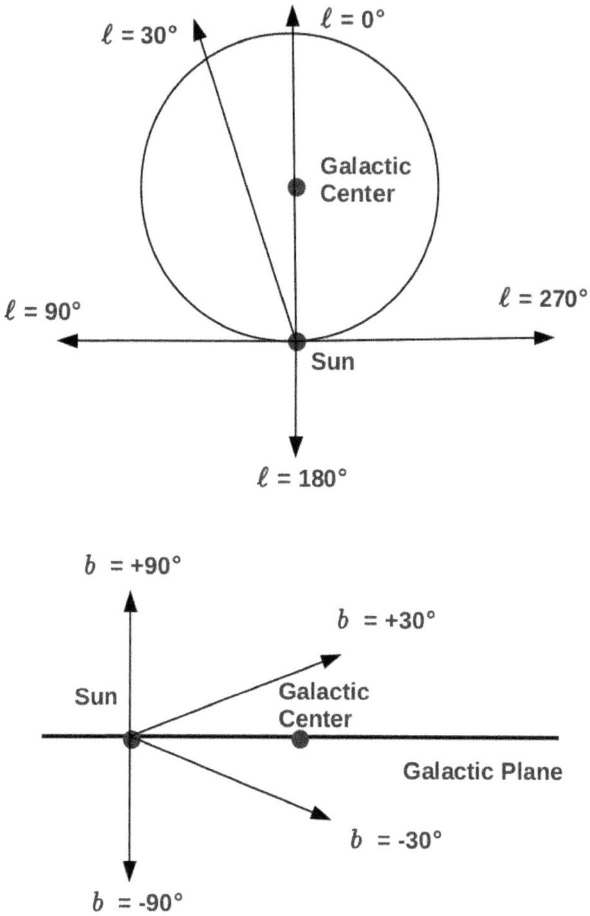

Figure 1.7: Any direction in the sky can be defined by its Galactic longitude and latitude. A face-on view of the Milky Way is shown in the upper figure and is shown as viewed from the north Galactic pole. We define Galactic longitude (ℓ) as the angle counter-clockwise relative to the Galactic center. Thus, the direction to the Galactic center is toward $\ell = 0°$ and the direction of the Galactic anti-center is toward $\ell = 180°$. The lower figure shows an edge-on view of the Miky Way. Galactic latitude is the angle out of the Galactic plane. A Galactic latitude of $b = 0°$ lies in the Galactic plane, while $b = +90°$ is perpendicular to the plane, towards the north Galactic pole.

application to light waves is not exactly the same as with sound waves, primarily because of the constancy of the speed of light. For velocities small with respect to the speed of light we can use what is called the *classical Doppler effect* (or *Doppler shift*). However, when velocities are an appreciable fraction of the speed of light there is an additional relativistic effect that also contributes to a shift in the observed frequency. In many applications in astronomy, especially the vast majority of studies of objects within the Galaxy and even in nearby galaxies, we can use the classical effect. There will be a number of cases, though, when the relative velocities are large and so we will need to use the more complicated relativistic formulation.

The Doppler effect provides an easy and useful tool for determining the radial velocity of an astronomical source by measuring the frequency or wavelength shift. Often, the velocities of objects we want to measure are relatively small and so the wavelength or frequency shifts are also small. The measurement of such small shifts in frequency or wavelength requires the observation of light in which the frequencies or wavelengths are well-defined and well-known. For this reason, the Doppler shift is usually measured for spectral lines.

1.4.1 Classical Doppler Effect

The classical Doppler effect is often explained in a beginning physics or astronomy class, so we will not give a derivation here. Instead we jump to the final equation that you need to know. In this section we discuss the classical Doppler effect, appropriate for velocities that are small relative to the speed of light, and in Section 1.4.2 we discuss the relativistic formulation. It is important to remember that only relative motion of the emitter towards or away from the observer causes a Doppler shift. Defining the velocity components from the perspective of the observer, we call the relative velocity along the line of sight the *radial velocity*, denoted by v_R, and the velocity perpendicular to the line of sight the *tangential velocity*, v_T. So, consider a light source emitting a light wave at frequency ν_e and moving directly away from the observer at speed v_R. Then, for velocities much less than the speed of light, the observed frequency, ν, can be approximated by

$$\nu \approx \frac{\nu_e}{1 + (v_R/c)}. \tag{1.10}$$

The equivalent equation expressed in wavelength is

$$\lambda \approx \lambda_e \, (1 + \frac{v_R}{c}), \tag{1.11}$$

where λ_e is the emitting wavelength and λ is the observed wavelength. These formulae are accurate to better than 1% for v/c less than 0.01, or for velocities less than about 3000 km s^{-1}. The sign of the velocity is also very important; if the light source is receding we define the velocity to be positive and if the light source is approaching we define the velocity to be negative. For sources that are receding, then, the wavelength observed is shifted to a longer wavelength than it was emitted and the frequency is shifted to a lower frequency. Since the astronomical use of the Doppler effect started with visible-wavelength astronomy, and a positive radial velocity shifts the spectrum toward longer wavelengths, and hence toward the redder end of the visible window, this is often called a *redshift*, regardless of whether we are observing at optical or radio wavelengths. For sources approaching us, the observed wavelength is shifted to shorter wavelengths or higher frequencies, and we call this a *blueshift*. To provide an equation for inferring the radial velocity (relative to the observer) of the emitter, we can rewrite the above two equations as

$$v_R \approx c\,\frac{\lambda - \lambda_e}{\lambda_e}, \quad \text{and} \quad v_R \approx c\,\frac{\nu_e - \nu}{\nu} \tag{1.12}$$

Note that these two equations are not symmetrical; in the first equation the change in wavelength is divided by the emitted wavelength, while in the second, the change in frequency is divided by the observed frequency.

Example 1.3:

A radio spectral line has an emitted frequency of 1420.4058 MHz. We observe an astronomical source and measure the frequency for this same spectral line to be 1419.9525 MHz. What is the radial velocity of this astronomical source and in what direction is it moving?

Answer:
We can use Equation 1.12 that relates the emitted frequency and the observed frequency. Because the frequency is shifted to lower frequencies, we know the object is moving away. Since astronomers usually express velocities in km s^{-1}, we use speed of light in these units, therefore

$$v_R \approx c\,\frac{\nu_e - \nu}{\nu} = 3.0 \times 10^5 \text{ km s}^{-1}\,\frac{(1420.4058 \text{ MHz} - 1419.9525 \text{ MHz})}{1419.9525 \text{ MHz}}$$

$$v_R = +95.8 \text{ km s}^{-1}.$$

Note that the velocity is positive, confirming that the object is moving away.

Unfortunately, there are two commonly used approximations for the classical Doppler effect. Equations 1.10 through 1.12 describe what is often called the "optical definition" of the Doppler effect, and is the one most commonly used and the one which is preferred by the International Astronomical Union. However, also in use is the so-called "radio definition", particularly by radio astronomers, which is given by

$$v_R \approx c\,\frac{\nu_e - \nu}{\nu_e}.$$

This approximation is as accurate as the previous expression; however the two results differ slightly and diverge for v/c larger than 0.01. Be careful to verify which definition for the Doppler velocity is being used.

1.4.2 Relativistic Doppler Effect

When the relative velocity between a light source and an observer is large, we need to properly account for special relativistic effects. In particular, we need to include time dilation between the two reference frames, which causes a decrease in the observed frequency by the Lorentz factor γ. Time dilation will occur regardless of the direction of the velocity, and so the full relativistic Doppler effect formulation involves a time dilation term, which contains the magnitude of the complete velocity vector, $\vec{v}$, and a term due to the classical Doppler effect which involves only the radial velocity, v_R. Again without derivation, the observed frequency, ν, is given by

$$\nu = \frac{1}{\gamma}\,\frac{\nu_e}{1 + (v_R/c)} = \frac{\nu_e\,\sqrt{1 - (v^2/c^2)}}{1 + (v_R/c)}, \tag{1.13}$$

where the emitted frequency is ν_e and γ is given by

$$\gamma = \frac{1}{\sqrt{1 - (v^2/c^2)}}.$$

As before, the sign of the radial velocity depends on whether the light source is approaching (negative) or receding (positive). Since the relativistic formulation depends on both the radial velocity and the total velocity, without further information about the direction of motion, it is impossible to infer the radial velocity from just the measured frequency shift.

If the relative motion between the light source and observer is only radial, then we can use a simplified expression involving only the radial velocity. In this special case, the observed frequency will be

$$\nu = \nu_e \frac{\sqrt{1-(v_R^2/c^2)}}{1+(v_R/c)}.$$

This expression is often rewritten in the form

$$\nu = \nu_e \sqrt{\frac{1-(v_R/c)}{1+(v_R/c)}}. \tag{1.14}$$

In wavelengths, the equivalent expression is

$$\lambda = \lambda_e \sqrt{\frac{1+(v_R/c)}{1-(v_R/c)}}. \tag{1.15}$$

Note that if v_R/c is small, then Equations 1.14 and 1.15 are equivalent to the classical Doppler shift expression. Once the velocity exceeds a few percent of the speed of light, it is important to use the relativistic formula to derive an accurate velocity. Also note that as v approaches c, the wavelength approaches infinity and the frequency approaches zero.

Equations 1.14 and 1.15 can be rearranged to solve for the radial velocity of the source from the observed frequency or wavelength. The radial velocity is

$$v_R = c\left(\frac{1-(\nu/\nu_e)^2}{1+(\nu/\nu_e)^2}\right) \tag{1.16}$$

or

$$v_R = c\left(\frac{(\lambda/\lambda_e)^2-1}{(\lambda/\lambda_e)^2+1}\right) \tag{1.17}$$

Example 1.4:

For the same radio spectral line in Example 1.3, if the source's motion is all radial and it is moving toward us at 30,000 km s^{-1} or 3.0×10^9 cm s^{-1} at what frequency would we observe this spectral line?

Answer:
In this case the velocity is 10% the speed of light, so we will need to use the full relativistic formulation of the Doppler effect. Applying Equation 1.14 with v_R/c = -0.1 (remember we define motion toward us as a negative velocity), we find

$$\nu = \nu_e \sqrt{\frac{1-(v_R/c)}{1+(v_R/c)}} = 1420.406 \text{ MHz} \sqrt{\frac{1-(-0.1)}{1+(-0.1)}}. = 1570.318 \text{ MHz}.$$

The observed frequency will be significantly larger than that emitted. Note that if you had used the classical formula, you would have gotten a frequency of 1578.229 MHz, too high by about 8 MHz.

If the relative motion is all perpendicular to the line of sight, there is no classical Doppler effect; however, since there is still time dilation, there is a transverse relativistic Doppler effect. Starting with Equation 1.13, if there is no radial component to the velocity, so that $\vec{v}$ is entirely perpendicular or transverse to the line of sight, the transverse Doppler effect is given by

$$\nu = \nu_e \sqrt{1 - \frac{v^2}{c^2}}. \tag{1.18}$$

The transverse Doppler effect always decreases the observed frequency relative to the emitted frequency.

Example 1.5:

If the motion of the source of the spectral line emission in Example 1.4 is all transverse to the line of sight, at what frequency would we observe this spectral line?

Answer:
In this case we will need to use the relativistic transverse Doppler formulation. Applying Equation 1.18 with $v/c = 0.1$, we find

$$\nu = \nu_e \sqrt{1 - \frac{v^2}{c^2}} = 1420.406 \text{ MHz} \sqrt{1 - (0.1)^2} = 1413.286 \text{ MHz}.$$

The observed frequency is significantly different than if the motion was all in the radial direction and is redshifted relative to the emitted frequency.

1.5 COSMOLOGICAL REDSHIFT AND THE EXPANDING UNIVERSE

One of the most important discoveries in the history of astronomy was that our Universe is expanding. Spectra of galaxies obtained by Vesto Slipher showed that most galaxies were redshifted, indicating that the galaxies were moving away. George Lemaître theorized in 1927 that the Universe was expanding and Edwin Hubble reported data in 1929 showing that the amount of a galaxy's redshift was directly proportional to its distance. This observational relationship between a galaxy's apparent recession velocity (v_{rec}) and distance (d) is called *the Hubble-Lemaître Law,* and is expressed as

$$v_{rec} = H_o\, d,$$

where the constant of proportionality, H_o, is called Hubble's constant. Hubble's constant is approximately 70 km s^{-1} Mpc^{-1}. Today we know that the value of Hubble's 'constant' varies in time, so we use the subscript 'o' to indicate the current value.

Our understanding of the origin of the Hubble-Lemaître Law is that we live in an expanding Universe. Galaxies' redshifts, in this case, are not actually due to the Doppler effect, or motion through space, but rather that space, itself, is expanding. The Doppler shift can produce either a blueshift or redshift depending on the direction of relative motion, but the expansion of space always produces a redshift. Since this redshift is not a result of a galaxy's motion through space, it is called a *cosmological redshift.* The cosmological redshift is fundamentally different from a Doppler shift. Light emitted by a distant galaxy takes time to reach the observer, and during this time, the space through which it travels expands, and as a consequence of traveling through an expanding space the wavelength of the light is stretched. The longer the light travel time the greater space has expanded and the greater wavelengths are stretched. Since the light travel time increases with distance, the light from the more distant galaxies suffers a greater stretching or greater redshift, and therefore it is not surprising that there is a relationship between distance and recessional velocity or redshift. Since space is expanding everywhere, observers in any galaxy will find the light from other galaxies to be redshifted, and that the redshift will be proportional to distance. Hence, the Hubble-Lemaître law does not indicate that we are at any preferred location or at the center of the expansion.

To be a little more quantitative, one can define a dimensionless *scale factor*, $R(t)$, which describes how the relative distance between objects changes with time. We usually define the scale factor so that now (time t_o) we have $R(t_o) = 1$. In this case, $R(t)$ gives the ratio of the distance between any two galaxies at time t to the distance today. $R(t)$ describes the expansion history of space. In an expanding universe the scale factor was less than unity in the past, as the galaxies were closer together, and will be greater than unity in the future, as the galaxies move farther apart. The wavelength of light emitted by a distant galaxy is stretched in proportion to the change in the scale factor of the Universe between the time when the light was emitted, t_e, and the time the light was detected, t_d. Therefore we have

$$\frac{\lambda_d}{\lambda_e} = \frac{R(t_d)}{R(t_e)},$$

where λ_e is the wavelength at which the light was emitted and λ_d is the wavelength of the light detected. Since we are making the detection now (at time t_o) and the scale factor is defined so that $R(t_o) = 1$, the ratio of the wavelength observed to that emitted is simply

$$\frac{\lambda_d}{\lambda_e} = \frac{1}{R(t_e)}. \tag{1.19}$$

Therefore the cosmological redshift is related solely to the change in the scale factor of the universe.

We often describe the observed wavelength shift of a galaxy's spectrum by a quantity called the redshift, z, which is defined by

$$z = \frac{\lambda_d - \lambda_e}{\lambda_e}, \tag{1.20}$$

where, as before, λ_d is the detected wavelength of a spectral line and λ_e is the emitted wavelength. The redshift defined in terms of frequency is given by

$$z = \frac{\nu_e - \nu_d}{\nu_d},$$

where ν_d is the detected frequency of a spectral line and ν_e is the frequency at which the spectral line was emitted. Note again, as was the case for the Doppler shift, that these two equations are not symmetrical as the denominator in the first equation is the *emitted* wavelength, while in the second equation the denominator is the *observed* frequency.

The measured redshift, z, has two contributions, the first due to the Doppler effect discussed in Section 1.4 and the second due to the expansion of the Universe, which is called the cosmological redshift (which we will denote z_C). We will first consider the cosmological redshift. We can combine Equations 1.19 and 1.20 to find a simple relationship between the redshift due to the expansion of the Universe and the scale factor,

$$1 + z_C = \frac{1}{R(t_e)}. \tag{1.21}$$

For any object, the light we observe was emitted in the past when $R(t_e) < 1$ and therefore the expansion of the universe causes all objects to be redshifted and z_C is always greater than 0.

In addition to the cosmological redshift, a galaxy's motion through space produces a Doppler shift. This motion through space is called the galaxy's *peculiar velocity*. An example of such motion occurs in galaxy clusters, where the galaxies are bound due to their mutual gravity, and thus orbit about the center of mass of the cluster. Observationally, we cannot tell the difference between the shift due to the expansion of the Universe and that due to a galaxy's peculiar velocity. Although a galaxy's peculiar motion can produce either blueshifts or redshifts depending on the direction of its relative motion, the expansion of the Universe always produces a redshift. For peculiar velocities that are non-relativistic (as is often the case), we can write the observed redshift as

$$1 + z \;=\; (1 + z_C)\,\left(1 + \frac{v_R}{c}\right),$$

where z_C is the cosmological redshift and v_R is the radial component of a galaxy's peculiar velocity. For nearby galaxies, the Doppler shift due to peculiar velocities often dominates the observed shift in wavelength or frequency, while for galaxies greater than a few tens of Mpc away the redshift is dominated by the cosmological redshift.

Suppose we measure the redshift of a galaxy to be $z = 1$, and in this case the observed redshift is dominated by the cosmological redshift. We can therefore conclude that when the light was emitted by this galaxy the scale factor of the Universe was one-half what it is today. Therefore the separation between us and the galaxy is twice as large today as compared to what it was at the time the light was emitted.

It is important to remember that when we observe a distant galaxy we see it as it was in the past; this is particularly important when we discuss the distant Universe. With a full understanding of the expansion history of the Universe we can convert the cosmological redshift into a *look-back time*, defined as the amount of time that has elapsed since the light was emitted. Observing galaxies at very high redshift gives us a glimpse of the Universe as it was in the very distant past.

Example 1.6:

The rest wavelength of the spectral line in Example 1.3 is at 21.107 cm. You observe this line in a very distant galaxy to be at a wavelength of 30.5 cm. What is the redshift of this galaxy? If the observed wavelength shift is entirely due to the expansion of the Universe, what was the scale factor of the Universe at the time the spectral line was emitted?

Answer:
The redshift is given by Equation 1.20. Therefore,

$$z \;=\; \frac{\lambda_d - \lambda_e}{\lambda_e} \;=\; \frac{30.5 \text{ cm} - 21.107 \text{ cm}}{21.107 \text{ cm}} \;=\; 0.445.$$

The scale factor at the time the light was emitted is given by Equation 1.21. Solving for $R(t)$, we find:

$$R(t) \;=\; \frac{1}{z+1} \;=\; \frac{1}{0.445+1} \;=\; 0.692.$$

Therefore when the light was emitted the scale factor of the Universe was 0.692.

In other words, the Universe has expanded by a factor of $1/0.692 = 1.45$ between the time the light was emitted and the time the light was received.

The Hubble-Lemaître law is also useful for inferring the distance to a galaxy when peculiar velocities are unimportant. A simple rearrangement of the Hubble-Lemaître law gives

$$d = \frac{v_{rec}}{H_o}.$$

This is an approximate relation only correct for relatively nearby galaxies because of the time dependence of the Hubble constant. Light received from very distant galaxies was emitted when the Hubble constant had a different value and in this case the above relation is not valid. Additionally, we must remember that the observed redshift contains some Doppler shift due to peculiar motions. An alternative form of the above equation uses the fact that the measurable quantity is the redshift. For small redshifts, combining Equations 1.12 and 1.20, the recessional velocity is simply cz, and therefore

$$d = \frac{cz}{H_o}. \tag{1.22}$$

Remember that this is only accurate for small redshifts; to have distances accurate to better than 10% requires the redshift to be less than 0.3. For larger redshifts, when a precise distance is needed, the correct calculation involves a more complicated equation that includes the specifics of how $R(t)$ changes with time, and this will be discussed in the next section.

Example 1.7:

We measure the redshift of a galaxy to be $z = 0.1$. What is the distance to this galaxy?

Answer:
We assume that the redshift is sufficiently large so that it is entirely due to the cosmological redshift, yet small enough that we can use Equation 1.22 to estimate the distance. The distance is then given by

$$d = \frac{cz}{H_o} = \frac{(3.0 \times 10^5 \text{ km s}^{-1})\ (0.1)}{70 \text{ km s}^{-1} \text{ Mpc}^{-1}} = 429 \text{ Mpc}.$$

1.6 DISTANCE AND AGE CALCULATIONS

In the previous section we discussed that Hubble-Lemaître law can be used to infer distances for objects with small redshifts. As we previously mentioned, the distance of a galaxy with $z < 0.3$ can be approximated to within 10% by using Equation 1.22. An accuracy of 1% requires that $z < 0.04$. There are many astronomical objects, though, that are at larger redshifts where this approximation is invalid, and so we need a method to infer the distance of an object with a large z. This calculation, as discussed earlier, depends on the expansion history of the Universe and does not have an analytical solution; a numerical integration must be performed.

The concept of an object's distance in an expanding universe is actually not so simple; during the time that photons travel from the object to the observer the distance changes. In one definition of distance, called the *proper distance* and denoted d_p, we imagine stopping the Universe's expansion and physically measuring the distance (with a cosmological tape measure, for example). The proper distance is a function of time and we denote the current proper distance as $d_p(t_0)$.

Although conceptually useful, we rarely use proper distance. In astronomy, there are two very common equations which use distance. The flux, F, of a source of radiation with luminosity L decreases with distance, d, as (see Equation 1.4)

$$F = \frac{L}{4\pi d^2},$$

and the apparent angular size, Θ, of a source of physical size, D, decreases with distance as

$$\Theta = \frac{D}{d}.$$

For objects at cosmological distances these equations are incorrect if we simply insert the current proper distance. For the relation between flux and luminosity, because of the expansion of space during their travel time, the photons are redshifted and the space between photons increases. Hence, the energies of the photons decrease by a factor of $1 + z$ and the rate that photons enter our telescope decreases by another factor of $1 + z$, and so the detected flux decreases by a factor of $(1 + z)^2$. If using the current proper distance, then, the relation between flux and luminosity becomes

$$F = \frac{L}{4\pi [d_p(t_0)]^2 (1 + z)^2}.$$

The angular size measurement is also altered by the expansion of space. Without explanation, when using a proper distance, an object's angular size relates to the object's physical size by

$$\Theta = \frac{D(1 + z)}{d_p(t_0)}.$$

Since the concept of distance is ambiguous, a common practice is to name and define a distance for each type of distance measurement. The distance that appears in the standard relation between flux and luminosity is called the *luminosity distance*, denoted d_L, and defined by

$$d_L = \sqrt{\frac{L}{4\pi F}}.$$

The luminosity distance is related to the current proper distance by

$$d_L = d_p(t_0)(1 + z). \tag{1.23}$$

The *angular size distance*, denoted by d_A, is defined by

$$d_A = \frac{D}{\Theta},$$

and is related to the current proper distance by

$$d_A = \frac{d_p(t_0)}{1 + z}. \tag{1.24}$$

Clearly, the luminosity distance and angular size distance are related by

$$d_L = d_A(1 + z)^2. \tag{1.25}$$

We now explain how to calculate the proper distance for a given z. We can then use either Equation 1.23 or 1.24 in our calculations involving distance, as needed. In this process, we will also show how to calculate the present age of the Universe, t_0, and the age when the light of redshift z was emitted, t_e. The difference of these ages gives how far into the past we are viewing, called the *look-back time*, $t_0 - t_e$.

The calculation of proper distance involves the light travel time while also accounting for the fact that all distances in the Universe were increasing while the light was in transit. The calculation, therefore, involves an integration over time of the light travel distance and including the shorter length scales in the past. In Section 1.5 we introduced the concept of the scale factor, $R(t)$, to describe the expansion of the Universe as a function of time. With the scale factor set equal to one today, Equation 1.21 shows a simple relation between the cosmological redshift of a distant object and the scale factor of the Universe at the time the light was emitted. The redshift, which is the measured quantity, provides information about the Universe's age when the light was emitted, and the values of cosmological parameters determine the time dependence of the scale factor.

Without explanation, the cosmological parameters needed are Hubble's constant, H_0, which describes the current expansion rate of the Universe, and the density parameters Ω_M and Ω_Λ, which are unitless measures of the relative amounts of matter (baryonic and dark) and dark energy. The density parameters are described in Chapter 10; for the discussion in this section, we only need to insert values of these parameters into the equations below. As of the writing of this book, precise values of Ω_M and Ω_Λ have not yet been firmly established; however all evidence suggests that the sum of these density parameters is unity (corresponding to a flat universe). Additionally, the exact form of dark energy and how its density depends on the scale factor are not well known. While the mass density depends on scale factor as $R(t)^{-3}$, the equations below assume that the dark energy density is constant with scale factor. Since the calculation of proper distance involves an integration over time, from the time of emission of the light to the time of observation (i.e., today), we need equations for these times. We also need the time dependence of the scale factor. Without derivation, these equations are as follows. The current age of the Universe, assuming both a non-zero value of Ω_Λ and that $\Omega_M + \Omega_\Lambda = 1$, is

$$t_0 = \frac{2}{3H_0\sqrt{\Omega_\Lambda}} \sinh^{-1}\left[\sqrt{\Omega_\Lambda/\Omega_M}\right], \tag{1.26}$$

and the age of the Universe when light at redshift z was emitted was

$$t_e = \frac{2}{3H_0\sqrt{\Omega_\Lambda}} \sinh^{-1}\left[\sqrt{\frac{\Omega_\Lambda}{\Omega_M}}\,\frac{1}{(1+z)^{3/2}}\right]. \tag{1.27}$$

The equation for the scale factor is

$$R(t) = \left[\sqrt{\frac{\Omega_M}{\Omega_\Lambda}}\sinh\left(1.5\sqrt{\Omega_\Lambda}H_0 t\right)\right]^{2/3}. \tag{1.28}$$

The proper distance, then, is given by

$$d_p = \int_{t_e}^{t_0} \frac{c}{R(t)}dt. \tag{1.29}$$

The value of H_0 for this calculation requires a simple conversion of units: $H_0 = 70$ km s^{-1} Mpc^{-1} $= 2.27 \times 10^{-18}$ s^{-1}. For the density parameters, we will use the approximately correct values of $\Omega_M = 0.3$ and $\Omega_\Lambda = 0.7$. Substituting these values into Equations 1.26-1.28 we get simpler expressions for t_0, t_e, and $R(t)$:

$$t_0 = 13.5 \text{ Gyr},$$

$$t_e = 11.1 \sinh^{-1}\left[\frac{1.53}{(1+z)^{3/2}}\right] \text{ Gyr},$$

$$R(t) = \left[0.655 \sinh\left(\frac{2.85 \times 10^{-18}}{\text{s}}\, t\right)\right]^{2/3},$$

with t in seconds. For t given in Gyr, we can rewrite $R(t)$ as

$$R(t) = \left[0.655 \sinh\left(\frac{0.0900}{\text{Gyr}}\, t\right)\right]^{2/3}.$$

In Appendix B we give a simple Mathematica[1] code for performing the integration in Equation 1.29.

Example 1.8:

(a) A distant astronomical object has a redshift $z = 1.0$. What was the age of the Universe when the light we receive from this object was emitted?

(b) What is the proper distance to this object?

Answer:
(a) The age of the Universe when light with redshift z was emitted is given by Equation 1.27. This gives us

$$t_e = 11.1 \sinh^{-1}\left[\frac{1.53}{(1+1.0)^{3/2}}\right] \text{ Gyr} = 5.7 \text{ Gyr}.$$

Since we inferred above the current age of the Universe to be about 13.5 Gyr, the light was emitted about 8 billion years ago.

(b) We obtain the proper distance from the numerical integration in Equation 1.29. Using the Mathematica code in Appendix B this yields

$$d_p = 3.30 \times 10^9 \text{ pc} = 3300 \text{ Mpc}.$$

Because calculating cosmological distances can be cumbersome, as it requires a numerical integration, a common temptation is to use the Hubble-Lemaître Law approximation to roughly estimate the distance. If we were to do that with the object in Example 1.8, we would get

$$d = \frac{cz}{H_0} = \frac{3 \times 10^5 \text{ km s}^{-1} \times 1.0}{70 \text{ km s}^{-1} \text{ Mpc}^{-1}} = 4286 \text{ Mpc},$$

which is 30% larger than the correctly calculated value. The error in this estimate based on the Hubble-Lemaître Law approximation increases with increasing redshift.

[1]Wolfram Research, Inc., Mathematica, Version 11.3, Champaign, IL (2018).

QUESTIONS AND PROBLEMS

1. An observation of an astronomical source of duration 100 seconds through a filter with a bandwidth of 1 MHz yields a total detected energy of 5×10^{-10} erg. If the observation were made with a telescope with a collecting area of 1000 cm^2, what is the flux density of the source?

2. A circular astronomical source has a radius of 1×10^{18} cm and is at a distance of 100 pc. If the source has a uniform intensity of $I_\nu = 1 \times 10^{-16}$ erg s^{-1} cm^{-2} Hz^{-1} sr^{-1}, what would be the flux density of this source?

3. The Sun has a total luminosity of 3.9×10^{33} erg s^{-1}. What is the flux of the Sun as measured on the Earth? Give the flux in both cgs and SI units.

4. We measure the flux density of an astronomical source to be 5 Jy at a frequency of 100 MHz. What is the equivalent flux density of the source if we measured it per unit wavelength interval instead of per unit frequency interval?

5. In Section 1.3.2, we provide the equatorial coordinates for the Galactic Center. The seconds of right ascension are given to two decimal places, while the seconds of declination are given to just one decimal place. Why do we need more precision in RA than we do in Dec?

6. In Section 1.3 we introduced the idea of expressing angular position (the right ascension) in terms of *time* rather than *angle*. However, there are some subtle issues involved when calculating *angular displacements* in right ascension. In particular, to convert from time differences to angular offsets, one must take the declination into account. The standard equivalence that 1 hour of time equals 15° of arc is only valid *on the equator*. To see how this works, calculate the angular displacement between positions 1 and 2, given by:

$$\text{Position 1} : \alpha_1 = 19^{\text{h}}25^{\text{m}}30.0^{\text{s}}; \quad \delta_1 = +60°00'00.''$$

$$\text{Position 2} : \alpha_2 = 19^{\text{h}}25^{\text{m}}50.0^{\text{s}}; \quad \delta_2 = +60°00'00.''$$

   What would the angular offset be if $\delta_1 = \delta_2 = +90°$?

7. The Earth is a moving platform and thus introduces a velocity relative to some astronomical source. First, compute the magnitude of the velocity of the Earth due to its rotation about its axis at the equator. Second, compute the magnitude of the velocity of the Earth in its orbit about the Sun. For a spectral line at 1420.4058 MHz, what is the maximum frequency shift due to the Doppler effect for each of these motions?

8. An important radio spectral line has a frequency of 1420.406 MHz. We observe two astronomical sources producing this spectral line. If we measure the frequency of the line in the first source to be 1420.845 MHz and in the second to be 1420.143 MHz, what can we infer about the velocities of these sources? What limitations do we face when dealing with their velocity vectors?

9. An astronomical source producing the radio spectral line at a frequency of 1420.406 MHz is moving purely tangentially with a value of 50,000 km s^{-1} or 5×10^9 cm s^{-1}. What frequency and wavelength would we observe for this spectral line?

10. We observe the same spectral line as in the previous problem in another source. For this source we measure a frequency of 500.0 MHz. If the motion is all in the radial direction, what is the radial velocity of this source? Is it moving toward or away from us?

11. We detect galaxies with redshifts of $z = 9$. What does that tell us about the change in the scale factor of the Universe between the time the light left one such galaxy and the time we detected the emission?

12. In a very distant galaxy we detect the spectral line emitted at 1420.406 MHz to be at a frequency of 1045.0 MHz. What is the redshift of the galaxy? What was the scale factor of the Universe when the light was emitted?

13. For the galaxy in Question 12, calculate the age of the Universe when the light we now see was emitted by this galaxy and the look-back time.

14. We observe the light from a distant galaxy which is at $z = 2$. Compute the proper distance of this galaxy and the age of the Universe when the light was emitted. Compare your answers with those for a $z = 1$ galaxy from Example 1.8.

CHAPTER 2

Propagation of Radiation

IN this chapter we describe the interactions between light and matter that are needed to understand how light propagates through matter. As we will discuss in later chapters, we can use the framework developed here to predict the emission produced by an astronomical source and to determine how that radiation might be altered by matter in space as it travels to our telescope. The process by which light propagates through a medium is called *radiative transfer*. We will discuss in Section 2.1 the transfer of radiation in general, and then in Section 2.2 discuss radiative transfer in an ionized and magnetic medium.

2.1 RADIATIVE TRANSFER

Although the interactions between light and matter occur on a microscopic scale, the photons pass through matter on a macroscopic scale. In this section we will produce a heuristic derivation of what is commonly called the equation of radiative transfer. We start this discussion by considering the ways in which matter can alter radiation that passes through it. Once electromagnetic radiation is emitted, if it passes through matter that matter can (1) absorb some of the radiation, (2) emit more radiation, (3) absorb, and then re-emit radiation at a new frequency, or (4) scatter some of the radiation into a new direction.

2.1.1 Absorption of Radiation

We first concentrate on a medium's ability to attenuate radiation by absorbing light. In this process the photons are destroyed and the photon energy is converted to internal energy of the absorbing medium. With this in mind, imagine a ray of light from a background source incident on an infinitesimally thin layer of a medium of thickness ds, and consider the attenuation of that radiation by the medium. This situation is illustrated in Figure 2.1. Let I_ν^o be the intensity of radiation incident on the medium at frequency ν, and dI_ν be the change in intensity of the radiation in passing through the layer. Since the medium removes radiation, dI_ν must be negative. Also, since the medium cannot remove more radiation than is incident on it, the amount of radiation absorbed must depend on the amount of incident radiation. In general, the layer of the medium will remove a certain fraction of the incident radiation. The fraction of incident radiation that is removed must depend on the distance that the radiation travels to pass through the layer, ds, and on the properties of the medium, most importantly its ability to absorb light. So, a reasonable and fairly simple equation for the change in intensity is

$$dI_\nu = -I_\nu^o \, \kappa_\nu \, ds, \tag{2.1}$$

where κ_ν is a macroscopic measure of the medium's ability to absorb the radiation at frequency ν per unit length of the medium. The units of κ_ν are inverse length, e.g. cm^{-1} or m^{-1}. The parameter κ_ν is often referred to as the *absorption coefficient*, or linear absorption coefficient, of the medium.

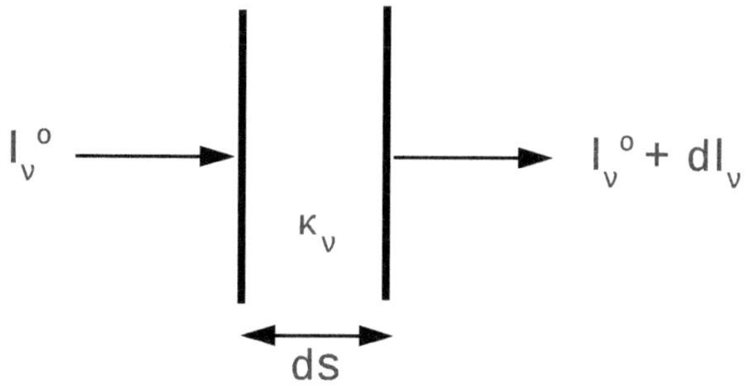

Figure 2.1: An infinitesimally thin layer of thickness ds of a medium that has an ability to absorb radiation characterized by an absorption coefficient κ_ν. The incident intensity of the radiation on the layer is I_ν^o and the emergent intensity is $I_\nu^o + dI_\nu$.

In addition to absorption, radiation can also be scattered by matter. In this scattering process, a photon interacts with the medium and emerges in a new direction and in some instances with slightly different wavelength or frequency. Therefore, like absorption, this process can reduce the intensity of light in the initial direction of propagation. Analogous to absorption, we can define a scattering coefficient that is a macroscopic measure of the medium's ability to scatter the radiation as a function of frequency. The combined effect of absorption and scattering in reducing the intensity of light can be described by an extinction coefficient, which is the sum of the absorption and scattering coefficients. For the sake of simplicity, in this section we will ignore scattering and consider only the absorption process.

A medium's ability to attenuate radiation will depend on the nature of the particles in the medium and also on the density of these particles. It is often convenient to modify Equation 2.1 so that the dependence on the density is shown explicitly. The density can be expressed in terms of a *number density* (the number of particles per unit volume of space), n, or a *mass density* (the mass of particles per unit volume of space), ρ. In the former case, we can define another parameter σ_ν which is related to κ_ν by

$$\kappa_\nu = \sigma_\nu \; n. \tag{2.2}$$

Then, the decrease in intensity is given by

$$dI_\nu = -I_\nu^o \; \sigma_\nu \; n \; ds. \tag{2.3}$$

Note that the units of σ_ν must be a length squared or an area (e.g. cm^2). In fact, in modeling the interactions between the photons and particles in the medium, the parameter σ_ν, which is called the *cross-section*, can be viewed as a measure of the cross-sectional area of a target, i.e., the particle, as viewed by the other interacting particle, in this case the photon, and is a function of frequency.

When it is the mass density that is shown explicitly, and not the number density, the ability of the medium to absorb or scatter the radiation is described by a parameter called the mass absorption coefficient or *opacity* and we will denote this by κ_ν^m, where the superscript m indicates this is a mass absorption coefficient. Note that the units of the absorption coefficient and mass absorption coefficient defined earlier are quite different. The opacity is a measure of the cross-section for interaction per unit mass of matter, and therefore is defined such that

$$\kappa_\nu^m = \sigma_\nu n/\rho = \kappa_\nu/\rho,$$

where ρ is the mass density (mass per unit volume) of the medium producing the absorption. The word opacity comes from the word opaque, which means that an object blocks light from passing through it. The units of opacity are cm^2 g^{-1}. Rewriting Equation 2.1 in terms of the mass absorption coefficient we have,

$$dI_\nu = -I_\nu^o\, \kappa_\nu^m \rho\, ds.$$

In the rest of this section, we will use only the absorption coefficient, κ_ν, to describe the absorption process.

There is a potential source of confusion about these two absorption coefficients. The symbols used to define these parameters are often switched in different textbooks. Some authors define the absorption coefficient by κ_ν, while others define the mass absorption coefficient or opacity by the same symbol. When consulting a textbook confirm which definition is used; the units (cm^{-1} or cm^2 g^{-1}) are a good indicator. For this book, we will use the definitions of absorption coefficient given above.

Now we return to our discussion of the attenuation of radiation as it passes through a medium. We obtain the amount of radiation that emerges from a medium with depth L, by considering the medium to be composed of many infinitesimal layers, each of thickness ds. Consider an arbitrary layer that has radiation I'_ν incident, and this layer produces a change of intensity of dI'_ν. Therefore, for this layer we have

$$dI'_\nu = -I'_\nu\, \kappa_\nu\, ds.$$

Rearranging, we get

$$\frac{dI'_\nu}{I'_\nu} = -\kappa_\nu\, ds.$$

To solve for the effect of the entire medium, we need to integrate the above equation. To obtain the limits of integration we assume that the medium extends from $s = 0$ to $s = L$ (the total depth of the medium) and that the incident intensity at $s = 0$ is I_ν^o and the emergent intensity (at $s = L$) is I_ν. Therefore,

$$\int_{I_\nu^o}^{I_\nu} \frac{dI'_\nu}{I'_\nu} = -\int_0^L \kappa_\nu\, ds.$$

Integrating the left-hand side, we find

$$\ln\left(\frac{I_\nu}{I_\nu^o}\right) = -\int_0^L \kappa_\nu\, ds.$$

An important quantity is the *optical depth*, denoted τ_ν, which is defined by

$$\tau_\nu = \int_0^L \kappa_\nu\, ds. \qquad (2.4)$$

It is helpful to remember the dependence of optical depth on the number density of the absorbers and so substituting in for κ_ν using Equation 2.2, we have

$$\tau_\nu = \int_0^L \sigma_\nu n \, ds. \tag{2.5}$$

If the absorption characteristics of the medium are uniform, then

$$\tau_\nu = \kappa_\nu L = \sigma_\nu n L. \tag{2.6}$$

The resultant intensity can then be written in terms of the optical depth as

$$I_\nu = I_\nu^o \, e^{-\tau_\nu}. \tag{2.7}$$

Example 2.1:

Assume the optical depths of three media are 0.1, 1.0 and 3.0 at frequency ν. Considering each medium separately, determine the fraction of the incident radiation at frequency ν that passes through each medium.

Answer:
The fraction of light at frequency ν that passes through the medium is I_ν / I_ν^o, which is given by $e^{-\tau_\nu}$. For $\tau_\nu = 0.1$, the fraction is $e^{-0.1} = 0.90$ (therefore 90% of the light passes through the medium), for $\tau_\nu = 1.0$, the fraction is $e^{-1.0} = 0.37$ (or 37%), and for $\tau_\nu = 3.0$, the fraction is $e^{-3.0} = 0.05$ (or 5%).

If τ_ν is small (i.e., less than about 0.1), then from Example 2.1 we see that the intensity exiting the medium is almost as large as the incident intensity and so most of the radiation passes through. In this case, we say that the medium is transparent or *optically thin.* If τ_ν is large (i.e., greater than about 3) then most of the radiation is blocked and the medium is considered opaque or *optically thick.* For values of τ_ν of order 1, the medium is foggy with some noticeable fraction of the background radiation passing through. In general, κ_ν, and hence τ_ν, depends on frequency, which is why we have written it with the subscript ν. A medium is often transparent at some frequencies and opaque at others.

2.1.2 Emission of Radiation

Any medium that absorbs light can also emit light. Therefore, to get a more general expression for the emergent intensity, we must account for any emission that was produced by the medium. We characterize the emission from the medium by a macroscopic quantity called the *emission coefficient* or sometimes just called the *emissivity.* The emission coefficient is a function of frequency and we denote this by j_ν. The quantity j_ν is the amount of intensity emitted at frequency ν per unit length of the medium. Therefore, the emission added by the medium in an infinitesimally thin layer, ds, is given by

$$dI'_\nu = j_\nu \, ds. \tag{2.8}$$

Since the emission coefficient is intensity per unit length, it will have units of energy per second per unit area per unit frequency per unit solid angle per unit length, and thus its cgs units are erg s^{-1} cm^{-2} Hz^{-1} sr^{-1} cm^{-1} or erg s^{-1} cm^{-3} Hz^{-1} sr^{-1} and its SI units are J s^{-1} m^{-3} Hz^{-1} sr^{-1}.

We first consider only emission from a medium assuming no incident radiation. However, we must still include absorption, as emission produced within the medium has to travel through part of the medium to escape, and will therefore be partially absorbed. Consider our medium to be composed of a large number of infinitesimally thin layers. One of the thin layers is located at a position such that the optical depth from the far side of the medium (relative to the observer) to the layer is given by τ_ν'. Therefore the emission from this layer must pass through an optical depth of $\tau_\nu - \tau_\nu'$ to escape, where τ_ν is the total optical depth of the medium as illustrated in Figure 2.2. Based on our discussion of absorption, we know that the emission from this layer will suffer an absorption by a factor of $e^{-(\tau_\nu - \tau_\nu')}$. Therefore the incremental intensity, dI_ν', that emerges from the medium due to this layer is given by

$$dI_\nu' = j_\nu \; ds \; e^{-(\tau_\nu - \tau_\nu')}.$$

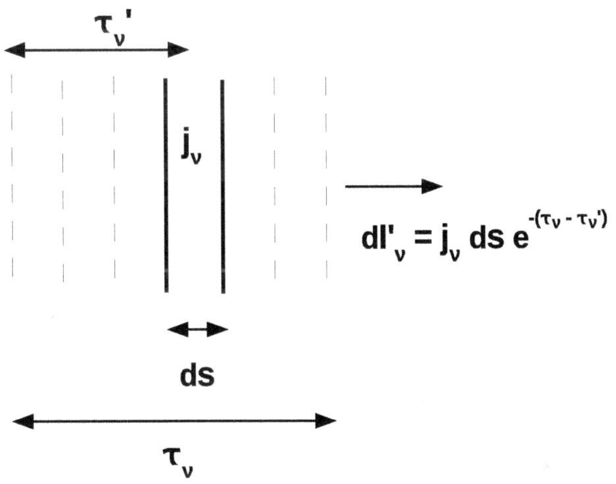

Figure 2.2: One of many thin layers of thickness ds in a medium characterized by an emission coefficient of j_ν. This thin layer is located at a position in the medium in which the optical depth from the far side of the medium to the layer is given by τ_ν'. The total optical depth of the medium is τ_ν; thus the emission produced in this thin layer must pass through an optical depth of $\tau_\nu - \tau_\nu'$ to escape.

We can find the total emission from the medium by summing the emission over all of these thin layers taking into account the subsequent absorption. The total emergent intensity is given by

$$I_\nu = \sum_i dI_\nu'(i) = \sum_i \; j_\nu(i) \; ds \; e^{-[\tau_\nu - \tau_\nu'(i)]},$$

where $j_\nu(i)$ is the emission coefficient of the ith layer and $\tau_\nu'(i)$ is the optical depth from the far side of the medium to the ith layer, and where we sum over all incremental layers. Since the incremental optical depth associated with a thin layer of thickness ds is just $d\tau_\nu'$ $= \kappa_\nu ds$, we can substitute for ds into the above sum to yield

$$I_\nu = \sum_i \; \frac{j_\nu(i)}{\kappa_\nu(i)} \; d\tau_\nu'(i) \; e^{-[\tau_\nu - \tau_\nu'(i)]}.$$

We can replace the summation with an integral, and rewrite the total emergent intensity as

$$I_\nu = \int_0^{\tau_\nu} \frac{j_\nu}{\kappa_\nu}\, d\tau'_\nu\, e^{-(\tau_\nu - \tau'_\nu)}, \tag{2.9}$$

where we are integrating over the thickness of the medium in optical depth units, and therefore the limits of integration go from 0 to the optical depth of the entire medium, τ_ν. Remember that j_ν and κ_ν may be functions of position (or, equivalently, of optical depth), although we do not show this dependence in Equation 2.9. If the quantity j_ν/κ_ν is independent of position (or optical depth), then this term can be taken outside of the integral, giving

$$I_\nu = \frac{j_\nu}{\kappa_\nu}\, e^{-\tau_\nu} \int_0^{\tau_\nu} e^{\tau'_\nu}\, d\tau'_\nu,$$

which upon integration gives

$$I_\nu = \frac{j_\nu}{\kappa_\nu}\, e^{-\tau_\nu}(e^{\tau_\nu} - 1).$$

Rearranging, we obtain

$$I_\nu = \frac{j_\nu}{\kappa_\nu}\,(1 - e^{-\tau_\nu}). \tag{2.10}$$

Equation 2.10 is not a general solution because of the assumption that $j_\nu/\kappa_\nu$is independent of position. We often lack information about how a medium changes along the line of sight, and in these cases, Equation 2.10 provides the emergent intensity in terms of the average values of j_ν and κ_ν. As we will see, even these averaged properties can provide valuable information about the emitting objects.

The ratio of the emission coefficient to the absorption coefficient is called the *source function* and given the symbol S_ν, and thus

$$S_\nu \;=\; j_\nu/\kappa_\nu. \tag{2.11}$$

The source function has units of intensity. Be careful not to confuse the source function with flux density, as radio astronomers often use the same symbol for both.

Note that if the optical depth of the medium is small (optically thin), then $e^{-\tau_\nu} \approx 1 - \tau_\nu$, and in this optically thin limit, we can approximate Equation 2.10 as

$$I_\nu \approx \frac{j_\nu}{\kappa_\nu}\,\tau_\nu \;\approx\; S_\nu\,\tau_\nu \;\text{(optically thin limit)}. \tag{2.12}$$

Since $\tau_\nu = \kappa_\nu L$, where L is the total thickness of the medium, the emission in this limit can be expressed simply as

$$I_\nu \;\approx\; j_\nu L \;\text{(optically thin limit)}. \tag{2.13}$$

In the optically thin limit, absorption by the medium is unimportant, and the intensity of the emission grows linearly with increasing thickness of the medium.

In the opposite limit, where the optical depth is very large (optically thick), then $e^{-\tau_\nu} \approx 0$, and the intensity approaches the source function:

$$I_\nu \;\approx\; \frac{j_\nu}{\kappa_\nu} \;\approx\; S_\nu \;\text{(optically thick limit)}. \tag{2.14}$$

In the optically thick case, emission and absorption in the medium compete, and the maximum emission is given by the source function.

To fully understand the emission from a medium we need to know the source function. In the next chapter we will discuss the important case of emission and absorption of radiation by thermal processes, which is referred to as thermal radiation. We will show that for thermal radiation the source function is equal to the *Planck function*, which is the intensity or brightness of a *blackbody*.

2.1.3 General Radiative Transfer Equation

We can now write the general equation for the emergent intensity from a medium, allowing for both absorption and emission within the medium and including incident radiation with intensity I_ν^o. This situation is illustrated in Figure 2.3. It should be obvious that the emergent intensity is given by the sum of two terms, one due to the attenuated incident radiation (Section 2.1.1) and the other due to the emission from the medium (Section 2.1.2). Thus, the general expression for the emergent intensity is

$$I_\nu = I_\nu^o\, e^{-\tau_\nu} + \int_0^{\tau_\nu} \frac{j_\nu}{\kappa_\nu}\, d\tau_\nu'\, e^{-(\tau_\nu - \tau_\nu')}. \tag{2.15}$$

The quantity j_ν/κ_ν is the source function and, as before, if it is independent of position we can bring it out of the integral. After integration the emergent intensity is then

$$I_\nu = I_\nu^o\, e^{-\tau_\nu} + S_\nu(1 - e^{-\tau_\nu}). \tag{2.16}$$

This equation is of fundamental importance and will be used often in coming chapters. It is worth noting that if the source function is larger than the incident intensity ($S_\nu > I_\nu^o$), then the intensity increases as the radiation passes through the source and so the emergent intensity will always be greater than the incident intensity. However, if the incident intensity is larger than the source function ($I_\nu^o > S_\nu$), then the radiation intensity decreases and the emergent intensity will always be less than the incident intensity.

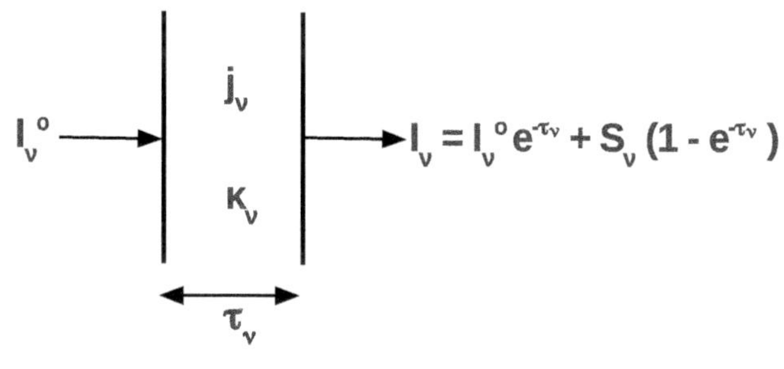

Figure 2.3: Radiation with intensity I_ν emanates from a medium characterized by an absorption coefficient, κ_ν, emission coefficient, j_ν, and with incident intensity of I_ν^o. The emergent intensity is composed of two terms; the first is the attenuated incident intensity and the second is the intensity emitted by the medium after correction for attenuation losses.

Example 2.2:

A spotlight beam of intensity of $I_\nu = 5.0 \times 10^{-8}$ erg s^{-1} cm^{-2} Hz^{-1} sr^{-1} at a frequency 5.0×10^{14} Hz is pointed at a cube of gas. On the opposite side of the cube, a detector, which responds only to light at 5.0×10^{14} Hz, measures the intensity of the light to be 3.0×10^{-8} erg s^{-1} cm^{-2} Hz^{-1} sr^{-1}. The spotlight is turned off and the intensity of the light emanating from the cube at the same frequency is then measured to be 2.0×10^{-8} ergs s^{-1} cm^{-2} Hz^{-1} sr^{-1}.

(a) What are the optical depth and source function of the cube of gas at this frequency?

(b) If the cross-section for interaction between the photons and gas particles is 5.7×10^{-18} cm^2, the cube is 50 cm in length and uniform in density, what is the number density of the gas particles?

(c) Compare the source function of the gas to the intensity of the spotlight beam and contrast these two objects as sources of light. What does the source function tell you about the foreground object?

Answer:
(a) We have two unknowns, the source function and the optical depth, and two different set-ups: spotlight *on* and spotlight *off*. For both situations, we use Equation 2.16. When the spotlight is *off* there is no incident intensity and so $I_\nu^o = 0$, yielding a fairly simple equation, i.e.,

$$I_\nu(off) = S_\nu \ (1 - e^{-\tau_\nu}).$$

When the spotlight is *on*, we have

$$I_\nu(on) = I_\nu^o \ e^{-\tau_\nu} \ + \ S_\nu \ (1 - e^{-\tau_\nu}).$$

By subtracting the first equation from the second we get an equation involving only the optical depth, i.e.,

$$I_\nu(on) - I_\nu(off) = I_\nu^o \ e^{-\tau_\nu}.$$

Solving for the optical depth and plugging in the observed intensities we have

$$\tau_\nu = -\ln[\frac{3.0 \times 10^{-8} - 2.0 \times 10^{-8}}{5.0 \times 10^{-8}}] = 1.61.$$

Now, we can substitute this value for the optical depth into the first equation and solve for the source function. We get

$$2.0 \times 10^{-8} \text{ erg s}^{-1} \text{ cm}^{-2} \text{ Hz}^{-1}\text{sr}^{-1} \ = \ S_\nu \ (1 - e^{-1.61}),$$

and therefore

$$S_\nu = 2.5 \times 10^{-8} \text{ erg s}^{-1} \text{ cm}^{-2} \text{ Hz}^{-1} \text{ sr}^{-1}.$$

(b) Using Equation 2.2, $\kappa_\nu = n\sigma_\nu$, where $\sigma_\nu = 5.70 \times 10^{-18}$ cm^2, Equation 2.4, assuming κ_ν is independent of position, $\tau_\nu = \kappa_\nu L$, and knowing that the number density is constant, we have $\tau_\nu = n\sigma_\nu L$. Using the optical depth from part (a), the number density is then

$$n = 1.61/(50.0 \text{ cm } \times 5.70 \times 10^{-18} \text{ cm}^2) \ = \ 5.65 \times 10^{15} \text{ gas particles per cm}^3.$$

(c) The source function of the gas is smaller than the intensity of the spotlight. This indicates that the gas is fundamentally a weaker light source at this frequency. The source function indicates the brightest that the gas could get on its own, and this is less than the intensity produced by the spotlight. As we will discuss in Chapter 3, a common determining factor in a source's radiation ability is its temperature. The gas, in this case, could be relatively cool, compared to the filament in the spotlight.

2.2 PROPAGATION IN AN IONIZED MEDIUM

As electromagnetic (EM) waves propagate through a medium with free electrons, those electrons will respond to the electric field of the wave. Similar to the manner in which the electric field due to the redistribution of mobile electrons in a conductor can cancel a static electric field inside the conductor, the movement of free electrons in an ionized medium in response to an electromagnetic wave alters the net fields. The medium must be electrically neutral, however, and so there must be as much positive charge as negative. Except in some special circumstance, the positive charge will be due to free protons and ions. Because these positive charges have much larger masses than the electrons (1840 times or greater), their accelerations in response to the electric field in the waves are much smaller. Their effect on the EM waves, therefore, is much less than that of the electrons. In comparison to the electrons, the effect of the protons is insignificant. Throughout the rest of this chapter, we will focus only on the free electrons and ignore the effect of the positive charges.

Free electrons pervade the interstellar medium of the Galaxy, and so correct analyses of observations of radio waves emitted by objects far away must include the effects of EM wave propagation through an ionized medium. The interpretations of some observations can be significantly altered if we do not account for these effects. These effects can also be used to provide additional probes for studying the interstellar medium itself. We will introduce and explain three concepts known as *plasma frequency*, *dispersion measure*, and *Faraday rotation*. In this section, we explain these phenomena conceptually, provide the relevant equations and give some examples. Discussion of the equations are provided in the following sections and the derivations are given in Appendix D.

2.2.1 Plasma Frequency

Imagine an electromagnetic wave approaching a region containing a number density of free electrons (number of electrons per unit volume), n_e . The electric field in the EM wave will accelerate these electrons, with charge $-e$, according to

$$\vec{a} = \frac{-e}{m_e}\vec{E}, \tag{2.17}$$

where m_e is the mass of the electron. There is also an interaction with the magnetic field in the wave but that acceleration is so much smaller that it can be ignored. Since the electric field in the wave oscillates sinusoidally, the accelerations of the electrons must also oscillate, changing direction periodically and varying in magnitude sinusoidally. The resulting velocities and positions of the electrons will also oscillate, with the same period as the EM wave. As the electrons' positions change, they will produce a net electric field which will also oscillate but in the opposite sense as the electric field of the incoming EM wave. If the frequency is low enough, the free electrons will be able to respond rapidly enough to the oscillating electric field of the electromagnetic wave to completely cancel the wave inside the medium. Outside the medium, at the incident surface, however, the motion of the electrons will produce an oscillating field propagating in the opposite direction, away from the surface. The wave, in this case, is completely reflected. The lowest frequency that can propagate through the ionized medium and not be reflected is called the *plasma frequency*.

At frequencies higher than the plasma frequency, the electric field changes direction so rapidly that the electrons cannot move enough during the short periods to cancel the electric field of the wave. However, a larger density of electrons will produce a larger total electric field such that, even with the short acceleration periods, the electric field of the wave can be canceled. Therefore, the plasma frequency must depend on the density of free electrons.

The plasma frequency is a measure of the maximum possible response rate of a particular density of free electrons to an applied oscillating electric field. A higher plasma frequency implies that there are more electrons per unit volume to counter an applied electric field. Likewise, higher-frequency electromagnetic waves need more electrons to completely reflect the incident wave.

We derive the plasma frequency equation in Appendix D, in which we show that the plasma frequency, in radians per second, is given in cgs units by

$$\omega_{\mathrm{p}} = \sqrt{\frac{4\pi n_e e^2}{\epsilon m_e}}, \tag{2.18}$$

where ϵ is the electric permittivity of the medium, described in Appendix D. In a tenuous plasma $\epsilon \sim 1$ and so the plasma frequency is often written without the ϵ in Equation 2.18. We can express ω_p in units of observing frequency by dividing by 2π. Inserting the values of the constants and assuming $\epsilon \sim 1$ we have

$$\nu_{\mathrm{p}} = 8.97\ \mathrm{kHz}\ \sqrt{\frac{n_e}{\mathrm{cm}^{-3}}}. \tag{2.19}$$

The plasma frequency of the Earth's ionosphere is of special interest because it determines the lowest frequency radiation that can reach ground-based telescopes. Additionally, the plasma frequency of the interstellar medium indicates the lowest frequency that can propagate through the Galaxy.

Example 2.3:

(a) What is the lowest frequency extraterrestrial EM radiation that can be detected at the Earth's surface? Although, the Earth's ionosphere is highly variable, a reasonable number density of free electrons is $n_e \approx 10^6\ \mathrm{cm}^{-3}$.

(b) The interstellar medium (the gas between stars) in the Galaxy contains free electrons of varying density. The majority of the interstellar medium has a free electron density less than 1 cm^{-3}. Use this density to calculate the plasma frequency of interstellar regions in general.

Answer:

(a) Radiation lower than the ionosphere's plasma frequency will be reflected and so this sets the lower frequency limit of detectable radio waves coming from space.

Using Equation 2.19 with $n_e = 10^6$ cm^{-3} we find

$$\nu_\mathrm{p} = 8.97 \text{ kHz} \sqrt{\frac{10^6 \text{ cm}^{-3}}{\text{cm}^{-3}}} \approx 9000 \text{ kHz} \approx 9 \text{ MHz}.$$

(b) Again using Equation 2.19, but this time with $n_e = 1$ cm^{-3}, we get

$$\nu_\mathrm{p} = 8.97 \text{ kHz} \sqrt{\frac{1 \text{ cm}^{-3}}{\text{cm}^{-3}}} \approx 9 \text{ kHz}.$$

The reflection of lower frequency waves by the free electrons in the Earth's ionosphere sets the lower frequency limit of the radio window, which is approximately 9 MHz, as calculated in Example 2.3. Additionally, the number density of electrons in the interstellar medium (ISM) is less than that in the Earth's ionosphere, and so the plasma frequency of the ISM is not relevant to Earth-based radio observations, since any frequency that will pass through the Earth's ionosphere can pass through the ISM. Of course, electromagnetic waves can also be reflected by the lower surface of the ionosphere, which permits long-distance radio communication on the Earth at frequencies below ν_p of the ionosphere.

2.2.2 Dispersion Measure

Waves of different frequencies will not propagate identically through a medium with free electrons. As we showed in Section 2.2.1, for example, waves below the plasma frequency will not propagate at all. Similarly, we show in Appendix D that for waves above the plasma frequency, those of higher frequency will propagate faster than those of lower frequency. This becomes important when observing a short pulse of radiation. A pulse has a very short extent in time and, by the Fourier relation between time and frequency, must have a large range of frequencies (see Appendix E). When a pulse travels a large distance through the interstellar medium, the lower frequencies of the pulse will travel more slowly and will arrive at the Earth later than the higher frequencies of the pulse. The pulse, as a packet in time, is *dispersed* during its travel.

The astrophysical application of the dispersion of a pulse is mostly relevant to observations of signals emitted by pulsars (compact remnants of massive stars that have extremely large magnetic fields). We discuss pulsars in Chapter 7, and will refer back to the equations and concepts presented here. As the name implies, pulsar signals are very regular bursts of extremely short pulses. These signals are also polarized, which will be relevant in Section 2.2.3.

In practice, the radio astronomer probes the dispersion by measuring the arrival times of a pulse as a function of frequency and infers from these times the average density of electrons along the line of sight. The time for an electromagnetic wave of a given frequency to travel a distance L is given by

$$t_\mathrm{tr} = \sqrt{\mu\epsilon}\,\frac{L}{c} + \frac{\sqrt{\mu\epsilon}\,e^2}{2\pi\epsilon m_e c}\,\frac{1}{\nu^2}\int_0^L n_e ds, \tag{2.20}$$

where μ is the magnetic permeability (described in Appendix D). For a tenuous plasma $\epsilon \approx 1$ and $\mu \approx 1$ and this is applicable for most of the discussion in this book. The other parameters appearing in Equation 2.20 are defined in Section 2.2.1. The mathematical derivation of Equation 2.20 is given in Appendix D. We see that the amount of delay depends on the frequency of the radiation. The lower the frequency, the more effectively the electrons respond to the electric field of the wave, and hence the slower the wave travels.

Now consider measuring the arrival time of a pulse at two different frequencies, ν_1 and ν_2. Since the first term in Equation 2.20 is independent of frequency, the difference in their travel times due to the response of free electrons along the line of sight is

$$t_{\rm tr}(\nu_2) - t_{\rm tr}(\nu_1) = \frac{\sqrt{\mu\epsilon}\, e^2}{2\pi\epsilon m_e c}\left(\frac{1}{\nu_2^2} - \frac{1}{\nu_1^2}\right)\int_0^L n_e\, ds. \tag{2.21}$$

We see that the difference in the arrival times depends on the integral

$$\int_0^L n_e\, ds.$$

This type of integral – of density over path length – is extremely common in astronomical measurements and so is given a name, called the *column density*, and commonly denoted by N. Astronomers frequently cannot separate the dependencies on number density and path length, and often have to settle for a combination of the two, and that is the column density. In the case at hand, the spread of arrival times of a pulse depends on the column density of free electrons, and so we will denote this as N_e,

$$N_e = \int_0^L n_e\, ds \tag{2.22}$$

One can think of N_e as a measure of the total number of free electrons in a column of unit area along the line of sight to the pulsar. In Chapter 4 we will encounter equations containing the column densities of atoms and molecules. Note that the dimensions of column density are cm^{-2}. However, since densities are given with units of cm^{-3} and distances in pc, column density is sometimes expressed in units of pc cm^{-3}; this helps to remind us that this is not an inverse area but a density times a distance.

The difference in arrival time of two different frequencies, we see, depends on three factors. The first factor is simply a combination of constants and the second factor depends solely on the two observing frequencies, which are known for any given observation. It is the third factor, the column density of free electrons, that varies and determines the amount of dispersion of a pulse from any particular pulsar. The column density of free electrons is also called *dispersion measure*, which we abbreviate as DM,

$$DM \equiv \int_0^L n_e\, ds = \langle n_e \rangle L, \tag{2.23}$$

where $\langle n_e \rangle$ represents the average electron density along the path length, L. We get an expression for DM in terms of the measured values by re-arranging Equation 2.21. Substituting in for the constants, approximating for a tenuous plasma we set $\mu \approx 1$ and $\epsilon \approx 1$, and converting to convenient measurement units, the dispersion measure is given by

$$DM = 2.41 \times 10^{-4}\ \text{cm}^{-3}\ \text{pc}\left(\frac{\Delta t_{\rm tr}}{\text{s}}\right)\left(\frac{1}{(\nu_2/\text{MHz})^2} - \frac{1}{(\nu_1/\text{MHz})^2}\right)^{-1}. \tag{2.24}$$

Example 2.4:

A pulsar is at a distance of 200 pc, and the average free electron density along the path from this pulsar to Earth is 8.00×10^{-2} cm^{-3}. This pulsar is observed at 500 and 600 MHz.

(a) At which frequency will the pulses arrive at Earth first?

(b) What is the dispersion measure to this pulsar?

(c) What is the difference in the arrival times of the pulse at the two frequencies?

Answers:
(a) As evident in Equation 2.20, the lower frequency waves have a longer time delay, so the higher frequency, 600 MHz, arrives first.

(b) Using Equation 2.23 we find that the dispersion measure is

$$DM = (0.0800\ \mathrm{cm}^{-3})\ (200\ \mathrm{pc}) = 16.0\ \mathrm{cm}^{-3}\ \mathrm{pc}.$$

By the definition of dispersion measure, this is equal to the column density of free electrons along the line of sight to the pulsar.

(c) Inverting Equation 2.24 we get

$$\Delta t_{\mathrm{tr}} = \left(\frac{16.0}{\mathrm{cm}^{-3}\ \mathrm{pc}}\right)\left(\frac{1}{(500\ \mathrm{MHz/MHz})^2} - \frac{1}{(600\ \mathrm{MHz/MHz})^2}\right)\frac{1}{2.41\times10^{-4}}\ \mathrm{s}$$

$$\Delta t_{\mathrm{tr}} = 0.0811\ \mathrm{s}.$$

2.2.3 Faraday Rotation

We now consider an electromagnetic wave incident on a region of space containing both free electrons and a magnetic field. The free electrons, with charge $-e$ and mass m_e, are accelerated by the electric field, $\vec{E}$, of the wave as given by Equation 2.17. The electrons gain a velocity, $\vec{v}_e$, in the direction opposite to $\vec{E}$, which is perpendicular to the wave propagation direction. In the presence of an external magnetic field, $\vec{B}$, with the newly gained velocity component, the electrons also experience acceleration given by

$$\vec{a}_{\mathrm{mag}} = \frac{-e}{m_e}\left(\frac{\vec{v}_e}{c}\times\vec{B}\right).$$

This equation is written in cgs units; you may have seen this equation before without the 'c,' which is the proper form in SI units. This situation is depicted in Figure 2.4.

Consider the component of $\vec{B}$ parallel to the wave's direction of propagation, which we label as $B_{||}$. If the magnetic field direction is such that $B_{||}$ is opposite the propagation direction of the EM wave, then $B_{||}$ is a negative quantity. The acceleration, then, is perpendicular to $B_{||}$, causing the electron to follow an arc around the direction of propagation. More specifically, applying the right-hand rule (remembering the negative sign of the electron), regardless of the direction of $\vec{E}$, the direction of the acceleration due to a positive $B_{||}$ is to turn the electron in the clockwise direction from the perspective of an observer looking in the direction of propagation of the wave (see right panel in Figure 2.4). The electron, therefore, cycles in the same direction as a right-circularly polarized (RCP) wave, in the radio astronomy convention for circular polarization (see Section 1.2.5). Conversely, if $B_{||}$ is negative, then the electron will cycle in the opposite direction. Therefore, because of the

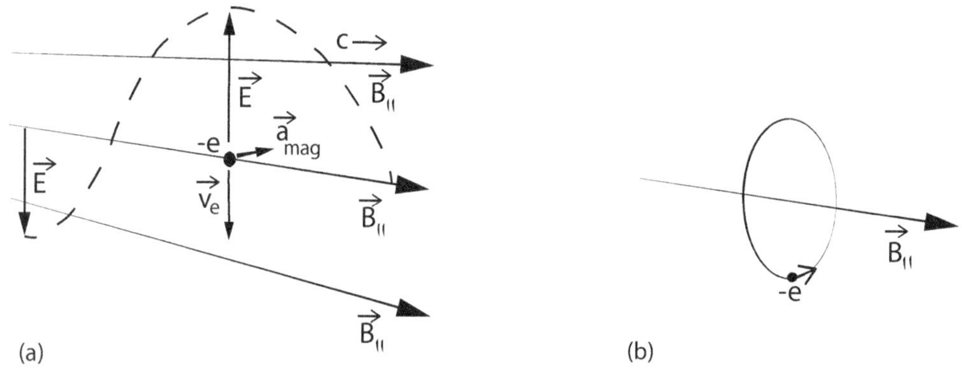

Figure 2.4: (a) An incident linearly-polarized electromagnetic wave accelerates a free electron in the direction opposite the electric field in the wave. The component of the magnetic field in the medium in the direction of wave propagation ($B_{||}$) then accelerates the electron in the direction perpendicular to both the electric field and direction of propagation. (b) For a wave propagating in the direction of $B_{||}$, the magnetic force causes the electron to curve around the magnetic field line in the same sense as a right-circularly polarized wave.

combined effect of the electric field of the wave and the $B_{||}$ in the medium, the electrons move in circles in the same direction that the electric field of the RCP waves cycle, and in the opposite direction of the LCP waves. Similarly, if $B_{||}$ is pointed in the opposite direction, the electrons cycle in the same sense as the LCP waves.

An interesting situation, that leads to a measurable effect, is the propagation of a linearly polarized electromagnetic wave in this medium. Recall that a linearly polarized wave can be described as the sum of right and left circularly polarized waves (see Section 1.2.5) and the relative phase of the RCP and LCP waves determines the angle of linear polarization. Since the electrons will respond to the combination of the wave's electric field and the medium's magnetic field with cyclical motion in the same sense as the RCP waves (if $B_{||}$ is in the direction of propagation), their response has a greater effect on the propagation of the RCP waves than on the propagation of the LCP waves. Following our discussion of the dispersion measure, we infer that the RCP waves propagate more slowly than the LCP waves and that there is a larger arrival delay of the RCP waves than the LCP waves.

The slower propagation speed of the RCP waves, along with the familiar relation between frequency and wavelength $c = \lambda\nu$, also means that they have a shorter wavelength in this medium than the LCP waves of the same frequency. The path length from source to observer, then, must contain a larger number of RCP wavelengths than LCP wavelengths. In other words, the RCP waves experience a greater phase shift per unit distance than the LCP waves and, hence, the polarization angle of the linearly polarized wave rotates with distance from the source. This is known as *Faraday rotation*. Note that if the medium's magnetic field is pointed opposite the direction of propagation of the EM wave, then it is the LCP wave that propagates more slowly.

The amount of rotation (in radians) of the linear polarization of an electromagnetic wave, derived in Appendix D, is given by

$$\Delta\theta = \frac{\sqrt{\mu}\ e^3}{2\pi\sqrt{\epsilon}\ m_e^2 c^2}\frac{1}{\nu^2}\int_0^L n_e B_{||} ds, \tag{2.25}$$

where $B_{||}$ is the component of the magnetic field along the direction of propagation of the radiation.

We see in Equation 2.25 that the amount of polarization rotation is smaller at higher frequencies. As we saw in our discussion of dispersion measure, the electrons are less able to respond to higher frequency electromagnetic waves. In the case of Faraday rotation, the frequency of the electrons' cycling motion will not match that of the circularly polarized wave. In fact, the frequency at which electrons cycle around in a given magnetic field is well-known and easily determined. This is known as the cyclotron frequency, discussed in Section 3.3.1, and is given by

$$\omega_c = \frac{eB}{m_e c}$$

where the units of the cyclotron frequency is radians per second. In hertz, the cyclotron frequency is

$$\nu_c = \frac{1}{2\pi}\frac{eB}{m_e c}.$$

For reasonable values of B in the interstellar medium (of order μgauss), the cyclotron frequency is quite small, of order tens of hertz, and so the frequencies of waves we observe will all be much greater than the cyclotron frequency. Furthermore, the difference between the cycling frequency of the electrons and that of the waves will be greater at the higher frequencies. Therefore, the difference between the propagation of the RCP and LCP waves will be smaller for higher frequency waves, and, hence, the rotation of the polarization angle of a linearly polarized wave will be smaller at higher frequencies.

One can measure this effect by observing linearly polarized radiation at two different frequencies, ν_1 and ν_2. The difference in the measured polarization angles will be

$$\theta(\nu_2) - \theta(\nu_1) = \frac{\sqrt{\mu}\, e^3}{2\pi\sqrt{\epsilon}\, m_e^2 c^2}\left(\frac{1}{\nu_2^2} - \frac{1}{\nu_1^2}\right)\int_0^L n_e B_{||} ds.$$

In terms of wavelengths, this is

$$\theta(\lambda_2) - \theta(\lambda_1) = \frac{\sqrt{\mu}\, e^3}{2\pi\sqrt{\epsilon}\, m_e^2 c^4}(\lambda_2^2 - \lambda_1^2)\int_0^L n_e B_{||} ds. \tag{2.26}$$

Putting the observables together on the left hand side, we get an expression known as the *rotation measure* (RM).

$$RM \equiv \frac{\theta(\lambda_2) - \theta(\lambda_1)}{\lambda_2^2 - \lambda_1^2} = \frac{\sqrt{\mu}\, e^3}{2\pi\sqrt{\epsilon}\, m_e^2 c^4}\int_0^L n_e B_{||} ds. \tag{2.27}$$

The integral is similar to that in the dispersion measure that we encountered in Equation 2.21, but now includes the line-of-sight magnetic field. As in the case of the dispersion measure, we generally do not know n_e or $B_{||}$ individually as a function of s, but must content ourselves with the integral of their product.

It is often convenient to approximate the integral in terms of the average free electron density, $\langle n_e \rangle$, and the average parallel magnetic field, $\langle B_{||} \rangle$. In these terms,

$$RM \approx \frac{\sqrt{\mu}\, e^3}{2\pi\sqrt{\epsilon}\, m_e^2 c^4}\langle B_{||}\rangle\langle n_e\rangle L. \tag{2.28}$$

Inserting the constants, setting $\epsilon \sim \mu \sim 1$, and converting to convenient units of measurement, the rotation measure can be written as

$$RM = 80.9 \text{ rad cm}^{-2}\left(\frac{\langle B_{||}\rangle}{\text{gauss}}\right)\left(\frac{\langle n_e\rangle}{\text{cm}^{-3}}\frac{L}{\text{pc}}\right). \tag{2.29}$$

The rotation measure can be useful for inferring magnetic fields. With multi-frequency, polarization observations, one determines the rotation measure, and if one can also obtain the dispersion measure, as can be done in pulsar observations, then by dividing the rotation measure (Equation 2.29) by the dispersion measure (Equation 2.23) one obtains the average magnetic field along the line of sight,

$$\langle B_{||} \rangle = 0.0124 \text{ gauss} \left(\frac{RM}{\text{rad cm}^{-2}} \right) \left(\frac{DM}{\text{cm}^{-3} \text{ pc}} \right)^{-1} . \tag{2.30}$$

Example 2.5:

A polarization-sensitive observation at 5000 MHz with a bandwidth of 250 MHz is made of the pulsar in Example 2.4, which is at a distance of 200 pc with an average free electron density along the path to Earth of 8.00×10^{-2} cm^{-3}. The polarization angle of the radiation across the bandpass is found to rotate by 5.0°. What is the average parallel magnetic field, which is often called the magnetic field along the line of sight, to this pulsar?

Answer:
The average magnetic field along the line of sight is given by calculating RM/DM and inserting into Equation 2.30. The DM was determined in Example 2.4 to be 16.0 cm^{-3} pc. So, we now only need to determine RM, as defined in 2.27. The wavelengths at the edges of the observed bandpass are

$$3.00 \times 10^{10} \text{ cm s}^{-1}/4.875 \times 10^{9} \text{ Hz} = 6.15 \text{ cm}$$

and

$$3.00 \times 10^{10} \text{ cm s}^{-1}/5.125 \times 10^{9} \text{ Hz} = 5.85 \text{ cm},$$

and the given $\Delta\theta_{12}$ is $5.00°(\pi/180°) = 8.73 \times 10^{-2}$ radians. With these data, the RM is

$$RM = \frac{8.73 \times 10^{-2} \text{ rad}}{(6.15 \text{ cm})^2 - (5.85 \text{ cm})^2} = 2.42 \times 10^{-2} \text{ rad cm}^{-2}.$$

So the average line-of-sight magnetic field, given by Equation 2.30, is

$$\langle B_{||} \rangle = 0.0124 \text{ gauss} \left(\frac{2.42 \times 10^{-2} \text{ rad cm}^{-2}}{\text{rad cm}^{-2}} \right) \left(\frac{16.0 \text{ cm}^{-3} \text{ pc}}{\text{cm}^{-3} \text{ pc}} \right)^{-1} ,$$

and so

$$\langle B_{||} \rangle = 1.9 \times 10^{-5} \text{ gauss}.$$

Note that this is only the component of B along the line of sight. The total magnetic field is almost certainly larger.

QUESTIONS AND PROBLEMS

1. Dust in the interstellar medium of the Galaxy can attenuate the visible light emission we observe from stars. Assume that at visible wavelengths the absorption coefficient due to dust in the interstellar medium is 6.3×10^{-22} cm^{-1}. If a star is 100 parsecs away, what is the optical depth due to the dust between us and the star? How much is the star's visible light attenuated due to dust?

2. In Equation 2.12 we make use of the optically thin approximation, where $1 - e^{-\tau} \approx \tau$. How small does τ have to be for this approximation to be accurate to 10%? to 1%?

3. At a wavelength of 100 cm an astronomical radio source is characterized by absorption coefficient $\kappa_\nu = 1.3 \times 10^{-20}$ cm^{-1} and source function $S_\nu = 7.7 \times 10^{-14}$ erg s^{-1} cm^{-2} Hz^{-1} sr^{-1}. If the depth of the radio source is 1 pc, what is the intensity of the radiation emitted by the source? How would the intensity change as the depth of the source is increased to 10 pc? to 100 pc? If the depth of the source were to increase beyond 100 pc would the intensity continue to increase?

4. Behind the source in Question 3 is a second astronomical source with an intensity at 100-cm wavelength of $I_\nu = 2.0 \times 10^{-14}$ erg s^{-1} cm^{-2} Hz^{-1} sr^{-1}. What is the total intensity if the depth of the first source were 1 pc, 10 pc, and 100 pc?

5. Imagine a spherically-symmetric radiating source as depicted in Figure 2.5. A core region is opaque at all wavelengths and emits radiation with a constant intensity I_0(core) at all frequencies. The surrounding shell of gas is cooler. The radii of the core and shell are R_{core} and R_{shell}, respectively. The gas in the shell has an emission coefficient $j_{\nu 1}$ and absorption coefficient $\kappa_{\nu 1}$ at frequency ν_1 and is an insignificant absorber at frequency ν_2.

   (a) What is the intensity of the radiation at ν_2 emanating from path A?

   (b) Derive an expression for the intensity at ν_1 emanating from path A as a function of the given parameters.

   (c) Derive an expression for the intensity at ν_1 emanating from path B. (Ignore the possibility of radiation emitted from the core being scattered by gas in the outer shell into path B.)

   (d) Of the intensities in the above questions, which do you expect to be the largest and which the smallest? Explain why.

6. Calculate the plasma frequency of the Sun's atmosphere just above the photosphere, using (and citing) any reliable reference source to obtain the values of any needed parameters. What are the implications for radio observations of the Sun?

7. A radio source is found to emit a continuum spectrum which suddenly drops to zero at frequencies below 200 MHz. Assuming this is due to blocking of radiation by a plasma screen, infer the free electron density in this screen.

8. (a) In the travel of electromagnetic waves of 1-GHz frequency, what column density of free electrons is needed to cause a delay of the arrival equal to one period of the wave?

   (b) If this travel occurs through a medium with 0.1 free electrons per cm^3, what path-length through this medium is needed?

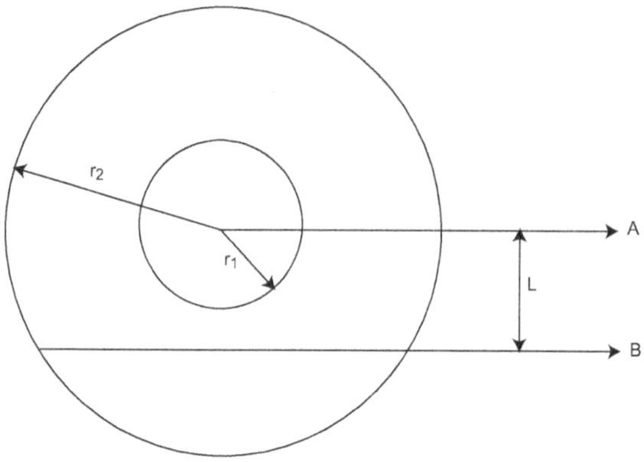

Figure 2.5: Radiation emanates along two separate paths, A and B, from a spherically-symmetric, but inhomogeneous source of radiation.

9. At ten solar radii (10 $R_\odot$), the Sun's corona has an average free electron density of 8000 cm^{-3} and magnetic field of $\approx$ 0.01 gauss. Assume that the electron density and the magnetic field strength are constant along the path length through the Sun's corona. If a pulsar were observed when the line of sight passed through the Sun's corona, would Faraday rotation due to the Sun's corona be detectable? What path length through a medium with these values of n_e and $B_{||}$ would yield a rotation measure of 1°?

10. A plane polarized pulse is observed over a range of wavelengths from 8.0 cm to 14.0 cm. The arrival time of the pulse at 14.0 cm is observed to occur 40 milliseconds later than at 8.0 cm. Also, the polarization position angle of the radiation changes with wavelength and is found to differ between 8 and 14 cm by 60°.

    (a) Estimate the free-electron column density along the path length of the pulse. Give your answer with units cm^{-3} pc and with cm^{-2}.

    (b) Calculate the average line-of-sight magnetic field along this path.

11. Explain why the plasma frequency is dependent on the column density of only free electrons, and not bound electrons? Why is the column density of free protons not important?

12. Why does Faraday rotation depend only on the component of B along the line of sight and not the entire B field?

13. If the charge of the electron were magically increased by a factor of two and the mass increased four-fold, how would that affect plasma frequency values, measurements of the dispersion of a pulse, and the Faraday rotation of a polarized signal?

14. Imagine a Universe in which protons are as low mass as electrons. What effect would this have on:

    (a) plasma frequency?

    (b) dispersion measure?

    (c) Faraday rotation?

CHAPTER 3

Continuum Emission Processes

In this chapter we discuss the physics of the radiation mechanisms that produce continuous spectra. The emission of a continuous spectrum (or continuum) usually involves a large number of emitting particles with a range of energies. As we explain shortly, the emitting particles are almost always electrons. However, the distribution of the electron energies as well as the emission mechanism can be either *thermal* or *non-thermal*. The energy distribution is called "thermal" if it can be described by a statistical function based on thermodynamic equilibrium, and is governed by one parameter — the temperature. Thermal emission occurs as a consequence of the random (or "stochastic") electromagnetic interactions, or "collisions," between charged particles with thermal motions. A thermal energy distribution of electrons emits radiation by these random processes producing *thermal radiation*. This emission can be directly linked to the temperature of the electrons. Other emission processes exist that are not governed by temperature, and these produce *non-thermal radiation*. A form of non-thermal continuum emission, called synchrotron emission, occurs when relativistic electrons are accelerated by a magnetic field. In this case the emission produced is not dependent on a temperature. In this chapter, we will model the production of radiation in these two cases. First, though, we outline the basic idea of the production of radiation at the particle scale.

3.1 RADIATION FROM ACCELERATED CHARGES

Classically, electromagnetic radiation is produced by accelerating charges. We will not go into the mathematical details of the physics here, but instead provide a conceptual presentation of this process. The mathematical details can be found in an upper level electricity and magnetism text.

To start, consider the electric field of a stationary isolated charge whose electric field lines are purely radial. Panel (a) in Figure 3.1 illustrates the field lines in this case. Now imagine this charge undergoes a brief acceleration upwards, and a short time later when the charge is moving at a constant velocity we re-examine the field lines. The field lines at this later time are illustrated in panel (b) of Figure 3.1. The acceleration of the charge must cause a change in its field, and that change propagates outward through space at the speed of light. The field seen by observers near the charge corresponds to the charge after acceleration; however, far away the field is that of the charge before acceleration. The field lines during the acceleration, which lie between these two regions, are kinked and in this transition region the electric field has both a radial and tangential component. We associate the tangential component with the electric field of a propagating electromagnetic wave.

The magnitude of the tangential component of the electric field determines the strength of the electromagnetic radiation. As can be seen in panel (b) in Figure 3.1, the tangential

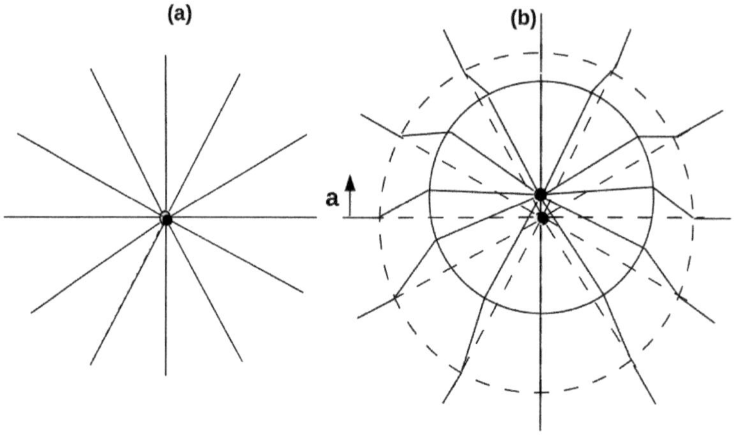

Figure 3.1: Panel (a) shows the electric field lines from an isolated stationary charge. We assume that this charge is briefly accelerated upwards, and in panel (b) the field lines are shown a short time later, after the acceleration ends. In panel (b), near the charge, the electric field lines correspond to the charge after acceleration while far away the field lines correspond to the charge before acceleration. The acceleration of the charge produces kinked electric field lines that are composed of both radial and tangential components. The tangential component is associated with the radiation produced by the accelerated charge and propagates outward at the speed of light.

component of the electric field varies with direction, and thus the strength of the radiation produced is not isotropic. Note that along the axis in which the charge was accelerated the tangential component of the electric field is zero; thus there is no radiation emitted in either direction along the acceleration axis. The largest tangential component of the electric field occurs perpendicular to the acceleration; the strongest emission is produced in these directions.

We can use the coordinate system shown in Figure 3.2 to define the tangential component of the electric field as a function of direction given by angles θ and ϕ. For a small acceleration, a, (and a non-relativistic particle), one can derive from geometrical arguments that the tangential component of the electric field in direction (θ, ϕ) and at a distance r, is given by

$$E_{\text{tangential}} = \frac{qa \sin\theta}{rc^2},$$

where q is the charge. Note that there is no dependence on the angle ϕ, so the tangential field for a given angle θ and distance r is the same for all angles ϕ. The power radiated per unit area in each direction is proportional to the square of $E_{\text{tangential}}$ and is called the Poynting flux (S). The Poynting flux in this case can be expressed as

$$S = \frac{1}{4\pi} \frac{q^2 a^2}{c^3} \frac{\sin^2\theta}{r^2}.$$

The pattern of emitted flux per unit area is a dipole pattern, which has a torus or doughnut shape. The Poynting flux is largest where $\theta = \pi/2$ radians and is zero in the directions where $\theta = 0$ or π radians. Finally, we can integrate the Poynting flux over the surface area of a sphere of radius r to give the total power radiated by an accelerated charge

$$P = \frac{2q^2 a^2}{3c^3}. \tag{3.1}$$

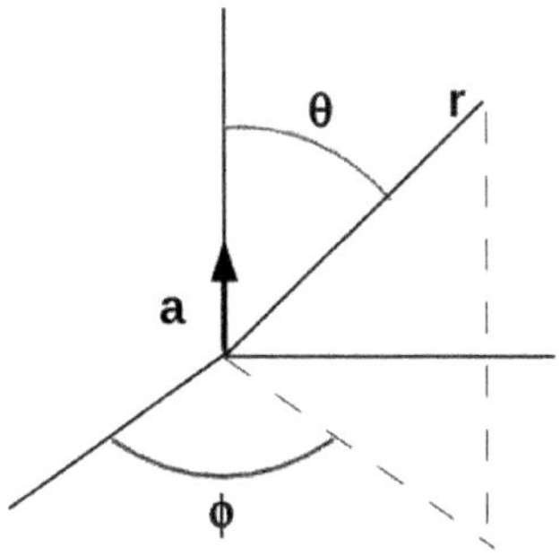

Figure 3.2: A charge is accelerated vertically in the coordinate system illustrated. The text discusses the radiated power per unit area in the direction (θ, ϕ) and at distance r.

This last expression is called Larmor's formula and is given here in cgs units.

So far we have ignored how charged particles are accelerated. Commonly, the acceleration of charged particles occurs due to Coulomb interactions. It is important to note that when these Coulomb interactions are either electron-electron or proton-proton interactions, the tangential fields produced by the equal and opposite acceleration of the two charged particles cancel in the far-field, and so these interactions do not produce electromagnetic radiation. The interactions that are important for producing electromagnetic radiation are those involving electrons and ions and most commonly electrons and protons. In these interactions, since the electron is much less massive than an ion, the electron undergoes a much larger acceleration. For this reason, we generally only need to consider radiation produced by electrons and can ignore that produced by ions.

Larmor's formula provides the amount of power radiated by an accelerated charged particle, but does not provide any information about the wavelength or frequency at which this radiation is emitted. Fortunately, from the Coulomb interaction between charged particles it is straightforward to determine the acceleration of a charged particle as a function of time, and thus how the tangential component of the electric field, produced by the acceleration, varies as a function of time, $E(t)$. What we would really like to know is the electric field of the emitted radiation as a function of frequency, $E(\nu)$. We state without derivation, that there is a mathematical relation between $E(t)$ and $E(\nu)$. These two functions are related by a *Fourier transform*, introduced in Appendix E. In short, the Fourier transform deconstructs the time variations of the electric field, $E(t)$, as a sum of many sinusoidal components, each with a specific frequency and amplitude. Thus, $E(\nu)$ represents the amplitudes of these sinusoidal components as a function of frequency. If the acceleration, and thus the tangential electric field, occurs over a short time, then a broad range of frequencies or wavelengths is needed for the sum of sinusoidal functions to fit $E(t)$. If the acceleration occurs over a much longer time, then a narrower range of frequencies or wavelengths is needed. Therefore, the spectrum of emitted radiation depends on the time dependence of the acceleration of the charged particle. Since many Coulomb encounters produce relatively brief accelerations, these interactions produce emission over a broad range of frequencies. Additionally, in regions of space containing free electrons and ions, the Coulomb interactions involve a continuous range of velocities and interaction distances. This will modify the spectrum produced in a single Coulomb encounter.

Example 3.1:

Consider an electron that undergoes simple harmonic motion in the y-direction. The y-position is given by $y = y_0 \sin(\omega t)$. What is the time-averaged power of the radiation produced by this oscillating electron?

Answer:
We first find the acceleration of the electron as a function of time by taking the second time derivative of the position:

$$d^2y/dt^2 \quad = \quad -\omega^2 y_0 \sin(\omega t).$$

We can now compute the power emitted as a function of time using the Larmor formula in Equation 3.1. Substituting into the equation, we have

$$P(t) \quad = \quad \frac{2e^2}{3c^3}\,[-\omega^2 y_0 \sin(\omega t)]^2 \quad = \quad \frac{2e^2\omega^4 y_0^2}{3c^3}\,\sin^2(\omega t).$$

In averaging over time, we use the fact that the time average of $\sin^2(\omega t)$ is 1/2. Thus, the time-averaged power is

$$\langle P \rangle \quad = \quad \frac{e^2\omega^4 y_0^2}{3c^3}.$$

We now apply these general principles to more specific cases.

3.2 THERMAL RADIATION

Thermal radiation is produced by the thermal motions of charged particles. Any body with a temperature greater than absolute zero has thermal motions. These thermal motions of gas particles or the thermal vibrations of atoms in a solid naturally lead to random Coulomb interactions that accelerate charged particles which produce electromagnetic radiation. Therefore, all bodies with temperatures above absolute zero produce thermal radiation. These thermal motions are characterized by a temperature and likewise so is the process of thermal emission; therefore, one requires a clear definition of temperature.

The definition of temperature is not as simple as one might expect. The complication is that there are many different forms in which microscopic energy can be stored. For example, there is the kinetic energy in the random motions of particles and there is energy in the internal excitations of the electronic states in atoms or the electronic, vibrational and rotational states in molecules. One can define a temperature by the average kinetic energy of the atoms and molecules or by the average energy in the excitations within the atoms or molecules. These two temperatures will be the same only if there is a good coupling (energy exchange) between the motions and the excitations of the atoms or molecules.

For thermal radiation, we need a temperature that characterizes the motions of the individual particles. For an ideal gas, in which all collisions are elastic, the particles achieve thermodynamic equilibrium and the distribution of particle speeds is given by the *Maxwell-Boltzmann distribution*, which is characterized by temperature. For real particles, collisions are not always elastic, as energy can be transferred between the various forms. For example, in the collision of two atoms, both in their ground states, some of the kinetic energy could be absorbed by exciting the electrons within the atoms. After the collision, then, the atoms travel with smaller kinetic energies but in higher electronic states. This is an example of an inelastic collision, in which the energy of motion (thermal energy) was transferred to internal energy of the colliding bodies. With enough collisions, the transfer of energy between the atoms' motions and the atoms' energy states achieves thermodynamic equilibrium. The temperature as defined by the kinetic energy of the atoms in known as the *gas temperature* or *gas kinetic temperature* while the amount of internal excitation defines what we call the *excitation temperature* and in thermodynamic equilibrium these two temperatures are the same. The excitation temperature is discussed more fully in Chapter 4.

However, different forms of energy are often *not* well-coupled in many regions in space, particularly where the densities of the atoms and particles are extremely small. The low densities cause low collision rates and collisions are what enable the energy to become distributed among the different possible forms. The basic point here is that we need to define and keep track of the forms of energy and be aware that temperatures used to define these energies can differ.

3.2.1 Blackbody Radiation

When a body absorbs all light incident upon it at all wavelengths or frequencies, it is referred to as a blackbody. Therefore, a blackbody must be opaque and non-reflective. Such a body will emit thermal radiation called *blackbody radiation*, and its intensity is described by the *Planck function*, $B_\nu(T)$, where T is the body's temperature. We will discuss the mathematical form of $B_\nu(T)$ shortly. Although no object is a perfect blackbody, there are many opaque thermal radiation sources where blackbody radiation is a good approximation of their thermal emission. In these cases, in the equation of radiative transfer, Equation 2.16, we can replace the source function with $B_\nu(T)$.

In the 19th century, the spectrum of a blackbody was measured; however the physics of the day was unable to describe why the emission had the distribution with wavelength that was observed. One of the major breakthroughs in physics at the start of the 20th century was the realization that light has characteristics of both waves and particles. It was well-known that light exhibits properties of waves, such as undergoing interference and diffraction, but the idea that the energy of the light comes in quantized packets, called photons, with the energy of each packet proportional to the frequency of the light ($E = h\nu$) was new. This realization started with Max Planck's discovery that the spectrum of blackbody radiation can be fit by a model that assumes that the energy of the light is in quantized units.

The light we typically see coming from objects usually involves reflected or transmitted light and not emitted light. A blue shirt, for example, looks blue because it reflects primarily the blue frequencies of all the light that hits it, while its emitted light is primarily in the infrared. To model only the light emitted by the body, we can imagine that the body absorbs all light that is incident, reflecting nothing. Such a body is opaque, or optically thick, and is considered black because no light is reflected. For this reason, it is called a blackbody. The radiation modeled by Planck is that emitted by a blackbody. The name blackbody can be confusing since black implies an absence of light while in discussions of blackbody radiation we are interested in the radiation that is emitted by the body. Remember, though, that with the vast majority of objects that we see, the color of the object is determined by the frequencies of light that the object reflects, except for light bulbs, computer screens, and any other object that glows.

One way of understanding blackbody radiation is as a description of the maximum amount of radiation that an ordinary body of a particular temperature will emit solely as an attempt to cool. The total radiation emitted by a cooling body is not correctly described merely as the sum of all the individual emission processes since the particles within the body can also absorb radiation. In the end, the statistics of a large number of emission and absorption events determine the resultant radiation. When the number of interactions between the photons and the particles is large enough, the photons achieve thermal equilibrium with the body. Just as the Maxwell-Boltzmann distribution describes the distribution of particle speeds in thermodynamic equilibrium, there is another statistical distribution function, Bose-Einstein, that describes the distribution of photon energies in thermodynamic equilibrium. The Planck function, denoted by B_ν or B_λ, describes the intensity of the light emitted by a blackbody and can be derived from the Bose-Einstein distribution. The derivation of Planck's function can be found in most modern physics textbooks and will not be repeated here.

Planck's function is given by

$$B_\nu(T) = \frac{2h\nu^3}{c^2} \frac{1}{\exp(h\nu/kT) - 1}, \tag{3.2}$$

where $h = 6.626 \times 10^{-27}$ erg s is Planck's constant; $k = 1.38 \times 10^{-16}$ erg K^{-1} is Boltzmann's constant; c is the speed of light; ν is the frequency of the observation; and T is the temperature of the radiating body. Intensity, defined in Chapter 1, is the amount of energy per unit time per unit area per unit spectral range per unit solid angle. The spectral range can be defined in either wavelength units or frequency units. The subscript "ν" in Equation 3.2 indicates that the spectral range is defined in units of frequency and thus the units of intensity in Equation 3.2 are erg s^{-1} cm^{-2} Hz^{-1} sr^{-1}. The spectral range can also be expressed in wavelength units, and this gives the Planck function in another form, denoted as $B_\lambda(T)$. The equation for B_λ is

$$B_\lambda(T) = \frac{2hc^2}{\lambda^5} \frac{1}{\exp(hc/\lambda kT) - 1}, \tag{3.3}$$

and the units are erg s^{-1} cm^{-2} cm^{-1} sr^{-1} or erg s^{-1} cm^{-3} sr^{-1}. Radio astronomers generally express the Planck function using $B_\nu(T)$.

It is important to understand that, even though $B_\nu(T)$ and $B_\lambda(T)$ represent the same concept, they are *not* the same numerical quantity or even the same function. We give more discussion about their differences later. We first discuss some important features of the Planck function.

We display in Figures 3.3 and 3.4 log-log plots of these functions. As shown in these figures, blackbody emission is a continuous spectrum which peaks at some frequency or wavelength, approaches zero at the spectral limits ($\nu = 0, \lambda = \infty$ and $\nu = \infty, \lambda = 0$), and has different shapes on either side of the peak — one side an exponential (the high frequency and short wavelength side) and the other a power law (ν^2 on the low frequency side in B_ν and λ^4 on the long wavelength side of B_λ). Note that the Planck function depends only on the body's temperature and the frequency of the radiation. No other characteristic of the body is relevant; the intensity of radiation that a blackbody emits at any given frequency depends *only* on its temperature.

A key feature of blackbody radiation, apparent in Figures 3.3 and 3.4, is that as the temperature of the blackbody increases, the value of $B_\nu(T)$ increases at every frequency or wavelength, meaning that a hotter body produces more energy at every frequency. In practical terms, this has the important implication that at any given frequency, any particular value of intensity corresponds to exactly one temperature. Note that the blackbody curves for different temperatures never cross. So, even if you know the intensity of an opaque object at only a single frequency (remember from Section 1.2.4 that we can determine intensity

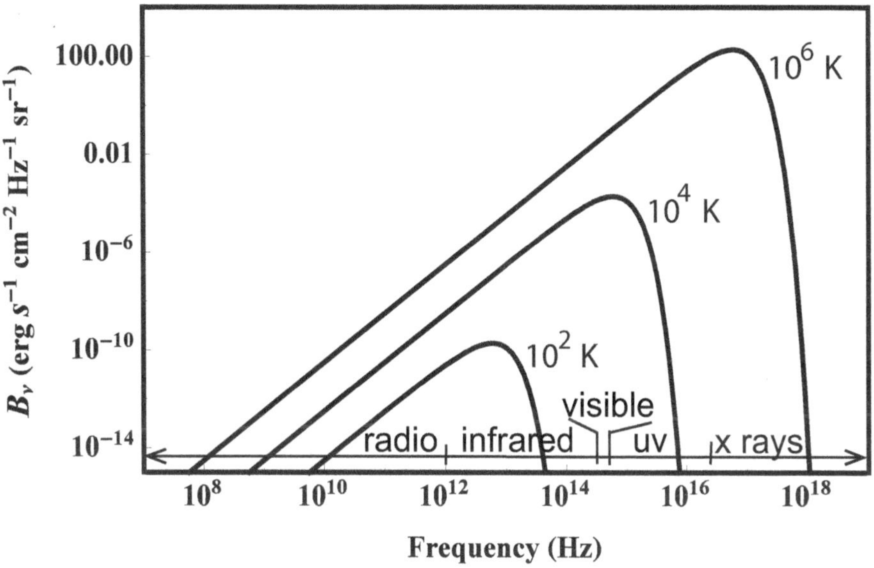

Figure 3.3: Log-log plot of $B_\nu(T)$ versus ν for three different temperatures.

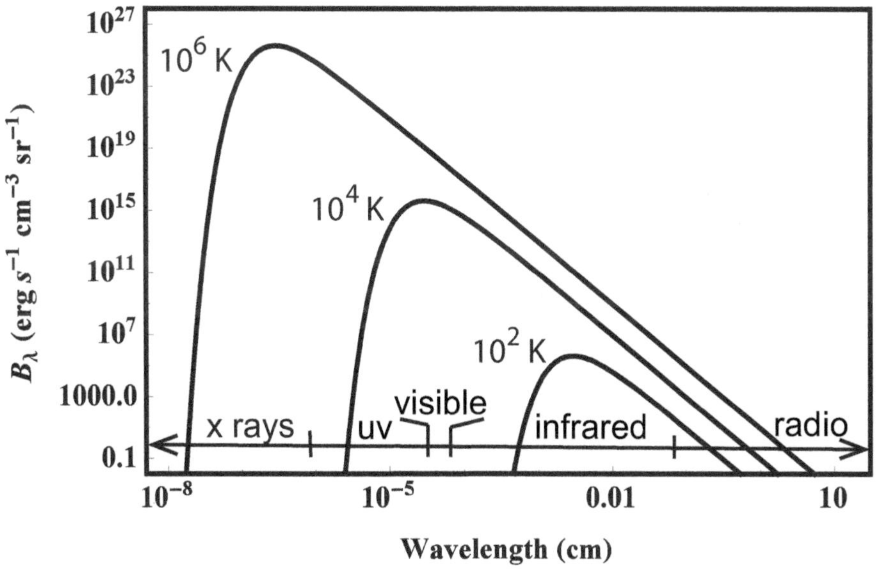

Figure 3.4: The Planck function, plotted as B_λ, for the same three temperatures as in Figure 3.3.

only for sources that are resolved), and if you know the radiating source is a blackbody then you can infer the temperature of the source from that single intensity measurement.

The total power emitted per unit area (erg s^{-1} cm^{-2}) by the surface of a blackbody can be obtained by integration of the Planck function over frequency or wavelength and solid angle. The result shows that the total power per unit area (or flux), F, is proportional to the fourth power of the body's temperature:

$$F = \sigma T^4. \tag{3.4}$$

This is known as the *Stefan-Boltzmann law* and $\sigma = 5.67 \times 10^{-5}$ erg s^{-1} cm^{-2} K^{-4} is the *Stefan-Boltzmann constant.*

Example 3.2:

An electric stove is turned off and the heating element (or burner) turns from red back to black, but the stove warning light indicates that the stove is still hot. An infrared sensor is aimed at the burner and the radiation at a frequency of 3.33×10^{14} Hz emanating from the burner is determined to have an intensity of 1.46×10^{-25} erg s^{-1} Hz^{-1} cm^{-2} sr^{-1}.

(a) Estimate the temperature of the stove burner.

(b) Estimate the total flux of electromagnetic radiation emitted by the stove burner.

Answers:
(a) The stove burner is solid and opaque and so the Planck function is a good approximation to the intensity emitted. Therefore, we can use Equation 3.2 to solve for the temperature. Rearranging, we have

$$\exp\left(\frac{h\nu}{kT}\right) - 1 = \frac{2h\nu^3}{c^2}\frac{1}{I_\nu}.$$

Substituting in the numeric values, this is

$$\exp\left(\frac{1.60 \times 10^4 \text{ K}}{T}\right) - 1 = \frac{4.90 \times 10^{17} \text{ erg s}^{-2}}{9.00 \times 10^{20} \text{ cm}^2 \text{ s}^{-2}} \times \frac{1}{1.46 \times 10^{-25} \text{ erg s}^{-1} \text{ Hz}^{-1} \text{ cm}^{-2} \text{ sr}^{-1}}$$
$$= 3.73 \times 10^{21}.$$

Further manipulation yields

$$T = \frac{1.60 \times 10^4 \text{ K}}{\ln(3.73 \times 10^{21})} = 322 \text{ K (or } 120^\circ\text{F).}$$

(b) Since the burner is opaque at almost all wavelengths, we can approximate the total flux emitted by the Stefan-Boltzmann law. Using Equation 3.4, we have

$$F = \sigma T^4 = \left(5.67 \times 10^{-5} \text{ erg s}^{-1} \text{ cm}^{-2} \text{ K}^{-4}\right)(322 \text{ K})^4$$
$$= 6.09 \times 10^5 \text{ erg s}^{-1} \text{ cm}^{-2}.$$

The average photon energy emitted by a blackbody can be calculated from the ratio of the total energy flux to the total photon flux, as given by

$$\langle E_{\rm ph}\rangle = \frac{\int B_\nu(T)d\nu}{\int (B_\nu(T)/h\nu)d\nu}.$$

The result of this calculation shows that the average photon energy is

$$\langle E_{\rm ph}\rangle = 2.70\ kT = (3.73 \times 10^{-16}\ {\rm erg\ K^{-1}})\ T. \tag{3.5}$$

Since this is a thermal distribution of photons, it should not be surprising that the average energy depends only on the temperature.

Figures 3.3 and 3.4 also show that the location of the peak of the Planck spectrum depends on the body's temperature. Just as the average photon energy increases linearly with temperature, the peak intensity of the blackbody spectrum from a hotter body occurs at a higher frequency. The peak of the curve, of course, is obtained by determining where the derivative of the Planck function equals zero, and the equation relating the location of the peak to the body's temperature is called *Wien's Displacement law.* For $B_\nu(T)$, the frequency of the peak of the curve is given by

$$\nu_{\rm peak} = (5.879 \times 10^{10}\ {\rm Hz\ K^{-1}})\ T \approx (60\ {\rm GHz\ K^{-1}})\ T \tag{3.6}$$

while for $B_\lambda(T)$, the wavelength of the peak of the curve is

$$\lambda_{\rm peak} = \frac{0.2898\ {\rm cm\ K}}{T}. \tag{3.7}$$

It is important to appreciate that Equations 3.6 and 3.7 do not give the same information. This relates to the fact that intensity can be defined per unit frequency or per unit wavelength but these are not the same. Equating I_ν and I_λ is a common misconception that often leads to mistakes. If you substitute $\nu = c/\lambda$ in the equation for $B_\nu(T)$ you will not get the correct equation for $B_\lambda(T)$. The functions $B_\nu(T)$ and $B_\lambda(T)$ cannot be the same quantity, for they have different dimensions. To really drive the point home, use Equations 3.6 and 3.7 to calculate the peak frequency of $B_\nu(T)$, for a given T, and the peak wavelength of $B_\lambda(T)$ for the same T, and then see if $\nu_{\rm peak}$ and $\lambda_{\rm peak}$ satisfy $c = \nu_{\rm peak}\lambda_{\rm peak}$. You will find they do not. In contrast, the average photon energy is independent of whether it is calculated using B_ν or B_λ.

The peaks of these curves for the same temperature might not even occur in the same spectral band. Let's apply Wien's displacement law to the Sun as an example. Where does the spectrum of a blackbody with the temperature of the Sun's surface ($T = 5800$ K) occur? In Figure 3.5 we display the Planck curves for B_ν and B_λ for $T = 5800$ K. According to Equation 3.7, the peak of B_λ for the Sun's temperature occurs at $\lambda_{\rm peak} = 5 \times 10^{-7}$ m $= 500$ nm which corresponds to green, in the middle of the visible band. If we use Equation 3.6 to do the same calculation for B_ν, we find that the peak of B_ν occurs at $\nu_{\rm peak} = 3.4 \times 10^{14}$ Hz, which corresponds to a wavelength of $\lambda_{\rm peak} = c/\nu_{\rm peak} = 880$ nm, which is in the infrared! The peaks of B_ν and B_λ for the Sun are in different regimes of the EM spectrum.

How can the peak of the Sun's spectrum occur at two different frequencies? How can we resolve this discrepancy between B_ν and B_λ? These both correctly describe a thermal spectrum, but they provide the information in different terms. Which function one uses to display a spectrum is really just a matter of choice. Radio astronomers routinely use B_ν, while optical astronomers more commonly use B_λ.

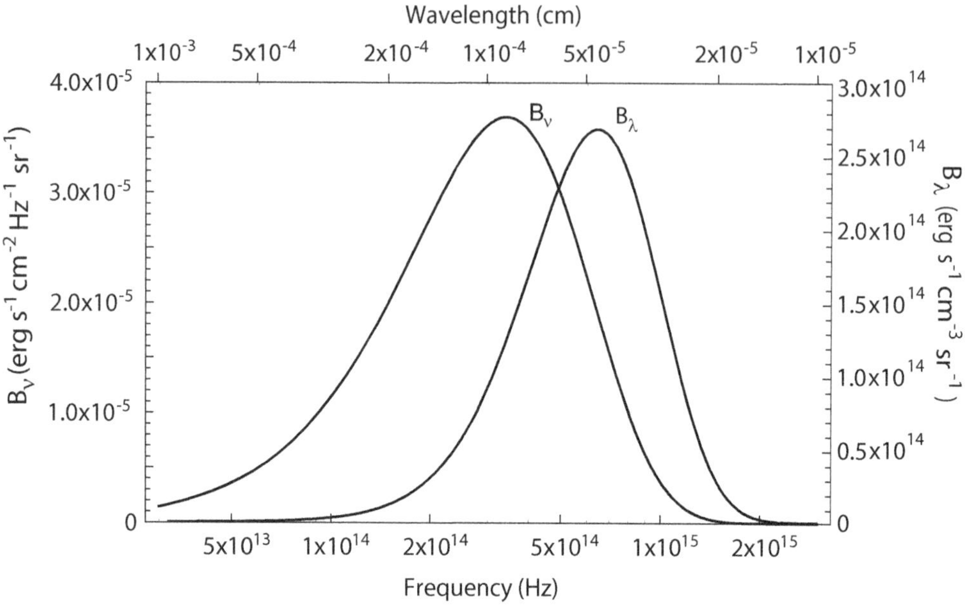

Figure 3.5: The intensity of a 5800 K blackbody is shown as both B_ν and B_λ.

What color the Sun appears to be is actually a very complex question that depends on the sensitivity of the eye to different colors and how the brain interprets the signals it receives from the eyes. So, it is reasonable to say that the Sun looks yellow, because that is how we perceive it. But, it is not correct to say that the Sun's blackbody spectrum peaks in the yellow. In fact, the spectrum of the Sun's radiation across the visible band is relatively flat. The entire visible band fits easily under the broad peak of the Sun's Planck spectrum; a better statement is to say that the Sun's spectrum is white, i.e., that it contains roughly an equal amount of all colors of the rainbow.

An alternative description of the "color" of a blackbody is to use the average energy of the photons, as described by Equation 3.5. By substituting in hc/λ for photon energy, we find that the wavelength of the average energy photon emitted by the Sun is given by

$$\left\langle \frac{hc}{\lambda} \right\rangle = 2.70\ k(5800\ \mathrm{K})$$

or

$$\lambda_{\langle E \rangle} = \frac{hc}{2.70\ k\ 5800\ \mathrm{K}} = 9.20 \times 10^{-5}\ \mathrm{cm}$$

or 920 nm, which is in the infrared.

What if you are given intensity in one form and you need to calculate it in the other? Fortunately, there is a straight-forward conversion. We start with the physical idea that the total power radiated over the whole spectrum must be equal. This means that the integral of intensity over the whole spectrum must be the same for both ways of expressing intensity, i.e.

$$\int_0^\infty I_\nu d\nu = \int_0^\infty I_\lambda d\lambda.$$

We can be even more specific and say that the total power carried by photons over a fixed part of the spectrum must be the same, so that the integrations will be equal when the limits are ν_1 and ν_2 and λ_1 and λ_2 as long as $\lambda_1 = c/\nu_1$ and $\lambda_2 = c/\nu_2$. We can choose the limits to be very close to each other and hence conclude that for each little piece of the spectrum,

$$I_\lambda d\lambda = -I_\nu d\nu.$$

The negative sign is needed because the λ and ν increase in opposite directions. And so,

$$I_\lambda = I_\nu \left(-\frac{d\nu}{d\lambda}\right).$$

Now to get the $d\nu/d\lambda$ we use

$$\nu = \frac{c}{\lambda},$$

from which we infer that

$$\frac{d\nu}{d\lambda} = \frac{d}{d\lambda}\left(\frac{c}{\lambda}\right) = -\frac{c}{\lambda^2}.$$

Substituting back in, we have

$$I_\lambda = \frac{c}{\lambda^2} I_\nu \tag{3.8}$$

and thus $B_\lambda = \frac{c}{\lambda^2} B_\nu$. If you use this conversion in either of the Planck functions you will find that one does convert to the other.

Example 3.3:

Imagine measuring the spectrum of the light emitted by a fellow human being, at a temperature of 98.6° F (or 37° C). Assume that there is no other light source in the room, so that there is no reflected light, and that the emitted spectrum fits the Planck function. (In reality the temperature of the outer layer of the skin will be less than 37° F, but we use this temperature for simplicity.)

(a) At what frequency will the intensity, I_ν, of the radiation emitted be greatest? Convert this to a wavelength.

(b) At what wavelength will the intensity, I_λ, be greatest? Convert this to a frequency.

(c) What is the average energy of the photons emitted? What wavelength and frequency does this correspond to?

(d) If the body's temperature were to increase, how would we detect this with a radiation detector that operates at a single frequency?

Answers:
(a) This question is really asking at what frequency is the peak of B_ν for this temperature; hence we use Equation 3.6. First, though, we must convert the temperature to units of kelvins,

$$T(\mathrm{K}) = 273 + 37 = 310 \text{ K}.$$

Now using Wien's Displacement Law in Equation 3.6, we have

$$\nu_{\text{peak}} = (5.879 \times 10^{10} \text{ Hz K}^{-1})\, 310 \text{ K} = 1.82 \times 10^{13} \text{ Hz}.$$

This corresponds to a wavelength of

$$\lambda = \frac{c}{\nu_{\rm peak}} = \frac{3.00 \times 10^{10} \text{ cm s}^{-1}}{1.82 \times 10^{13} \text{ Hz}} = 1.65 \times 10^{-3} \text{ cm} = 16.5 \ \mu\text{m},$$

which is in the infrared band. Although a 310 K object has its peak emission in the infrared, it also is easily detectable at radio wavelengths.

(b) Now we want the peak of I_λ, which requires Equation 3.7. We have then

$$\lambda_{\rm peak} = \frac{0.2898 \text{ cm K}}{310 \text{ K}} = 9.33 \times 10^{-4} \text{ cm} = 9.33 \ \mu\text{m}.$$

This corresponds to a frequency of $\nu = c/\lambda_{\rm peak} = 3.22 \times 10^{13}$ Hz.

(c) The average energy of the photons is given by Equation 3.5, and so

$$\langle E_{ph} \rangle = (3.73 \times 10^{-16} \text{ erg K}^{-1}) \ 310 \text{ K} = 1.16 \times 10^{-13} \text{ erg},$$

which corresponds to a wavelength

$$\lambda = \frac{hc}{E} = 1.72 \times 10^{-3} \text{ cm} = 17.2 \ \mu\text{m}$$

and a frequency $\nu = E/h = 1.75 \times 10^{13}$ Hz.

(d) If the temperature of the radiating surface increases, we will see an increase in intensity at all frequencies.

Example 3.4:

In an article about a visible-wavelength spectral observation of another galaxy, the measured flux density centered at a wavelength of 425.0 nm (1 nm $= 10^{-9}$ m $= 10^{-7}$ cm), with a bandwidth of 50.0 nm, is reported to be $F_\lambda = 2.30 \times 10^{-14}$ erg s^{-1} cm^{-2} nm^{-1}. Meanwhile, a radio observation of the same galaxy at 22.2 GHz with a bandwidth of 2.00 GHz yields a flux density of $F_\nu = 42.0$ Jy. (As explained in Section 1.2.3, 1 Jy $= 10^{-23}$ erg s^{-1} Hz^{-1} cm^{-2}.)

(a) In which band does this galaxy have a larger flux density?

(b) Compare the detected fluxes in the two bands.

Answers:
(a) The conversion between F_λ and F_ν is the same as between I_λ and I_ν, which is given in Equation 3.8. So we first convert the F_λ from the visible-wavelength observation to F_ν and then to Jy. Before applying Equation 3.8, though, we must calculate F_λ in purely cgs units by converting the nm^{-1} to cm^{-1} (a factor of 10^7). This is $F_\lambda = 2.30 \times 10^{-7}$ erg s^{-1} cm^{-3}. By Equation 3.8, then, the flux per Hz at 425 nm is

$$F_\nu = \frac{\lambda^2}{c} F_\lambda = \frac{(4.25 \times 10^{-5} \text{ cm})^2}{3.00 \times 10^{10} \text{ cm s}^{-1}} \ 2.30 \times 10^{-7} \text{ erg s}^{-1} \text{ cm}^{-3}$$

$$= 1.38 \times 10^{-26} \text{ erg s}^{-1} \text{ cm}^{-2} \text{ Hz}^{-1},$$

which equals 1.38×10^{-3} Jy. This is 4 orders of magnitude smaller than the radio-frequency flux density.

Let's now convert the radio-frequency flux density measurement to the same units as that in the visible. We get

$$F_\lambda = \frac{c}{\lambda^2} F_\nu = \frac{\nu^2}{c} F_\nu = \frac{(22.2 \times 10^9 \text{ Hz})^2}{3.00 \times 10^{10} \text{ cm s}^{-1}} 42.0 \times 10^{-23} \text{ erg s}^{-1} \text{ cm}^{-2} \text{ Hz}^{-1}$$

$$= 6.90 \times 10^{-12} \text{ erg s}^{-1} \text{ cm}^{-3}.$$

This is 5 orders of magnitude smaller than the visible-wavelength flux density!

How can we find that each measurement is many orders of magnitude smaller than the other one? It is rational to expect that one of the two is greater than the other, regardless of how they are compared. The resolution of this paradox is the difference between the quantities F_ν and F_λ. The amount of the spectrum contained in one Hz is miniscule at visible wavelengths (about one part in 10^{14}) and so a large value for F_ν at visible wavelengths would imply an incredible amount of total flux. Similarly, each nm at radio wavelengths is a very small sliver (one part in 10^7) and so the values of F_λ will naturally be very small at radio wavelengths. Although it is good to know how to convert between these quantities, the question asked in (a) is ambiguous and not really of much value. Question (b) is more interesting.

(b) The detected fluxes are given by multiplying the measured flux densities by the bandwidths. For the radio, this is

$$F = F_\nu \times \Delta\nu = 4.20 \times 10^{-22} \text{ erg s}^{-1} \text{ cm}^{-2} \text{ Hz}^{-1} \times 2.00 \times 10^9 \text{ Hz}$$

$$= 8.40 \times 10^{-13} \text{ erg s}^{-1} \text{ cm}^{-2},$$

while for the visible, we find

$$F = F_\lambda \times \Delta\lambda = 2.30 \times 10^{-14} \text{ erg s}^{-1} \text{ cm}^{-2} \text{ nm}^{-1} \times 50.0 \text{ nm}$$

$$= 1.15 \times 10^{-12} \text{ erg s}^{-1} \text{ cm}^{-2}.$$

We see that the flux detected in this example is a little larger at visible wavelengths. Note, though, that the amount of flux detected depends on the bandwidth.

In Section 2.1.2 we discussed emitted intensity as a function of optical depth in general. In the optically thick limit, given by Equation 2.14, we found that the intensity equals the source function, S_ν. For thermal radiation and a non-reflecting body, the source function equals the Planck function, $S_\nu = B_\nu$. For the astronomical sources we discuss in this book, when dealing with thermal radiation we can replace S_ν with B_ν. Making this substitution in Equation 2.11, we get that

$$B_\nu = j_\nu / \kappa_\nu. \tag{3.9}$$

We can then use Equation 3.9 in Equation 2.10 to get an expression for the emitted intensity of thermal radiation as a function of optical depth,

$$I_\nu = B_\nu(T)\,(1 - e^{-\tau_\nu}). \tag{3.10}$$

This is a useful equation. Equation 3.10 shows that B_ν is the maximum emitted intensity. The equation for the optically thin limit, as with that for I_ν in general, is

$$I_\nu = \tau_\nu \; B_\nu(T).$$

We conclude that an opaque body emits more intense radiation than a transparent body and that the Planck function describes the maximum intensity that a radiating body of given temperature will emit by thermal processes at each frequency.

Example 3.5:

A volume of ionized gas in interstellar space is known to emit thermal radiation. At a frequency of 400 MHz, a flux density of 200 Jy is determined to come from an area on the sky with solid angle of 2×10^{-5} sr. Use these data to infer a lower limit to the temperature of this gas.

Answer:
We are told that this gas emits thermal radiation, but not whether the volume of gas is opaque or transparent. But, since this radiation is thermal, we know that the maximum intensity it could have is given by the Planck function. Therefore, we set the observed intensity to be less than or equal to the Planck function. Since the observed intensity cannot be larger than the Planck function this will give us a lower limit to the temperature of the gas.

First, following the approach in Example 1.2, we calculate the intensity from the flux density and source solid angle:

$$I_\nu = \frac{F_\nu}{\Omega} = \left(\frac{200 \text{ Jy}}{2 \times 10^{-5} \text{ sr}}\right) \left(\frac{10^{-23} \text{ erg s}^{-1} \text{ cm}^{-2} \text{ Hz}^{-1}}{1 \text{ Jy}}\right)$$

$$= 1 \times 10^{-16} \text{ erg s}^{-1} \text{ cm}^{-2} \text{ Hz}^{-1} \text{ sr}^{-1}.$$

Setting the Planck function as the upper limit to this intensity and substituting in the values for h, ν, c, and k and rearranging, as in Example 3.2, we find

$$\exp\left(\frac{0.0192 \text{ K}}{T}\right) - 1 \leq 9.42 \times 10^{-6},$$

which requires that

$$T \geq 2040 \text{ K}.$$

3.2.2 Rayleigh-Jeans Approximation

At radio wavelengths the Planck function can be approximated by a much simpler expression. This approximation also leads to an extremely important definition (presented in Section 3.2.3). At radio wavelengths the frequency, ν, is so small that $(h\nu)/(kT) \ll 1$ for most reasonable temperatures (see Question 2 in the Questions and Problems list at the end of the chapter). The exponential in the denominator of the Planck function, then, can be approximated by a Taylor series expansion (the end-of-chapter Question 2 guides you through this expansion), yielding

$$\frac{1}{\exp(h\nu/kT) - 1} \approx \frac{kT}{h\nu} \;\; (\text{for } h\nu/kT \ll 1)$$

and so

$$B_\nu(T) \approx \left(\frac{2h\nu^3}{c^2}\right)\frac{kT}{h\nu} = \frac{2k\nu^2}{c^2}T \tag{3.11}$$

or

$$B_\nu(T) \approx \frac{2k}{\lambda^2}T. \tag{3.12}$$

Note how simple these expressions are in comparison to the Planck function (Equation 3.2). This approximation is very useful, provided you are in the realm where $h\nu/kT \ll 1$. This approximation is known as the *Rayleigh-Jeans approximation* and can almost always be used at centimeter wavelengths and longer. As demonstrated in end-of-chapter Question 2, only at the highest radio frequencies and with observations of cold objects does the approximation start to differ from the full expression. Equations 3.11 and 3.12 are so important in radio astronomy that you will want to memorize them.

Example 3.6:

A radio observation at a wavelength of 6.00 cm yields the determination that a particular radio source has a solid angle of 7.18×10^{-6} sr, is opaque and thermal, and has a flux density of 350 Jy.

(a) What is the temperature of the radio source?

(b) What is the intensity of this source at a wavelength of 2.70 cm? What is the flux density of the source at 2.70 cm?

(c) An observation is made of another opaque, thermal radio source that is twice as hot as the first source. What is its intensity at 2.70 cm?

Answers:
(a) Since the flux density is measured at 6.00 cm, which is still in the realm where the Rayleigh-Jeans approximation works well, we can solve for the temperature using Equation 3.12 along with Equation 1.9, which relates flux density and intensity,

$$F_\nu = \frac{2kT}{\lambda^2}\Omega.$$

Substituting $F_\nu = 3.50 \times 10^{-21}$ erg s^{-1} cm^{-2} Hz^{-1}, $\Omega = 7.18 \times 10^{-6}$ sr, and solving for T gives us

$$T = 63.6 \text{ K}.$$

(b) Again using Equation 3.12, we find that the intensity at 2.70 cm, which corresponds to frequency 11.1 GHz, is:

$$I_\nu(11.1 \text{ GHz}) = \frac{2(1.38 \times 10^{-16} \text{ erg K}^{-1})\ 63.6 \text{ K}}{(2.70 \text{ cm})^2}$$

$$= 2.41 \times 10^{-15} \text{ erg s}^{-1} \text{ cm}^{-2} \text{ Hz}^{-1} \text{ sr}^{-1},$$

and the flux density is

$$F_\nu(11.1 \text{ GHz}) = I_\nu(11.1 \text{ GHz})\ \Omega = 1.73 \times 10^{-20} \text{ erg s}^{-1} \text{ cm}^{-2} \text{ Hz}^{-1},$$

which is 1730 Jy.

(c) As Equation 3.12 shows, in the Rayleigh-Jeans part of the blackbody spectrum, the intensity is directly proportional to the temperature of the radiating body. So, if the second source is twice as hot as the first, then its intensity at the same wavelength will be twice as great. The intensity of the second source at 11.1 GHz, then, is 4.82×10^{-15} erg s^{-1} cm^{-2} Hz^{-1} sr^{-1}.

3.2.3 Brightness Temperature

Examine Equations 3.11 and 3.12 and note that, with the Rayleigh-Jeans approximation, the intensity of blackbody radiation is directly proportional to its temperature. This approximation, then, provides radio astronomers with an extremely convenient, alternative way of describing the intensity of radiation. At radio wavelengths, the intensity and the temperature of blackbody sources can be used interchangeably: they are linearly related by the proportionality constant $2k/\lambda^2$. Moreover, since intensity is a measure of the radiation emitted while temperature refers to a physical condition of the source, temperature is often the more interesting parameter as it could add to our understanding of the physics of the source. In the Rayleigh-Jeans approximation, we define *brightness temperature*, T_B, as

$$T_B = \left(\frac{\lambda^2}{2k}\right) I_\nu. \tag{3.13}$$

This describes the radiation intensity at a given wavelength in terms of the temperature of a blackbody that produces the same intensity at that wavelength.

We stress that T_B is a measure of the *radiation*, not necessarily a property of the emitting *object*. However, in the special case of a blackbody radiation source then the brightness temperature is the same as the temperature of the source. At higher frequencies and/or for lower temperature objects where $h\nu$ is not much smaller than kT, the Rayleigh-Jeans approximation is not applicable. In fact, at the high frequency end of the radio window and when observing relatively low temperature thermal sources, the Rayleigh-Jeans approximation may be inaccurate (see end-of-chapter problems). In these cases, the brightness temperature as defined by Equation 3.13 is not equal to the source temperature.

In summary, brightness temperature is a measure of intensity, and is equal to the temperature that the source would have if it were a blackbody. If the Rayleigh-Jeans approximation applies, the brightness temperature is directly proportional to intensity.

It should be clear that for non-thermal radiation (such as synchrotron radiation, in which relativistic electrons emit photons while accelerated by magnetic fields), brightness temperature is not related to the source temperature. However, it does still describe the radiation intensity and therefore is still a useful quantity.

Example 3.7:

What is the brightness temperature of the source discussed in Example 3.6? (At a wavelength of 6.00 cm, the source is opaque with thermal radiation and has a temperature equal to 63.6 K).

Answer:
Since the source is opaque and thermal and has no reflected light, its intensity equals the Planck curve intensity for the source's temperature. Although we could

solve for the brightness temperature using Equation 3.13, we should realize that because the source is a blackbody its brightness temperature is its physical temperature, that is $T_B = T = 63.6$ K.

3.2.4 Thermal Bremsstrahlung Radiation (or Free-Free Emission)

A free electron is allowed any energy and so can experience a change of energy of any amount. In a cloud of ionized gas, the free electrons will be accelerated by Coulomb interactions with the ions and will naturally emit photons of all frequencies and so produce a continuous spectrum. Since this emission results from electrons undergoing energy transitions from a free state to another free state, this is sometimes called *free-free emission*, in contrast to the emission of radiation from atoms or molecules which occur in *bound-bound* or *bound-free* transitions. This free-free emission is also called *bremsstrahlung radiation*, which is German for "braking radiation" to describe the radiation emitted when a free electron is decelerated.

To model the emitted intensity one starts by using Larmor's equation (Equation 3.1) to derive an expression for the emission coefficient, j_ν. Bypassing the long and tedious derivation, we outline the initial considerations and skip to the final result. First, one considers a single electron with velocity v on a path that will pass by an ion at a distance b; b is called the impact parameter. The acceleration of the electron is then calculated as a function of time due to its Coulomb interaction with the ion. From the time-dependent acceleration, Larmor's equation gives the power radiated by this electron as a function of time. The frequency dependence of the radiation, then, follows from the time dependence (see Appendix E). For an ensemble of electrons, the result is summed over all possible interactions, which involves an integration over both the thermal velocity distribution of the electrons, given by the Maxwell-Boltzmann distribution for a given temperature, and over all possible impact parameters, b. From such calculations, the power radiated by an ionized gas at temperature T is obtained.

The emission coefficient of bremsstrahlung radiation from a thermal distribution of electrons and ions is found to have a complicated dependence on temperature and density. For most temperatures and densities observed in radio astronomy, the emission coefficient (taken from *Physical Processes in the Interstellar Medium* by Lyman Spitzer[1]) is

$$j_\nu^{\text{ff}} = \frac{8}{3}\left(\frac{2\pi}{3}\right)^{1/2} \frac{Z^2\, e^6\, n_e\, n_i\, g_{\text{ff}}}{m_e^{3/2}\, c^3\, (kT)^{1/2}}\, e^{-h\nu/kT},$$

where n_e and n_i are the number densities of electrons and ions, Ze is the charge of the ions, T is the gas temperature, and g_{ff} is the *Gaunt factor*, which takes into account quantum mechanical corrections to the classical analysis. If we substitute for the constants using cgs units in the above equation, we find

$$j_\nu^{\text{ff}} = \left(5.44 \times 10^{-39} \text{ erg s}^{-1} \text{ cm}^3 \text{ Hz}^{-1} \text{ sr}^{-1} \text{ K}^{1/2}\right) \frac{Z^2 n_e n_i}{T^{1/2}}\, e^{-h\nu/kT}\, g_{\text{ff}}. \tag{3.14}$$

Recall from Chapter 2 that the units for emission coefficient are erg s^{-1} cm^{-3} Hz^{-1} sr^{-1}. The dependencies of j_ν^{ff} in Equation 3.14 can be understood in terms of basic principles. Since the emission requires the interaction of an electron with an ion, there is a dependence on the density of both. The time period over which the interaction occurs is inversely proportional

[1]John Wiley & Sons, Inc., 1978.

to the electron's velocity, which produces the $T^{-1/2}$ dependence, and the exponential term appears because of the integration over the Maxwell-Boltzmann distribution of the electron velocities.

The Gaunt factor in Equation 3.14 can be approximated at radio wavelengths and has the following dependence on temperature and frequency:

$$g_{\rm ff} \propto T^{0.15} \, \nu^{-0.1}.$$

Since at radio wavelengths $h\nu/kT \ll 1$, the exponential in Equation 3.14 is essentially equal to one. Setting the exponential to unity and multiplying by the Gaunt factor, the emission coefficient has the following dependence on the density, frequency and temperature

$$j_\nu^{\rm ff} \propto \frac{Z^2 n_e n_i}{\nu^{0.1} \, T^{0.35}}.$$

If the ionized gas is optically thin and has a uniform temperature and density, then the intensity of the emission is given by $I_\nu = j_\nu^{\rm ff} \, L$ (see Equation 2.13). Thus, in optically thin situations, the frequency dependence of the free-free intensity is

$$I_\nu^{\rm ff}(\text{thin}) \propto \nu^{-0.1}. \tag{3.15}$$

Therefore optically thin free-free radiation has an intensity that is relatively flat with frequency, decreasing very slowly with increasing frequency. In the calculation of the radiation emitted by a single electron-ion interaction, the short time duration of the interaction leads to the emission of a large range of frequencies (see Appendix E). The almost flat continuum described by Equation 3.15 is a consequence of there being an ensemble of electrons interacting with ions with a continuum of velocities and ion-electron impact parameters.

To find a general expression for the emergent intensity from an ionized gas region, we not only need to include the free-free emission process but also free-free absorption — the inverse of the emission process. Therefore we need an expression for the free-free absorption coefficient, which we can obtain by use of Equation 3.9, giving us

$$\kappa_\nu = \frac{j_\nu}{B_\nu(T)}.$$

This relation between κ_ν, j_ν, and B_ν can also be obtained from the concept that, when in thermodynamic equilibrium, the amount of energy per length absorbed, $\kappa_\nu B_\nu$, must equal the amount emitted per length, j_ν. At radio wavelengths the absorption coefficient (see review by Mark Gordon[2]) is found to be approximately

$$\kappa_\nu^{\rm ff} \approx 2.67 \times 10^{-20} \ {\rm cm}^{-1} \left(\frac{T}{\rm K}\right)^{-1.35} \left(\frac{\nu}{\rm GHz}\right)^{-2.1} \left(\frac{n_e}{{\rm cm}^{-3}}\right) \left(\frac{n_i}{{\rm cm}^{-3}}\right). \tag{3.16}$$

The frequency dependence in this equation arises from the frequency dependence of the emission coefficient ($\nu^{-0.1}$) and the frequency dependence of the Planck function in the Rayleigh-Jeans limit (ν^2).

A general expression for the emitted intensity as a function of optical depth is given in Equation 2.10. For free-free emission we can replace the source function with the Planck function and thus

$$I_\nu^{\rm ff} = B_\nu(T) \, (1 - e^{-\tau_\nu^{\rm ff}}). \tag{3.17}$$

[2]in Galactic and Extragalactic Radio Astronomy (2nd edition), Berlin and New York, Springer-Verlag, 1988.

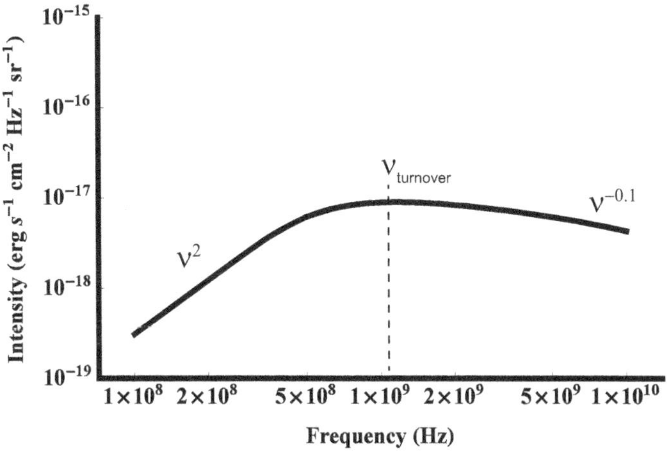

Figure 3.6: Log-log plot of the radio-frequency spectrum of a cloud of plasma. The cloud is transparent at higher frequencies and opaque at lower frequencies. The peak in the spectrum, at $\nu_{turnover}$, occurs where the optical depth of the ionized gas transitions from being optically thin to optically thick, which is approximately where $\tau_\nu = 1$.

Recalling, from Chapter 2, that $\tau_\nu = \int \kappa_\nu \, ds$, we find that the optical depth of the ionized gas strongly depends on frequency, $\tau_\nu^{\rm ff} \propto \nu^{-2.1}$, decreasing rapidly with increasing frequency. Because of the steep dependence of the optical depth on frequency, at very low frequencies the optical depth of the ionized gas will likely be large, and so the intensity will be equal to the Planck function. But at these low radio frequencies we can use the Rayleigh-Jeans approximation for the Planck function, and therefore in the optically thick regime, the intensity will be given by

$$I_\nu^{\rm ff}(\text{thick}) \;\propto\; \nu^2. \tag{3.18}$$

At higher frequencies, the optical depth becomes $\ll$1 and the emitting region is optically thin. In this regime the intensity is given by Equation 3.15. The behavior of the intensity of the free-free emission with frequency is illustrated in Figure 3.6. We stated earlier that the Planck function gives the maximum intensity of the thermal radiation from any source, and this is true for the free-free emission from ionized gas.

The frequency of the peak of the thermal emission spectrum, called the *turnover frequency*, occurs at the transition between the optically thin and optically thick frequency regimes. This transition occurs where the optical depth is of order unity and depends on the density, temperature and spatial extent of the ionized gas. Inserting the expression for $\kappa_\nu^{\rm ff}$ from Equation 3.16, the free-free optical depth is

$$\tau_\nu^{\rm ff} \approx \int_0^L \kappa_\nu^{\rm ff} ds \approx 2.67 \times 10^{-20} \left(\frac{\nu}{\rm GHz}\right)^{-2.1} \int_0^L \left(\frac{T}{\rm K}\right)^{-1.35} \left(\frac{n_e}{\rm cm^{-3}}\right)\left(\frac{n_i}{\rm cm^{-3}}\right)\left(\frac{ds}{\rm cm}\right),$$

where L is the extent of the ionized gas along the line of sight. If the temperature of the gas is roughly constant along the line of sight, we can rewrite this as

$$\tau_\nu^{\rm ff} \approx 2.67 \times 10^{-20} \left(\frac{\nu}{\rm GHz}\right)^{-2.1} \left(\frac{T}{\rm K}\right)^{-1.35} \int_0^L \left(\frac{n_e}{\rm cm^{-3}}\right)\left(\frac{n_i}{\rm cm^{-3}}\right)\left(\frac{ds}{\rm cm}\right).$$

The integral of $n_e n_i ds$ is an important quantity and is referred to as the *emission measure* (EM). Since lengths are commonly measured in parsecs, the units of EM are usually

cm^{-6} pc. Rewriting the optical depth in terms of the emission measure gives

$$\tau_\nu^{\text{ff}} \approx 0.0824 \left(\frac{T}{\text{K}}\right)^{-1.35} \left(\frac{\nu}{\text{GHz}}\right)^{-2.1} \left(\frac{EM}{\text{cm}^{-6}\ \text{pc}}\right). \tag{3.19}$$

By setting the optical depth to unity, we can write an expression that relates the turnover frequency to the physical properties of the source

$$\nu_{\text{turnover}} \approx 0.305\ \text{GHz} \left(\frac{T}{\text{K}}\right)^{-0.643} \left(\frac{EM}{\text{cm}^{-6}\ \text{pc}}\right)^{0.476} \tag{3.20}$$

The turnover frequency, once identified in a spectrum, can provide valuable information about the physical conditions of the free-free radiating gas. It may seem that the turnover frequency and the intensity in the optically thin part of the spectrum provide the same information (they both depend on $n_e n_i L$), and so it may not be obvious why the turnover frequency is at all useful. Recall, though, that if the source is unresolved, we do not measure intensity, but rather the source's flux density. Inferring intensity requires knowledge of the source size and whether there are any small-scale variations below the resolution of our radio telescope. The turnover frequency, on the other hand, is easily determined, provided that the spectral coverage is sufficient to identify the turnover.

Example 3.8:

A spherical region containing only ionized hydrogen gas, 10 pc in diameter, is observed to have a spectrum that fits ν^2 at lower frequencies and $\nu^{-0.1}$ at higher frequencies. The turnover frequency occurs at 5.0 GHz. At 1.7 GHz, the measured intensity of the radiation is $I_\nu = 9.09 \times 10^{-15}$ erg s^{-1} cm^{-2} Hz^{-1} sr^{-1}. Assuming that the gas is uniform in density and temperature, infer the temperature and density of the hydrogen gas.

Answer:
We can use the intensity at a frequency well below the turnover (where the emission is optically thick) to infer the temperature. At 1.7 GHz, which is below the turnover, the intensity is related to the gas temperature by the Rayleigh-Jeans approximation, Equation 3.11,

$$T = \frac{c^2}{2k\nu^2}\, I_\nu$$

$$= \frac{(3 \times 10^{10}\ \text{cm s}^{-1})^2}{2 \times 1.38 \times 10^{-16}\ \text{erg K}^{-1}\ (1.7 \times 10^9\ \text{Hz})^2}\ 9.09 \times 10^{-15}\ \text{erg s}^{-1}\ \text{cm}^{-2}\ \text{Hz}^{-1}\ \text{sr}^{-1}.$$

The gas temperature, then, is

$$T = 10{,}300\ \text{K}.$$

Now we can use Equation 3.20 for the turnover frequency, or Equation 3.19 with $\tau_\nu^{\text{ff}} = 1$, to determine the emission measure. We find

$$EM \approx 0.0824\ \text{cm}^{-6}\ \text{pc} \left(\frac{5\ \text{GHz}}{\text{GHz}}\right)^{2.1} \left(\frac{10300\ \text{K}}{\text{K}}\right)^{-1.35}.$$

The emission measure, therefore, is 9.3×10^7 cm^{-6} pc. The depth of the region of ionized gas is 10 pc, and since the plasma is pure hydrogen, we know that there are equal numbers of ions as electrons, so $n_i = n_e$; thus we derive an electron density of $n_e = 3000$ cm^{-3}.

3.3 NON-THERMAL RADIATION

The most important type of non-thermal emission at radio wavelengths results from electrons accelerated by magnetic fields. The emission from non-relativistic free electrons in a magnetic field is called *cyclotron radiation*. If the electrons are relativistic, on the other hand, the emission is called *synchrotron radiation*, and can extend to the highest frequencies of the electromagnetic spectrum. Synchrotron radiation, in contrast to cyclotron radiation, is the most common non-thermal emission mechanism in radio astronomy. Other than the Sun, the brightest objects in the sky at long radio wavelengths, e.g. Cygnus A, Cas A, Taurus A (a.k.a. the Crab Nebula), and Virgo A, are synchrotron sources. The 'A' in these objects' names indicates they are the brightest radio sources in those constellations. Figure 3.7 displays the spectra, on a log-log plot, of three of these objects. The shapes of the spectra emitted by these objects are clearly quite different from either blackbody or free-free emission.

Any explanation of synchrotron radiation must include special relativity, to properly address the dynamics of the radiating electrons and their interactions with the magnetic field. Combining relativity with magnetism is not an easy undertaking. With that in mind, we will avoid many of the details in derivations and focus primarily on the conceptual points.

The names of these radiation mechanisms come from particle physics. "Cyclotron" and "synchrotron" are the names of two types of particle accelerators in which charged particles are steered in a circle by way of a properly designed magnetic field. In synchrotron accelerators the particles reach relativistic speeds while those in the earlier cyclotron accelerators do not. The particle physicists realized that when charged particles are deflected by magnetic fields the particles would naturally radiate away some of their energy; this needed to be accounted for in the design of the accelerators. They developed the theories of these radiation processes, and now radio astronomers reap the benefits by using their results to understand the phenomena of the synchrotron-emitting objects in space.

3.3.1 Cyclotron Radiation

We start by considering a non-relativistic electron moving with velocity $\vec{v}$ in the presence of a magnetic field $\vec{B}$. The magnetic field exerts a force on the moving electron given in cgs units by

$$\vec{F} = -e\frac{\vec{v} \times \vec{B}}{c}.$$

Since $\vec{F}$ is proportional to the cross product of $\vec{v}$ and $\vec{B}$, the acceleration is always perpendicular to the velocity vector and circular motion must result. Because this is a cross product, only the component of $\vec{v}$ that is perpendicular to $\vec{B}$, which we'll denote as $v_\perp$, matters. The circular motion is determined by $v_\perp$, while the linear drifting motion is determined by the component of $\vec{v}$ parallel to $\vec{B}$, or $v_\parallel$. The electron circles around a $\vec{B}$-field line, while continuing along the field line at speed $v_\parallel$, as shown in Figure 3.8.

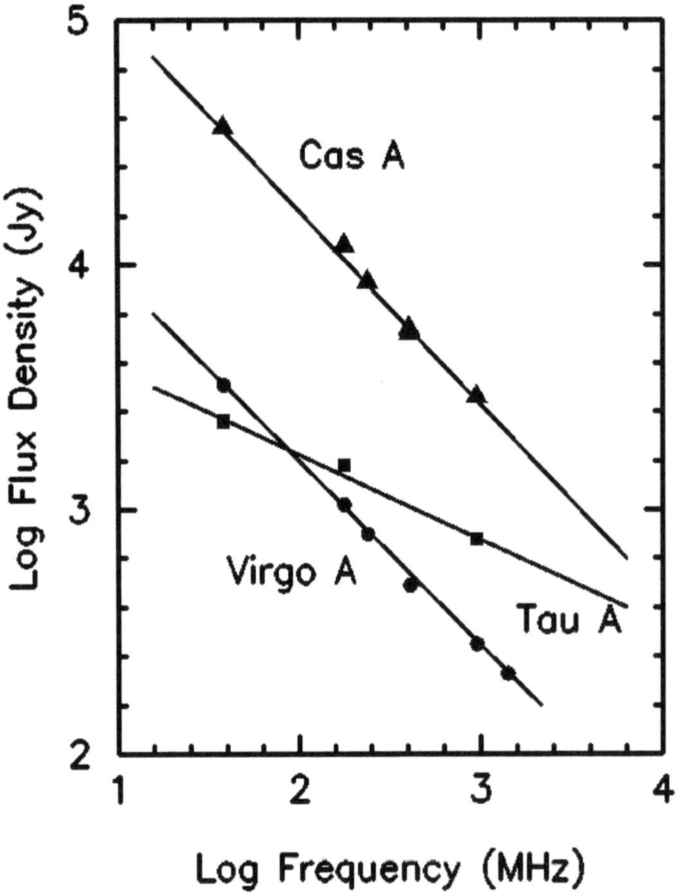

Figure 3.7: Log-log plot of spectra of three bright synchrotron radio sources. With the axes plotted logarithmically, the straight-line indicates a power-law relation, $F_\nu \propto \nu^\alpha$. The flux densities decrease with increasing frequency and so $\alpha < 0$. The data are taken from a paper by Conway, Kellermann, & Long (1963, Monthly Notices of the Royal Astronomical Society, vol. 125, p. 261).

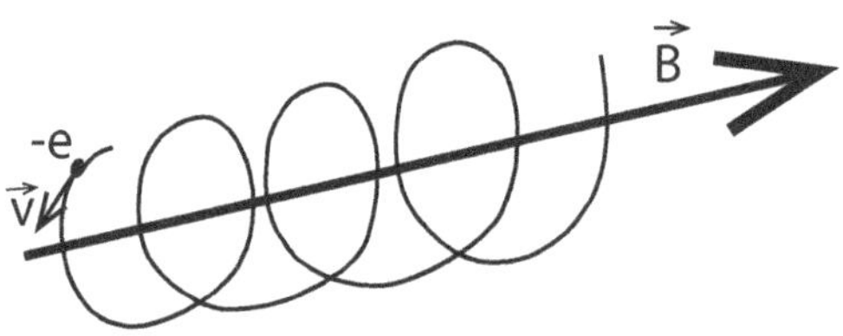

Figure 3.8: Motion of an electron in a uniform magnetic field.

We can set F equal to the centripetal force. For the classical case of a non-relativistic electron, this becomes

$$F_{\rm cent} = \frac{mv_\perp^2}{r} = e\frac{v_\perp B}{c} \tag{3.21}$$

The angular frequency of the circular motion is given by

$$\omega = v_\perp / r,$$

where ω has units of radians s^{-1}. This equation comes from the definition of a radian, in which θ(radians) $= s/r$, where s is the arclength, and so $\omega = d\theta/dt = (ds/dt)/r = v_\perp/r$. By dividing both sides of Equation 3.21 by $mv_\perp$, we get the angular frequency of the electron's circular motion. This is known as the electron's *cyclotron frequency* and is given by

$$\omega_{\rm c} = \frac{v_\perp}{r} = \frac{eB}{mc}. \tag{3.22}$$

Since the direction of motion of the electron is constantly changing, there is a continuous acceleration of the electron and hence radiation must be emitted. This is *cyclotron radiation*. The first order frequency of the radiation emitted will match the oscillation frequency of the electron. This is a consequence of the Fourier transform relation between the time and frequency domains of the radiation (see Appendix E). The primary frequency of cyclotron radiation, then, is the cyclotron frequency. The cyclotron frequency in Equation 3.22, though, has units of radians per second, and we generally discuss the frequency of radiation in hertz (cycles per second). In hertz, the emitted cyclotron radiation frequency is

$$\nu_{\rm c} = \frac{1}{2\pi}\frac{eB}{mc}. \tag{3.23}$$

Note that the only variable in the cyclotron frequency expression is the magnetic field. All electrons in the same magnetic field, regardless of their velocities, will emit at the same frequency. Those moving at larger speeds will cycle around the magnetic field in larger circles, but complete the circle in the same amount of time.

Because of this simple relation, when we detect cyclotron radiation, we get a direct measure of the strength of the magnetic field solely by knowing the frequency of the radiation. Since all electrons will experience the same cyclotron frequency, if this radiation is produced from a region of uniform magnetic field, for example, we will see a single frequency in the spectrum. Of course, any real region will have a range of magnetic fields, but if we can be certain that the detected radiation is cyclotron emission, then we get direct measures of the magnetic fields in the region.

Example 3.9:

The strength of the magnetic field in a sunspot can be as large as 1000 gauss.

(a) What is the cyclotron frequency of electrons in a sunspot?

(b) What frequency of radiation do these electrons emit?

Answer:
(a) By Equation 3.22, the electrons have a cyclotron frequency of

$$\omega_{\rm c} = \frac{(4.8 \times 10^{-10}\ {\rm esu})(1000\ {\rm gauss})}{(9.11 \times 10^{-28}\ {\rm g})(3.00 \times 10^{10}\ {\rm cm\ s^{-1}})} = 1.76 \times 10^{10}\ {\rm radians\ s^{-1}}.$$

The conversion of units in this calculation may be confusing. To help with this, recall from Chapter 1 that the cgs units of magnetic field are force per charge and so 1 gauss = 1 g cm s^{-2} esu^{-1}.

(b) The emitted radiation occurs at frequency

$$\nu_c = \frac{\omega_c}{2\pi} = 2.8 \text{ GHz}.$$

Example 3.10:

There is a magnetic field threading through the disk of the Milky Way Galaxy. Its magnitude is typically of order 2.0 μgauss. There is also an interstellar medium (occupying the space between the stars) that contains ions and free electrons. At what frequency do these electrons emit radiation due to their interaction with the Galactic magnetic field?

Answer:
By Equation 3.23,

$$\nu_c = \frac{1}{2\pi} \frac{(4.80 \times 10^{-10} \text{ esu})(2.0 \times 10^{-6} \text{ gauss})}{(9.11 \times 10^{-28} \text{ g})(3.00 \times 10^{10} \text{ cm s}^{-1})} = 5.6 \text{ Hz}.$$

Note that this is an extremely low frequency and is well below the lowest frequency that can propagate through the Galaxy, as calculated in Chapter 2 (see Example 2.3).

3.3.2 Synchrotron Radiation by a Single Relativistic Electron

We now consider the spectrum emitted by relativistic electrons in a magnetic field. This spectrum is, surprisingly, quite different from that of cyclotron radiation. It occurs over a large range of frequencies and is dependent on the electron's energy.

We first take a moment to remind readers of the Lorentz relativistic factor γ. This is defined mathematically as

$$\gamma = \frac{1}{\sqrt{(1 - (v/c)^2)}},$$

where v is the particle's velocity and c is the speed of light. This factor appears repeatedly throughout the equations of special relativity. It is the unitless multiplicative factor in the equations for time dilation, length contraction, and apparent mass. For the discussion here, γ must be included when describing the total energy and momentum of a particle, in which

$$E = \gamma m_0 c^2,$$

and

$$p = \gamma m_0 v,$$

where m_0 is the rest mass of the particle and γm_0 is the relativistic mass.

With regards to a relativistic electron cycling around a magnetic field line, we need to replace the mass in Equation 3.22 with its relativistic mass. Therefore the angular frequency of the electron's motion, which we call the electron's gyration frequency, is

$$\omega_{gyr} = \frac{eB}{\gamma m_e c}, \tag{3.24}$$

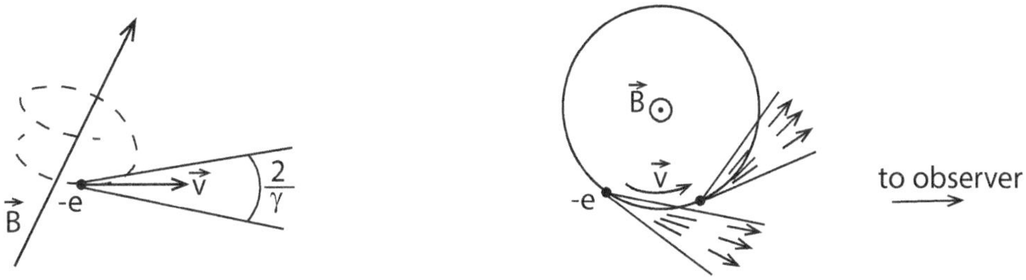

Figure 3.9: A relativistic electron moving in a helix about a magnetic field line emits radiation predominantly distributed over a cone of width inversely proportional to the electron's energy (left). For a magnetic field pointing into the page, the cone of emission passes over the observer causing an apparent pulse of emission (right).

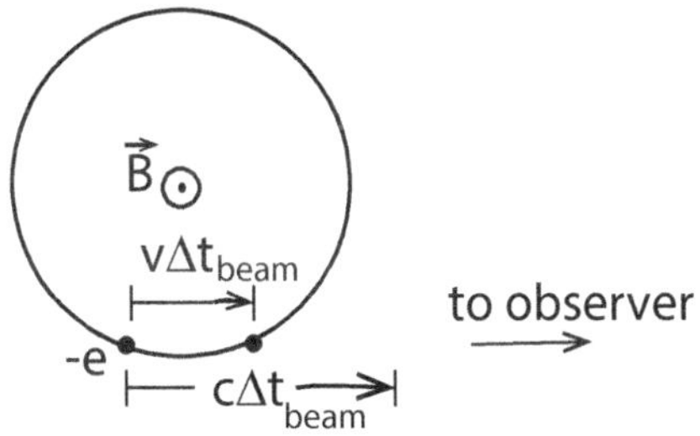

Figure 3.10: As the electron moves along the part of its gyration that emits radiation toward the observer, which occurs for a time period $\Delta t_{\rm beam}$ in the electron's frame, the radiation emitted at the start travels a distance $c\Delta t_{\rm beam}$ while the electron travels a distance $v\Delta t_{\rm beam}$. Therefore, the radiation at the back of the pulse is only a distance $(c - v)\Delta t_{\rm beam}$ behind the front of the pulse.

where m_e is the electron rest mass. Note that the gyration frequency is a factor of γ smaller than the electron's cyclotron frequency.

Because of two additional relativistic effects, the frequency of the emitted radiation in the observer's reference frame does not equal the electron's gyration frequency. First, in the reference frame of the observer, the radiation is beamed in the electron's direction of motion into a cone of angle $2/\gamma$ (see Figure 3.9 left). To the observer, then, the radiation will appear as a narrow pulse in time, as the radiation cone passes through the observer's line of sight (see Figure 3.9 right). Second, because the electron is traveling close to the speed of light, it almost keeps up with its own radiation. The distance between the beginning and end of the pulse is therefore shortened; see Figure 3.10. From the observer's perspective, this causes a shortening of the time width of the detected pulse.

The frequency of the peak of the emission is related to the time width of the pulse. Therefore we need to model the time width of the pulse in the observer's frame and use the inverse relation between frequency and time as independent variables of the intensity (see Appendix E). A thorough derivation is non-trivial and involves Fourier transforms of complex functions. However, we can obtain the salient aspects of the spectrum by focusing on the pulse width and its dependencies on the magnetic field strength, B, and the electron energy, E.

To model the pulse width, because of the beaming of the radiation into a cone of angle $2/\gamma$, the emission will reach the observer for a fraction $(2/\gamma)/(2\pi)$ of the gyration period

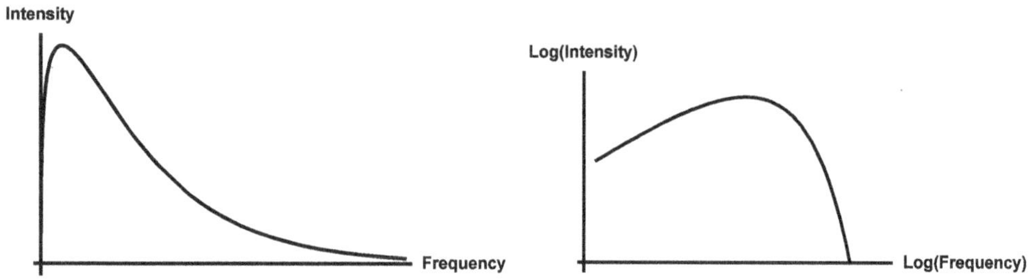

Figure 3.11: The shape of the spectrum of synchrotron radiation from a single electron. The left panel displays a linear plot which shows that the spectrum, although a continuum, is strongly peaked. The right panel displays a log-log plot, which is used in Section 3.3.3.

$(2\pi/\omega_{\rm gyr})$. Hence, the beam passes over the observer for a time period

$$\Delta t_{\rm beam} = \frac{2}{\gamma \omega_{\rm gyr}}.$$

Furthermore, by the end of the pulse, the electron has moved a distance $v\Delta t_{\rm beam}$ while the leading edge of the pulse has moved $c\Delta t_{\rm beam}$. This leads to an apparent time width of the pulse of

$$\Delta t_{\rm pulse} = \frac{c\Delta t_{\rm beam} - v\Delta t_{\rm beam}}{c} = \Delta t_{\rm beam}(1 - \frac{v}{c}).$$

The high speed of the electron, therefore, causes a shortening of the pulse width by a factor of $(1 - v/c)$. For velocities very close to c this factor is approximately equal to $1/2\gamma^2$ (see Question 15). The observed pulse width, then, is

$$\Delta t_{\rm pulse} = \frac{1}{2\gamma^2}\Delta t_{\rm beam} = \frac{1}{\gamma^3 \omega_{\rm gyr}}.$$

Substituting in the expression for the relativistic gyration frequency from Equation 3.24 we have

$$\Delta t_{\rm pulse} = \frac{1}{\gamma^2}\frac{m_e c}{eB}. \tag{3.25}$$

The inverse of the observed pulse width, which we call the *characteristic synchrotron frequency*, is then

$$\nu_{\rm char} = (\Delta t_{\rm pulse})^{-1} = \frac{e}{m_e c}\gamma^2 B.$$

A complete analysis shows that the actual spectrum covers a wide range of frequencies with a significant peak at a frequency smaller than the characteristic synchrotron frequency, as depicted in the left panel of Figure 3.11. We call the frequency of the peak intensity the *synchrotron peak frequency* of the individual electron, which is given by

$$\nu_{\rm peak} = 0.0692\frac{e}{m_e c}\gamma^2 B. \tag{3.26}$$

Expressed in terms of the relativistic energy, $E = \gamma m_e c^2$, we can rewrite this as

$$\nu_{\rm peak} = 0.0692\frac{e}{m_e^3 c^5}E^2 B. \tag{3.27}$$

The important aspect of this equation is that the peak frequency of the power emission is proportional to the electron's energy squared and to the magnetic field strength.

We will also want an expression for the total power of the radiation emitted by a single electron. We will, again, give an heuristic explanation to avoid a complicated derivation. We need to take care, though, about the different frames of reference. The acceleration of the electron in Larmor's equation must be calculated in the electron's frame of reference since it is doing the radiating.

Because of the beaming, the radiation that we detect is emitted in the electron's direction of motion and so is tangent to the electron's gyration motion. The acceleration of the electron, then, is perpendicular to the line of sight. Converting the acceleration from the electron's frame of reference to the observer's frame involves two time dilation factors and so two factors of γ are needed. Hence,

$$a_{\text{electron frame}} = \gamma^2 a_{\text{observer frame}}.$$

This can be shown formally by a Lorentz transformation of the acceleration. The acceleration of the electron in the observer's frame is given by

$$\gamma m_e \vec{a} = \frac{e\vec{v} \times \vec{B}}{c},$$

and so, in the electron frame,

$$a_{\text{electron frame}} = \gamma \frac{evB \sin\theta}{m_e c},$$

where θ is the angle between $\vec{v}$ and $\vec{B}$. Putting this into Larmor's equation, we have

$$P = \frac{2e^2}{3c^3} \left(\gamma \frac{eB \sin\theta}{m_e} \frac{v}{c} \right)^2.$$

Since $v/c \approx 1$ for a relativistic particle and $E = \gamma m_e c^2$, we can write this as

$$P(\text{single e}^-) \approx \frac{2e^4}{3m_e^4 c^7} \sin^2\theta \; E^2 B^2. \tag{3.28}$$

The key point to note here is that $P_{\text{synchr}} \propto E^2 B^2$, and so the power radiated is larger for higher energy electrons and for stronger magnetic fields.

3.3.3 Radiation by an Ensemble of Relativistic Electrons

Of course, we never actually detect the radiation from a single electron, so we need to determine the characteristics of the radiation emitted by a population of electrons of varying energies. As there is a broad range of electron energies, there will be a broad range of emitted frequencies. We showed the observed spectra of some luminous synchrotron sources in Figure 3.7. Since this figure is a log-log plot, the decreasing straight-lines indicate that the relations between the flux densities and frequencies are declining power-laws, which are described mathematically as

$$F_\nu \propto \nu^\alpha, \tag{3.29}$$

where $\alpha < 0$. In a typical synchrotron spectrum, the *spectral index*, $\alpha \approx -0.7$. We therefore need a model of these synchrotron sources that yields declining power-law spectra.

We obtain the shape of the spectrum by determining the total power of radiation emitted at each frequency. A single electron's spectrum is narrowly peaked at its synchrotron peak

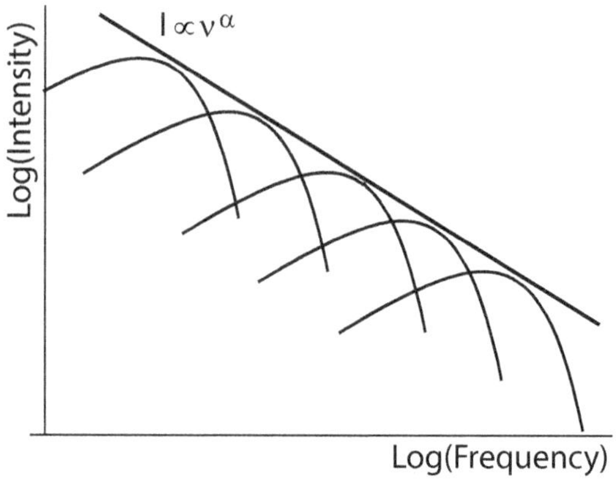

Figure 3.12: The creation of a declining power-law synchrotron spectrum by a declining power-law energy distribution of electrons. The spectrum emitted by a single relativistic electron, in a log-log plot, is shown in the right-panel of Figure 3.11. For each E, the spectrum produced by a single electron is multiplied by the number of electrons, $N(E)dE$, at that energy. Since the single-electron spectrum is highly peaked (for a linear plot, see Figure 3.11), the total spectrum shape, indicated by the straight, decreasing line, is dictated by the energy distribution of electrons.

frequency, given in Equation 3.27. We let $N(E)dE$ equal the number of electrons with energy in the range from E to $E + dE$, and recall that the power emitted by each electron of energy E is proportional to E^2B^2. Hence, the total power emitted between frequencies ν and $\nu + d\nu$ is roughly given by

$$P(\nu)d\nu \;\propto\; N(E)dE\; E^2B^2.$$

where E is the energy of the electron whose peak frequency is ν. The function $N(E)dE$ is called the *energy distribution* of the radiating electrons. The resultant spectrum shape is determined by the dependence of $N(E)dE$ on E. The declining power-law spectra of synchrotron sources, therefore, requires that $N(E)$ also be a declining power law. In Figure 3.12 we show how a declining power-law spectrum is created by adding up the synchrotron emission from a distribution of electrons with a declining energy distribution.

To represent the electron energy distribution as a declining power law we write

$$N(E)dE = N_0E^{-p}dE$$

where p is the *power-law index* of the electron energy distribution and is related to α, the spectral index. Note that we defined p with a negative sign explicitly shown in front of p, meaning that a positive p implies an energy distribution which *decreases* with larger energy, but we defined α, without the negative sign, meaning that a positive α indicates a spectrum whose intensity *increases* with frequency. Some astronomers choose to define α with the negative sign, so be sure to check which definition they are using. To avoid confusion, whenever astronomers discuss the spectral index they *should* be clear about which definition of α they are using.

Inserting the power-law proportionalities for $P(\nu)$ and $N(E)$ above, and ignoring the B dependence, we require that

$$\nu^\alpha d\nu \propto (E^{-p}dE)E^{\;2} \propto E^{2-p}dE.$$

By Equation 3.27 we have that $\nu \propto E^2$, and so $d\nu \propto 2E\ dE$. Substituting these into the left-hand side we have

$$E^{2\alpha}2E\ dE \propto E^{2-p}dE.$$

Regardless of all the other factors in the equation, these power indices must be equal, and so we require that

$$\alpha = \frac{1-p}{2}. \tag{3.30}$$

Keep in mind when using Equation 3.30 that α is defined as in Equation 3.29.

This is a useful result. It provides a means of inferring physical properties of a synchrotron source from its spectrum. By making a log-log plot of the spectrum, $\log(I_\nu) \propto \alpha \log \nu$, and measuring the slope, we can use Equation 3.30 to infer the power index of the energy distribution of the synchrotron electrons.

Example 3.11:

The birth of radio astronomy, marked by the first detection of extraterrestrial radio-frequency signals, was an observation of radio emission from the Galaxy by Karl Jansky in 1932. This emission we now know is due to synchrotron radiation, indicating that the plane of the Galaxy is infused with relativistic charged particles (cosmic rays) in a Galactic magnetic field. The spectrum of the Galactic emission has been determined to have a spectral index of $\alpha = -0.8$. According to these data, what is the power-law index of the energy distribution of the interstellar cosmic rays?

Answer:
Using Equation 3.30, the power-law index of the energy distribution is

$$p = 1 - 2\alpha = 2.6,$$

meaning that the energy distribution of the cosmic rays is

$$N(E)dE \propto E^{-2.6}.$$

Typically, we see synchrotron sources with $\alpha \sim -0.4$ to -1.0 (the most common spectral indices are from -0.7 to -0.8). Therefore, the electrons responsible for the synchrotron radiation must have energy power law indices of $p \sim 1.8$ to 3.0 (with the most common values of 2.4 to 2.6) . Theoretical models for the acceleration of electrons to relativistic energies, therefore, must be able to explain this energy distribution. An additional measure of the energy distribution of relativistic particles is provided by studies of cosmic rays entering the Earth's upper atmosphere. Cosmic rays with energies up to 10^{18} eV have been found to have a power-law energy distribution with $p = 2.8$. This is in good agreement with $p = 2.6$ found in Example 3.11.

3.3.4 Polarization of Synchrotron Radiation

Unlike thermal radiation, in which the processes are completely random, there is some order to the synchrotron emission process. There is a direction to the magnetic field and the electron's motion is directed toward the observer when emitting the radiation detected by the observer. Not surprisingly, then, synchrotron radiation can have a net polarization. This can be important and useful for a number of reasons. In Section 2.2.3 we showed that the polarization of radiation can be Faraday rotated as it travels through the interstellar medium and the amount of Faraday rotation can be used (in combination with the dispersion measure) to infer the average magnetic field along the line of sight. The amount of net polarization in a synchrotron source can also be used to estimate the amount of order in the source's magnetic field (see below). Without derivation, we briefly discuss the polarization of synchrotron radiation.

From an ensemble of electrons in a uniform magnetic field the received synchrotron emission will have a net linear polarization. As shown in the text *Radiative Processes in Astrophysics* by George Rybicki and Alan Lightman,[3] the fractional polarization (or "degree of polarization") in this case is

$$\frac{p+1}{p+7/3}$$

where p is the power-law index of the electron energy distribution, which is equivalent to

$$\frac{1-\alpha}{5/3-\alpha},$$

where α is the spectral index. Since α is typically of order -0.7, a source with uniform magnetic field can have fractional linear polarizations as large as 0.72. There are, in fact, some synchrotron sources which contain regions of relatively uniform magnetic fields.

If, on the other hand, the magnetic field direction in the source varies, including all orientations, the angle of polarization will also vary and will average to zero. We see, therefore, that determination of the net polarization of a synchrotron source provides information about the level of uniformity of the magnetic field, or lack thereof.

3.3.5 Optical Depth Effects: Synchrotron Self-Absorption

Our discussion concerning synchrotron emission so far has assumed that the emission is optically thin. However, as the intensity of the synchrotron radiation increases, absorption can become important. Absorption is the reverse process of emission, and in this case a relativistic electron in a magnetic field can absorb a photon of the appropriate energy. In a synchrotron radiation source, when an electron emits a photon there is some probability that the photon can then be absorbed by another electron. Since the absorptions are due to the same population of electrons that do the emitting, this process is called *synchrotron self-absorption*, often abbreviated *SSA*. The "self" in this definition emphasizes the difference from when the synchrotron emission is absorbed by other gas along the line of sight. In particular, another relevant mechanism of absorption of synchrotron emission is free-free absorption by non-relativistic, free electrons along the line of sight, which we address in Chapter 9.

Deriving either the emission coefficient or the absorption coefficient for synchrotron radiation is quite complex. Here we will only address the frequency dependence where self-absorption dominates, that is, where the emission is optically thick. We discussed earlier that for thermal radiation the competition between the emission and absorption processes leads to a limit to the intensity of thermal radiation given by the Planck function. A similar competition between emission and absorption processes for the synchrotron mechanism also leads to an upper limit on the intensity of the synchrotron emission. In this case, synchrotron self-absorption limits the brightness temperature of the radiation to be less than the effective temperature of the electrons.

---

[3] John Wiley & Sons, 1979.

So what is the effective temperature of relativistic electrons? Since the electrons do not have a thermal energy distribution, we cannot use the conventional definition of temperature; instead we define the effective temperature of the electrons to be $T_e \sim E/k$. However, there is a broad range of electron energies, and thus the electrons have no single effective temperature. Recall that synchrotron electrons of a given energy will emit radiation that is steeply peaked at a particular frequency, which we called the synchrotron peak frequency. Similarly, a relativistic electron is most likely to absorb a synchrotron photon of frequency equal to (or very close to) its synchrotron frequency. Therefore, we can think of the electrons of any given energy as interacting with photons of only the corresponding frequency. Hence, relativistic electrons of a particular energy or effective temperature only affect the emission at a particular frequency. We will indicate the effective temperature of the electrons by $T_{e,\nu}$ to indicate that the radiation at different frequencies interacts with electrons of different effective temperatures.

Let's now consider the brightness temperature of the radiation. At the extremely high brightness temperatures of synchrotron sources at radio frequencies, the Rayleigh-Jeans approximation always applies, and so the T_B in Equation 3.13,

$$T_B = \frac{c^2}{2k\nu^2} I_\nu.$$

provides a measure of the effective temperature of the radiation at any particular frequency.

We first determine at which frequencies the self-absorption is more important. We do this by finding where the optically thin spectrum would have a brightness temperature larger than the effective temperature of the electrons that the radiation interacts with. Inserting the optically-thin power-law dependence of the intensity, then, and focusing on just the frequency dependence, the brightness temperature would depend on frequency as

$$T_B \propto \nu^{-2} \nu^{\alpha} \propto \nu^{\alpha - 2}.$$

Since α is a negative number, and typically ≈ -0.7, the brightness temperature of the optically thin radiation increases dramatically toward lower frequencies. Considering the effective temperature of the electrons that interact with the radiation of frequency ν, we use the relation between the synchrotron peak frequency and the electron energy given by Equation 3.27, which shows that $\nu_{peak} \propto E^2$, and since the effective temperature of the electrons is proportional to the electron energy, we have

$$T_{e,\nu} \propto \nu^{1/2},$$

which increases toward higher frequencies. We see, therefore, that the ratio of the radiation brightness temperature to the effective temperature of the radiating electrons would be greater at lower frequencies and, hence, the self-absorption will be more important at lower frequencies.

When self-absorption does occur, the resultant radiation will have a brightness temperature equal to the effective temperature of the interacting electrons. We can, then, model the shape of the self-absorbed synchrotron spectrum by equating these temperatures. By substituting the electron effective temperature in for brightness temperature in the Rayleigh-Jeans approximation, and keeping only the frequency dependencies we have

$$I_\nu(\text{optically thick}) = \frac{2k\nu^2}{c^2} T_{e,\nu} \propto \nu^2 \, \nu^{1/2} \propto \nu^{5/2}. \tag{3.31}$$

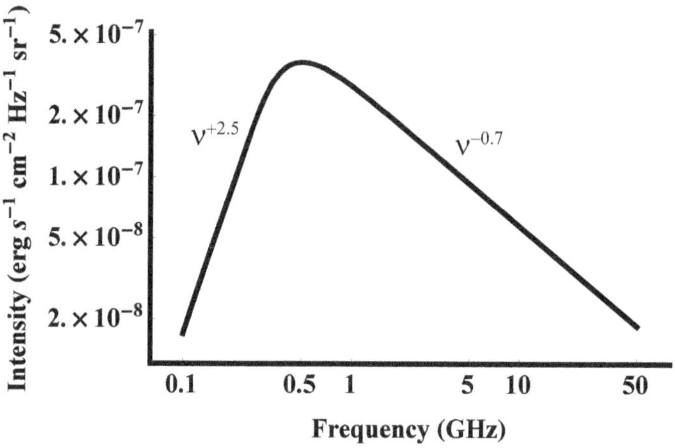

Figure 3.13: Composite synchrotron spectrum from a homogeneous source with significant synchrotron self-absorption at lower frequencies.

This is an intensity that *increases* with increasing ν. Since synchrotron self-absorption is greatest at very low frequencies, the total synchrotron spectrum for relativistic electrons with energy distribution $N(E) = N_0 E^{-p}$, and uniform density, in a uniform magnetic field B, will look like that shown in Figure 3.13.

As with bremsstrahlung radiation, the peak of the spectrum occurs at the frequency where the optical depth is of order unity and we, again, refer to this frequency as the *turnover frequency.* This occurs at the frequency where the brightness temperature of the emission in the absence of absorption is equal to the effective temperature of the electrons emitting at that frequency. The SSA turnover frequency depends on the magnetic field and the intensity as

$$\nu_{\mathrm{SSAturnover}} \propto B^{1/5} I_\nu^{2/5}. \tag{3.32}$$

Let's take a moment to examine this dependence and make sense of it. If the radiation intensity is large, then there is more energy in the photons and so absorption of the photons should occur at more frequencies, and hence the turnover frequency will occur at a higher frequency; this is consistent with Equation 3.32. And, if the magnetic field is strong, then for any electron energy the emission will occur at a higher frequency and so all effects, including the turnover of the spectrum, should shift upward in frequency; this also agrees with Equation 3.32.

Example 3.12:

Imagine a spherical synchrotron source with a uniform magnetic field of 10^{-2} gauss, angular diameter of 3 milli-arcseconds, and a spectral peak flux density of 10 Jy occurring at 1 GHz. If, instead, the source's diameter were 5 times larger, what would its flux density have to be for its spectral turnover to still occur at 1 GHz?

Answer:

The dependence of the SSA turnover frequency on the given parameters is shown in Equation 3.32. With the magnetic field and turnover frequency both staying constant, the intensity must also stay constant. Therefore, since $F_\nu = I_\nu \times \Omega = I_\nu \pi \Theta^2/4$, keeping the turnover frequency constant with a five-fold increase in Θ requires an increase in F_ν of a factor of 25.

The new flux density, then, is 250 Jy. The production of a given flux density of synchrotron emission requires a certain number of relativistic electrons, but the presence of an SSA turnover requires that the source be relatively opaque to the synchrotron photons. The ensemble of electrons will have a larger optical depth if those electrons are contained in a dense, and hence more compact volume. We see, then, that an SSA turnover at a frequency as high as 1 GHz requires either a very large flux density or that the source be compact.

Although we have given only an heuristic derivation, more careful and thorough efforts yield the same spectral shape. However, a spectrum as steep as $I_\nu \propto \nu^{2.5}$ is never observed. This was once considered a paradox. The reason is, in fact, quite simple. There is nothing wrong with the theory. The issue involves the structure of real sources. The discussion above applies only to homogeneous sources, and astronomical objects are never really homogeneous (for example, the electron densities are usually larger closer to the object centers). We discuss the effect of non-homogeneous sources in Chapter 9, in the context of Active Galactic Nuclei.

QUESTIONS AND PROBLEMS

1. Figure 3.14 displays a spectrum, as I_ν vs. ν, at radio frequencies of a source.

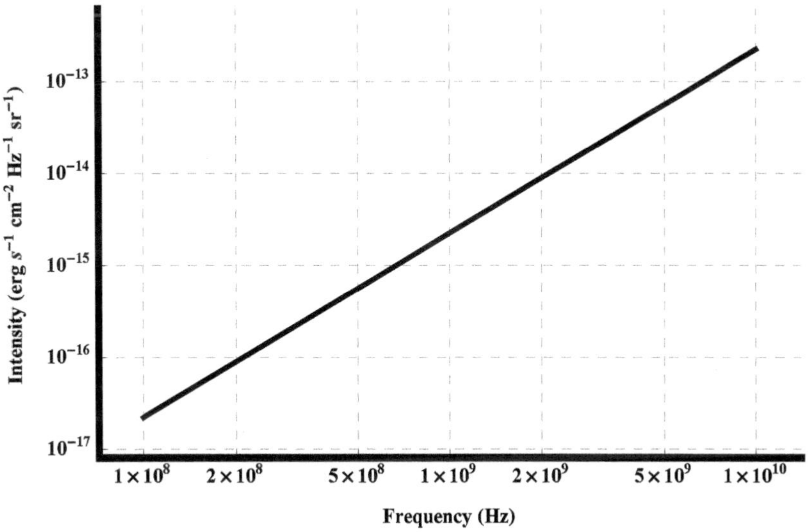

Figure 3.14: Radio-frequency spectrum of a particular source.

   (a) Calculate the brightness temperature at 2.0 GHz.

   (b) Can the plotted spectrum be thermal? Explain; give arguments why or why not.

   (c) Is the emission object optically thick or thin?

   (d) What is the gas temperature of the radiation source?

2. (a) Derive the Rayleigh-Jeans approximation by doing a Taylor-series expansion of $e^{h\nu/kT}$, where $h\nu/kT \ll 1$, and substitute into the Planck function.

(b) Considering that the high frequency edge of the radio window is at 1000 GHz, determine the minimum temperature that a source can have for the Rayleigh-Jeans Law to apply at all radio wavelengths. Estimate the minimum temperature, first by using the assumption implicit in the Rayleigh-Jeans Law by requiring $h\nu/kT \leq 0.1$, and then check your answer by solving for $B_\nu(T)$ using both the Planck function and the Rayleigh-Jeans Law and requiring that the two agree to better than 10%.

(c) Considering that the entire Universe bathes in the Cosmic Microwave Background, which is at a temperature of 2.73 K, determine the observing frequency range in which the Rayleigh-Jeans Law is always a safe approximation.

3. In Section 3.2.2 the Rayleigh-Jeans approximation for B_ν at small frequencies ($h\nu/kT \ll 1$) was presented.

   (a) Convert the Rayleigh-Jeans approximation for B_ν to B_λ. At what wavelengths does this approximation hold?

   (b) Perform a Taylor series approximation on the exponential in B_λ to derive the Rayleigh-Jeans approximation for B_λ. Does your answer agree with that in (a)?

4. Radiation from an opaque, thermal emission source has an intensity of 5.75×10^{-12} erg s^{-1} cm^{-2} Hz^{-1} sr^{-1} at 500 MHz. What is the temperature of the source?

5. An opaque, thermal emission source of 5000 K is spherical with a radius of 7.00×10^{10} cm.

   (a) What is the average energy of the photons emitted?

   (b) At what frequency is the peak in I_ν?

   (c) What is the total power emitted, including all frequencies?

6. The Cosmic Microwave Background (CMB) radiation fits a blackbody spectrum of T = 2.73 K.

   (a) At what frequency is the peak of I_ν of this radiation?

   (b) What is the average energy of the CMB photons?

7. An electron starting from rest is accelerated across a capacitor gap containing a uniform electric field E and a gap w.

   (a) What is the power radiated by the electron during this process?

   (b) What is the total energy emitted by the electron?

8. What is bremsstrahlung radiation? Why does its detection necessarily indicate a region of ionized gas?

9. The spectrum shown in Figure 3.6 is for a 10,000 K bremsstrahlung source. What is the source's emission measure?

10. A particular substance has an emission coefficient for a thermal radiation process given by

$$j_\nu = 5 \times 10^{-15} \text{ erg s}^{-1} \text{ cm}^{-3} \text{ Hz}^{-1} \text{ sr}^{-1} \frac{T^{0.5}}{\nu}.$$

   (a) What is the absorption coefficient, κ_ν, of this substance?

(b) If a cloud of this substance has a uniform temperature of 3000 K and a linear depth of 1000 km, what is its optical depth at $\nu = 10$ GHz along the direction of its length?

(c) What is the intensity of the radiation, as a function of frequency, emitted by this medium (in the direction of its length)?

11. The continuum radiation from a uniform, spherical ionized gas region is observed at frequencies of 2 and 20 GHz. The cloud contains only hydrogen, and has uniform structure with $T_e = 9{,}000$ K, n_e=10,000 cm^{-3}, and diameter = 1 pc.

    (a) What is the turnover frequency of the radiation from this cloud?

    (b) What is the free-free optical depth at each of the observed frequencies?

12. An ionized gas cloud is determined to have a free-free optical depth at $\nu = 8$ GHz of 0.3.

    (a) What is the free-free optical depth of this same cloud at $\nu = 2$ GHz?

    (b) If the brightness temperature of the radiation from the cloud at 8 GHz is 500 K, what is the brightness temperature at 2 GHz?

    (c) Write an approximate expression for the brightness temperature as a function of frequency for frequencies greater than 8 GHz.

    (d) Write an approximate expression for the brightness temperature as a function of frequency for frequencies less than 2 GHz.

13. Cyclotron radiation has been detected from the area around Jupiter at frequencies from to 5 to 30 MHz. What magnetic field strengths can we conclude must exist in Jupiter's magnetosphere to explain this radiation?

14. What is synchrotron radiation? What is it due to? What does the optically thin synchrotron spectrum look like? What does this spectrum tell us about the radiating particles? Comment on the connection to cosmic ray particles.

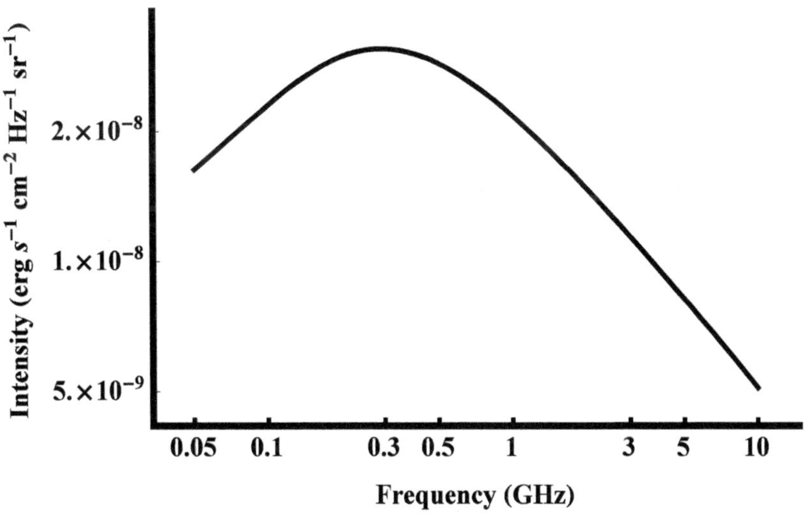

Figure 3.15: Log-log plot of the spectrum of a particular synchrotron source. The peak of the spectrum occurs at 300 MHz where the intensity is 3.11×10^{-8} erg s^{-1} cm^{-2} Hz^{-1} sr^{-1}.

15. In the derivation of Equation 3.25 in Section 3.3.2, we stated that for velocities very close to c, the factor $(1 - \frac{v}{c})$ is approximately equal to $1/2\gamma^2$. Show that this is true.

16. What is the frequency of the peak synchrotron emission from electrons with $\gamma = 400$ immersed in a magnetic field of 300 milligauss?

17. Figure 3.15 displays the spectrum as $\log(I_\nu)$ vs $\log(\nu)$ of a particular synchrotron source. Explain why the slope of the spectrum differs at lower frequencies from that at higher frequencies; why does the spectrum increase with frequency at the lower frequencies and decrease with frequency at the higher frequencies? What physical process is responsible for causing this difference?

CHAPTER 4

Spectral Lines

SPECTRAL lines result from the interaction of light with atoms, ions and molecules. The development of quantum mechanics, describing the behavior of matter and its interaction with light on atomic scales, led to the realization that certain properties (such as angular momentum or energy) of an atom or molecule are restricted to a discrete or quantized set of values. Changes of an atom's energy, for example, can only occur between these discrete sets of values. Light is also quantized into discrete particles of energy or photons, and unlike the processes we described in Chapter 3, the interaction of light with electrons bound to atoms and molecules usually involves only a single photon. In the following discussion we assume the reader has some background in modern physics.

Let us start our discussion by reviewing the electronic states of the hydrogen atom and its possible photon interactions. With regards to the electron's orbitals, the allowed *energy states* of the hydrogen atom, labeled by their principal quantum number n, are given by

$$E_n = -13.60 \left(\frac{1}{n^2}\right) \text{ eV}, \tag{4.1}$$

where 1 eV (electron volt) $= 1.60 \times 10^{-12}$ ergs. The lowest energy state (or the *ground state*) is denoted by $n = 1$, while higher values of n indicate higher energy states, often referred to as *excited states*. Note that the total electronic energy (kinetic and electric potential energy) is negative, as one would expect if the electron is bound to the nucleus. Often the energies of the excited states are measured relative to the energy in the ground state. In this case the relative energy levels of hydrogen are given by

$$E_n = 13.60 \left(1 - \frac{1}{n^2}\right) \text{ eV}.$$

A useful way to depict the allowed energy states of an atom or molecule is an *energy level diagram* that has energy on the vertical axis and the allowed energy states indicated as horizontal lines. The energy level diagram for the orbital energy of the electron in the hydrogen atom is shown in Figure 4.1. Note that the electronic energy levels for hydrogen get closer and closer together as the principal quantum number grows larger. If the energy added to an electron in the ground state of the hydrogen atom exceeds 13.60 eV, then the electron will not be bound to the nucleus and the atom is said to be ionized. For this reason, 13.60 eV is called the *ionization energy* of hydrogen.

The electron in a hydrogen atom can either gain energy or lose energy (if not in the ground state) in interactions with other particles or with radiation. In all cases, energy is conserved. Collisions between atoms and between atoms and free electrons can exchange electron orbital energy with the kinetic energy of the colliding particle. The electron can

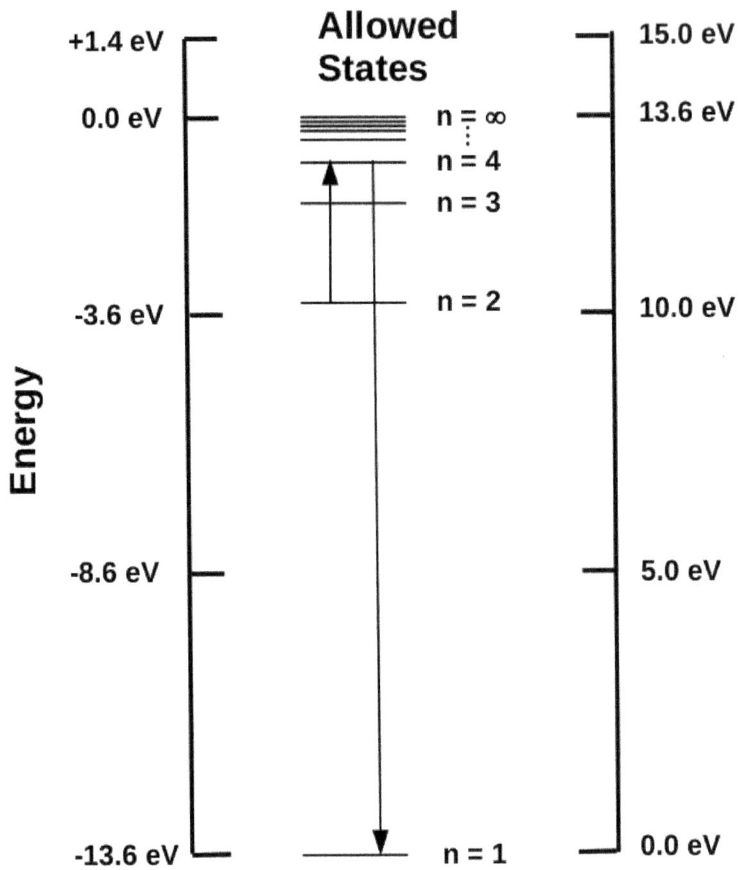

Figure 4.1: An energy level diagram for the electronic energy levels of the hydrogen atom. The allowed states are denoted by the principal quantum number n. The vertical axis on the left shows the total kinetic and electric potential energy, which is negative for the allowed bound states. If the energy is positive then the electron is unbound. The vertical axis on the right indicates the energy relative to the ground state.

gain energy at the expense of the colliding particles and move from a lower energy state to a higher energy state. Alternatively, collisions can transfer orbital energy of an excited electron to the colliding atom, moving the electron from a higher energy state to a lower energy state.

The electron in the hydrogen atom can also absorb or emit individual photons provided the photons have precisely the correct amount of energy to change the electron from its current energy state to another allowed energy state. Thus, the photon must have the precise wavelength or frequency that corresponds to that energy. For example, a hydrogen atom with its electron in the $n = 2$ state can absorb a photon that has precisely the amount of energy needed to put the electron into one of the excited states with $n \geq 3$. In Figure 4.1, the upward arrow shows a hydrogen atom changing from the $n = 2$ state to the $n = 4$ state. The photon absorbed by the atom will need an energy that is the difference between the energies of the $n = 4$ and $n = 2$ states, or about 2.55 eV. Remember that the energy of a photon is $E = h\nu$ or $E = hc/\lambda$; therefore the corresponding wavelength of this photon is about 486.1 nm (a visible wavelength). A hydrogen atom with its electron in the $n = 4$ state can spontaneously emit a photon and drop to any of the lower states. For instance, if the electron drops to the $n = 1$ state, as represented by the downward arrow in Figure 4.1, the photon carries away the energy difference (about 12.75 eV) between the two states. This photon has a wavelength of about 97.2 nm (a ultraviolet wavelength).

Changes of energy states involving the absorption or emission of a photon are called *radiative transitions*, while energy changes involving interactions with other particles are called *collisional transitions*. We will return to the hydrogen atom later in this chapter as it is responsible for producing some important radio wavelength spectral lines, but first we discuss radiative transitions in general.

Example 4.1:

What is the wavelength of a photon emitted during a radiative transition when an electron in a hydrogen atom changes from the $n = 3$ state to the $n = 2$ state ?

Answer:
First we compute the energies of the $n = 2$ and $n = 3$ states using Equation 4.1.

$$E(n=2) \;=\; -13.60\left(\frac{1}{2^2}\right)\,\text{eV} \;=\; -3.40\ \text{eV}.$$

$$E(n=3) \;=\; -13.60\left(\frac{1}{3^2}\right)\,\text{eV} \;=\; -1.51\ \text{eV}.$$

The difference in energy is 1.89 eV or $3.02{\times}10^{-12}$ erg. Since $E = hc/\lambda$, the wavelength is given by:

$$\lambda \;=\; \frac{hc}{E} \;=\; \frac{(6.63\times10^{-27}\ \text{erg s})\ (3.00\times10^{10}\ \text{cm s}^{-1})}{3.02\times10^{-12}\ \text{erg}}$$

$$= \; 6.56\times10^{-5}\ \text{cm} \;=\; 656\ \text{nm}.$$

This visible wavelength emission line is the well-known Balmer α line, usually denoted as Hα.

4.1 EMISSION AND ABSORPTION LINES

In order to infer any information about the conditions of an astronomical source from its observed spectral lines, we need an understanding of the physics that determines the intrinsic strength of a particular emission or absorption line. We can characterize the intrinsic line strengths using parameters called the *Einstein coefficients*, while a second formulation, which we will only mention briefly, is to use a parameter called the *oscillator strength*. As you read the following sections, keep in mind that both atoms and molecules can undergo radiative transitions, emitting and absorbing photons. For the sake of brevity we often only mention atoms; however the discussion that follows applies equally well to molecules. In fact, at radio frequencies, most of the spectral lines detected are due to molecules not atoms. Later in this chapter we will give a more detailed discussion of the molecular spectral lines that are commonly observed at radio frequencies.

4.1.1 Einstein Coefficients

The Einstein coefficients provide a measure of the probability that a given atom or molecule undergoes a given transition by the emission or absorption of radiation. These coefficients can be computed using quantum mechanics or they can be measured in the laboratory; however most radio astronomers use tabulated values of these rate coefficients. Radiative transitions can occur via either electric or magnetic interactions. The most common type of interaction is called an *electric dipole transition* in which the radiation interacts with the atom's electric dipole. However, other interactions are possible, leading to magnetic dipole transitions or electric quadrupole transitions (even higher moments are also possible, but not very important). The electric dipole transitions have the largest line strengths, followed by magnetic dipole transitions and finally electric quadrupole transitions. Each type of transition has rules, called selection rules, that determine whether such a radiative transition between two states can occur. Most radiative transitions detected are electric dipole transitions and are often referred to as *permitted transitions.* However, some changes of energy states do not involve permitted electric dipole transitions, and in these cases magnetic dipole or electric quadrupole transitions may connect the states. These magnetic and higher-dipole transitions occur at much slower rates than electric dipole transitions. Because these transitions do not obey the electric dipole selection rules and proceed slowly, they are often referred to as *forbidden transitions.*

Radiative transitions can occur by three different processes: (1) *spontaneous emission,* (2) *absorption* and (3) *stimulated emission.* There is an Einstein coefficient defined for each of these processes. Unfortunately, the exact definitions of the Einstein coefficients for absorption and stimulated emission, in terms of their units, vary throughout the literature. The conceptual meanings are the same in all systems, but the exact way to use them differs mathematically. For our discussion, we will introduce the coefficients following only one convention.

4.1.1.1 Spontaneous Emission

An electron in an excited energy state can, on its own, drop to a lower allowed energy state, while emitting a photon that carries away the energy lost by the electron. This process is illustrated in Figure 4.2 and this type of radiative transition is called spontaneous emission. The atom does not undergo spontaneous emission immediately upon entering the excited state, but waits some period of time that follows probability statistics and can be computed from quantum mechanics. The inverse of the time the electron spends in an excited state is related to the rate of spontaneous emission. One way of describing the average rate at which an atom undergoes a particular spontaneous emission transition is by using the *Einstein A coefficient,* which is defined as the transition probability per unit time. If an atom is in an upper state, u, the probability per unit time that it will spontaneously drop to a lower state, l, is given by A_{ul}. Because A_{ul} is a rate, it has units of s^{-1}.

One can also view the inverse of A_{ul} as the average time an atom in state u will take to transition to state l. The atom may also have other lower energy states that it can drop to, and each of these transitions has its own rate, described by its own Einstein A coefficient. The average time an individual atom spends in the upper state, then, before undergoing spontaneous emission, is the inverse of the sum of all allowed spontaneous transition rates out of the upper level. The average time in the state, or the *lifetime,* is therefore given by

$$t_{\mathrm{life}} = \left(\sum_{l} A_{ul} \right)^{-1}, \text{ for all } l < u. \tag{4.2}$$

For example, the A coefficient for the electronic transition in hydrogen from the $n = 4$ to $n = 1$ level (see Figure 4.1) has a value of $6.8 \times 10^{7}\ \mathrm{s}^{-1}$. Including other allowed transitions from the $n = 4$ level, the lifetime of this excited state is only about 8×10^{-9} s.

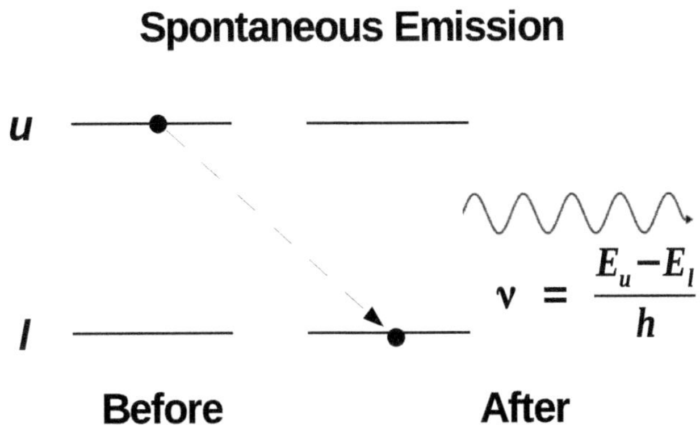

Figure 4.2: The spontaneous emission process is illustrated. An atom starts in an upper energy state and makes a spontaneous change to a lower energy state. The energy difference between the upper and lower states is carried away by a single photon.

In order for the Einstein rate coefficients to be useful we need to explain how they relate to observations of emission and absorption lines. In Section 2.1 we discussed the relation between the observed intensity and the emission and absorption coefficients of the radiative medium. We now derive the connection between the Einstein A_{ul} coefficient and the emission coefficient, j_ν, which you may recall has units of intensity per unit length, or erg s^{-1} cm^{-3} Hz^{-1} sr^{-1}.

For atoms undergoing transitions between a given upper state, u, to a lower state, l. with an energy difference, $\Delta E_{ul} = E_u - E_l$, we describe the rate of spontaneous emission of photons of frequency $\nu = \Delta E_{ul}/h$. For a large number of atoms, we can get the rate of spontaneous emissions per unit volume by multiplying the rate, A_{ul}, by the number density of atoms (number of atoms per unit volume) in the upper state of this transition, n_u. Therefore the rate of spontaneous emission per unit volume is $n_u A_{ul}$. For an ensemble of randomly oriented atoms, the photons are emitted in all directions so the emission is isotropic. Since a sphere contains 4π steradians of solid angle, dividing by 4π steradians gives the emission per unit solid angle. Putting this together, we find that the energy, dE, emitted per second, dt, per unit volume, dV, per unit solid angle, $d\Omega$, is given by

$$\frac{dE}{dt\; dV\; d\Omega} = \frac{\Delta E_{ul}\; n_u\; A_{ul}}{4\pi}.$$

Due to a number of important effects, collectively called *line broadening*, the photons produced by spontaneous emission are not all precisely at the same frequency, but are spread over a small frequency range that we will denote by $\delta\nu$. If the spontaneous emission is spread uniformly over a frequency width $\delta\nu$, then the emission coefficient, defined in Section 2.1, of a spectral line is given by

$$j_\nu = \frac{h\nu\; n_u\; A_{ul}}{4\pi\; \delta\nu}, \tag{4.3}$$

where we substituted $h\nu$ for ΔE_{ul}. In reality the emitted photons are not spread uniformly over the frequency width $\delta\nu$, and the distribution of emitted photons is characterized by a quantity called the *line profile function*, denoted by ϕ_ν. Thus, for an emitted photon, the line profile function describes the probability per unit frequency that a photon will be emitted at a particular frequency. The line profile function is normalized so that the integral over all frequencies is $\int \phi_\nu \, d\nu = 1$, and thus the units of ϕ_ν are inverse frequency. Including the line profile function, the line emission coefficient is

$$j_\nu = \frac{h\nu \, n_u \, A_{ul}}{4\pi} \, \phi_\nu. \tag{4.4}$$

In Section 4.1.2 we will discuss the effects that produce line broadening and the functional form of the line profile function.

4.1.1.2 Absorption

An atom in a lower energy state can absorb a passing photon of just the right energy (and therefore just the right frequency) to excite the atom into a higher allowed energy state. In this process the energy of the photon is entirely consumed and is converted into energy of the electron. This process is illustrated in Figure 4.3. As with spontaneous emission, the atom has a particular probability of undergoing this radiative transition, and this is, in part, described by the *Einstein B coefficient.* However, there is a difference between spontaneous emission and absorption. With absorption an input photon is needed, so the rate of absorption must also depend on the rate of incident photons and therefore the intensity of the radiation at the appropriate frequency. The Einstein B coefficient for absorption is labeled B_{lu}, and the probability per unit time, or rate, that an atom in a lower state, l, will absorb a photon and move to the upper state, u, is given by $B_{lu} I_\nu$, where I_ν is the intensity of the radiation field at the transition frequency. Therefore the units of B_{lu} must be erg^{-1} cm^2 Hz sr. Be careful, as there are other definitions for the B coefficient.

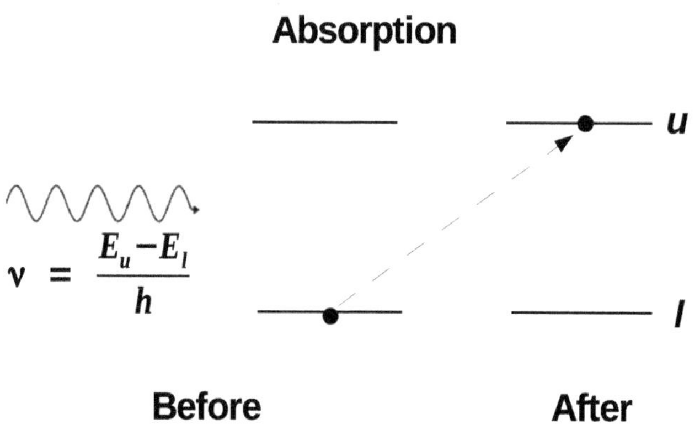

Figure 4.3: The absorption of radiation is illustrated. An atom starts in a lower energy state and absorbs a single photon to end up in an upper energy state. For this process, the photon must have an energy equal to the difference in energy between the upper and lower states.

The rate of photon absorption per unit volume is given by $n_l B_{lu} I_\nu$, where n_l is the number density of atoms in the lower energy state of this transition. If we want to express this as the rate at which photon energy is absorbed per unit volume per unit solid angle, we need to multiply by the photon energy and, since the absorption process for an ensemble of randomly oriented atoms is independent of the direction of incident photons, we need to divide by the number of steradians in a sphere (4π). Therefore the change in energy in the radiation field, dE, per unit time, dt, per unit volume, dV, and per unit solid angle, $d\Omega$, is

$$\frac{dE}{dt\ dV\ d\Omega} = -\frac{h\nu}{4\pi}\, n_l\, B_{lu}\, I_\nu.$$

Finally, as in the case of spontaneous emission, the absorption of photons is not at a single frequency but spread over a range of frequencies due to line broadening. Therefore the amount of light energy absorbed per unit time, per unit volume, per solid angle, and per unit frequency, $d\nu$, is given by

$$\frac{dE}{dt\ dV\ d\Omega\ d\nu} = -\frac{h\nu}{4\pi}\, n_l\, B_{lu}\, I_\nu\, \phi_\nu.$$

Before we can relate the Einstein B coefficient to the absorption coefficient that we defined in Chapter 2, we first need to consider the process called stimulated emission.

4.1.1.3 Stimulated Emission

The third type of radiative process is called stimulated emission. This type of radiative process occurs when an atom in an excited state interacts with a photon of energy exactly equal to the energy that the atom would emit were it to undergo spontaneous emission (see Figure 4.4). The incident photon essentially provides a stimulus that induces the atom to drop down to the lower energy level immediately, emitting a photon. A special aspect of the emitted photon is that it has exactly the same energy and the same *phase* as the incident photon, so the photon that stimulated this process and the emitted photon are *coherent*. This property of stimulated emission is a key aspect in the functioning of lasers and can also occur naturally in some astronomical sources, which we call masers.

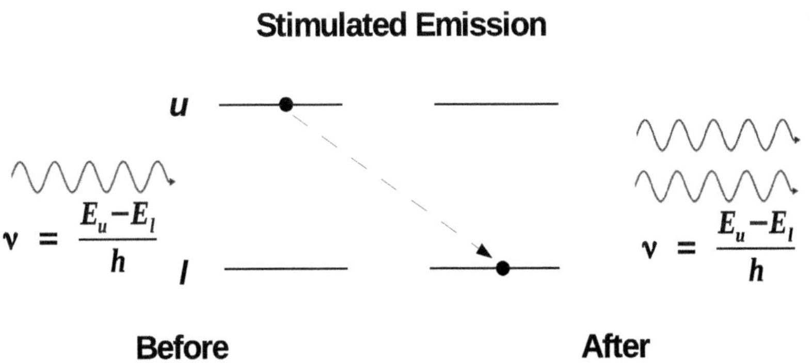

Figure 4.4: The process of stimulated emission is illustrated. An atom starts in an upper energy state and a photon of energy equal to the energy difference between the upper and lower states stimulates the atom to emit a photon and drop down to a lower energy state. The photon emitted has the same frequency as the photon providing the stimulus. The stimulating photon and the emitted photon are coherent with one another.

As in absorption, the rate of stimulated emission not only depends on the properties of the atom but also on the intensity of the incident radiation. The Einstein coefficient for stimulated emission from u to l is given by B_{ul}; it is labeled with a B because it involves an incident photon, but is different from the Einstein coefficient for absorption in the order of the subscripted states. Note that there is a significant difference between B_{lu} and B_{ul}, so be careful when writing these subscripts. The rate of stimulated emission is given by $B_{ul}I_\nu$. Following the same line of reasoning as for absorption, the rate that energy is emitted by stimulated emission per unit time, per unit volume, per unit solid angle and per unit frequency is given by

$$\frac{dE}{dt\ dV\ d\Omega\ d\nu} = \frac{h\nu}{4\pi}\ n_u\ B_{ul}\ I_\nu\ \phi_\nu.$$

The stimulated emission process adds to the intensity of the radiation field, while absorption removes intensity from the radiation field.

4.1.1.4 Absorption Coefficient

Remember from Section 2.1.1 that the absorption coefficient is the macroscopic measure of a medium's ability to absorb radiation at a particular frequency. Our goal is to relate the B coefficients to the absorption coefficient. We can view stimulated emission as negative absorption, and incorporate both absorption and stimulated emission processes in our formulation of the absorption coefficient. Combining these processes we find that the total energy change in the radiation per unit time per unit volume per unit solid angle and per unit frequency is given by:

$$\frac{dE}{dt\ dV\ d\Omega\ d\nu} = -\frac{h\nu}{4\pi}\ [n_l\ B_{lu} - n_u\ B_{ul}]\ I_\nu\ \phi_\nu.$$

In thermodynamic equilibrium, the absorption process always removes more energy than is added by stimulated emission, and therefore the change in energy is negative. However, in special cases, such as in a laser and in astronomical masers, which are out of thermodynamic equilibrium, the stimulated emission can be larger than the absorption and the medium will amplify an incoming signal. We will discuss astronomical masers later in this chapter.

From the definition of intensity (see Section 1.2.4), the change in intensity, dI_ν, is a change of energy per unit time, per unit area, per unit solid angle, and per unit frequency. However the expression derived above gives the change per unit volume and not per unit area. The volume element, dV, is equivalent to $dA\ ds$, where dA is an area element and ds is a path element. Using this relation, we can rewrite the above expression for the change in intensity as

$$dI_\nu = \frac{dE}{dt\ dA\ d\Omega\ d\nu} = -\frac{h\nu}{4\pi}\ [n_l\ B_{lu} - n_u\ B_{ul}]\ I_\nu\ \phi_\nu\ ds.$$

In Section 2.1.1 we defined the change in intensity as light passes through a medium of infinitesimal width ds in terms of the absorption coefficient as

$$dI_\nu = -\ I_\nu\ \kappa_\nu\ ds.$$

From these two relations we can express the absorption coefficient in terms of the B coefficients as

$$\kappa_\nu = \frac{h\nu}{4\pi}\ [n_l\ B_{lu} - n_u\ B_{ul}]\ \phi_\nu. \tag{4.5}$$

We now have expressions that give the emission and absorption coefficients in terms of the A and B coefficients that define the strength of spectral lines.

4.1.1.5 Relations between the Einstein A and B Coefficients

The Einstein A and B coefficients are also related to each other. We can consider the case of thermodynamic equilibrium, in which the rates of radiative excitations and de-excitations between two levels must be equal, to determine the relationship between these coefficients. We equate j_ν and $I_\nu \kappa_\nu$, where $I_\nu = B_\nu(T)$, and find that the relation between the spontaneous emission coefficient and the stimulated emission coefficient is

$$A_{ul} = \frac{2\,h\,\nu^3}{c^2}\,B_{ul}. \tag{4.6}$$

Second, the relation between the B coefficients for absorption and stimulated emission is

$$g_l\,B_{lu} = g_u\,B_{ul}, \tag{4.7}$$

where the quantities g_l and g_u are called *statistical weights* and are a measure of the degeneracy of each energy state. In our discussion concerning the electronic energy levels of the hydrogen atom we only discussed the principal quantum number; in reality one needs four quantum numbers to completely describe the electronic energy states. In addition to the principal quantum number, one also needs an orbital, a magnetic and a spin quantum number. Often, levels with different orbital, magnetic and spin quantum numbers share the same energy and are therefore called degenerate. The statistical weights measure the degree of degeneracy of each energy level. As is the case for the Einstein coefficients, the statistical weights are often tabulated. We can conclude that these relations, Equations 4.6 and 4.7, must also hold true in non-equilibrium conditions, since they do not depend on temperature, density or any other thermodynamic parameter.

An alternative method of characterizing the probability for emission or absorption in a spectral line is using a quantity called the oscillator strength, usually denoted by f. The oscillator strength and the Einstein B coefficient for absorption are related by

$$f_{lu} = \frac{h\nu\,m_e\,c}{4\pi^2\,e^2}\,B_{lu},$$

where m_e is the electron mass and e is the magnitude of the electron charge. Like the Einstein coefficients, the oscillator strengths are often tabulated for spectral transitions.

4.1.2 Line Broadening

As mentioned earlier, the emission or absorption in a spectral line is not at a single frequency, but has a finite frequency width due to line broadening. The line profile function, defined in Section 4.1.1.1, describes the frequency dependence of the emission or absorption across a spectral line. A spectral line can be broadened by various effects; the main two are *natural broadening* and *Doppler broadening.*

Natural broadening is a quantum mechanical effect that results from the *Heisenberg uncertainty principle* and describes the minimum width of a spectral line. A consequence of the uncertainty principle is that the energy of a quantum state can never be known precisely if the lifetime of that state is finite. This is expressed as

$$\Delta E\,\Delta t \geq \frac{h}{2\pi},$$

where ΔE is the uncertainty in the energy of a quantum mechanical state and Δt is the lifetime of the state. The lifetimes of the upper and lower states are given by Equation 4.2.

Due to the finite lifetimes, the Heisenberg uncertainty principle requires that the energy states each have a range of possible values. This, in turn, requires that an emitted or absorbed photon has a range of possible frequencies. From the uncertainty principle we see that ΔE depends inversely on the lifetimes of the states, which depend inversely on the Einstein A coefficients. The amount of natural broadening of a spectral line, therefore, is directly proportional to the sum of the Einstein A coefficients for transitions out of the upper and lower states. For the electronic transitions of hydrogen that were discussed earlier, the A coefficients are very large; therefore natural broadening for these ultraviolet and optical wavelength transitions can be important. However for radio wavelength spectral lines, the A coefficients for both atomic and molecular transitions are very small, so natural broadening is rarely important. For this reason, we will not discuss this broadening mechanism any further.

The most important broadening mechanism for radio wavelength spectral lines is Doppler broadening. This line broadening arises due to thermal, turbulent or macroscopic motions which result in shifts in the frequency due to the Doppler effect. The Doppler effect, discussed in Section 1.4, is produced by the relative radial motion of the observer and a light source. With *thermal broadening*, for instance, the random motions of the atoms and molecules contain a range of velocities, and consequently there will be a distribution of frequencies or wavelengths at which an ensemble of atoms or molecules will absorb or emit. The Doppler shift in frequency (see Equation 1.12) is given by

$$v_R \approx c\,\frac{\nu_e - \nu}{\nu},$$

where ν_e is the frequency at which the line was emitted in the rest frame of the atom (often called the rest frequency), ν is the observed line frequency and v_R the radial velocity of the atom or molecule relative to the observer. As discussed in Section 1.4, astronomers define the radial velocity to be positive if the source and observer are moving away from each other, and in this case the emitted photons will have slightly lower frequency than if there were no relative motion.

For pure thermal motion, the atoms have a Maxwell-Boltzmann velocity distribution and the line profile function has a Gaussian shape. In this case we define the line profile function due to Doppler broadening by

$$\phi_\nu = \pi^{-1/2}\,\Delta\nu_D^{-1}\,e^{-[(\nu-\nu_o)/\Delta\nu_D]^2}, \tag{4.8}$$

where $\Delta\nu_D$ is the Doppler broadening parameter and ν_o is the observed line center frequency. Note that ν_o may not equal ν_e if the entire ensemble of emitting or absorbing atoms is moving relative to the observer. For a gas with only thermal motions, the Doppler broadening parameter is given by

$$\Delta\nu_D = \frac{\nu_o}{c}\left[\frac{2kT_K}{m}\right]^{1/2}, \tag{4.9}$$

where T_K is the gas kinetic temperature and m is the mass of the atom or molecule that is emitting or absorbing. Since the thermal motions increase with increasing gas temperature, the Doppler broadening also increases with increasing temperature. Note that at line center ($\nu = \nu_o$) the value of the line profile function is just $\phi_\nu = \pi^{-1/2}\,\Delta\nu_D^{-1}$.

In addition to thermal motions, the gas atoms and molecules can also have turbulent motions. For instance, the gas might have internal flows, and in some cases these flows, instead of being steady, can become chaotic. The chaotic behavior of flows is called turbulent motion. Such turbulent motions also produce Doppler shifts and can contribute to line broadening. On even larger scales, the gas might have an ordered rotating, expanding or collapsing motion that will produce a range of shifts in the line center frequency. These large-scale motions can also contribute to line broadening. Another line-broadening process, called *pressure broadening*, is discussed later in this book, in Section 6.2.3. We don't discuss it further here, as it is usually unimportant in radio astronomy studies.

A measure of a spectral line width that is often used by radio astronomers is the full width of the spectral line at half of its maximum intensity, often called full-width-half-maximum (FWHM). For a thermally broadened spectral line characterized by the line profile function given in Equation 4.8, it is straightforward to show that the FWHM line width is given by

$$\Delta\nu_{\rm FWHM} = 2\ \sqrt{\ln 2}\ \Delta\nu_D \ = \ 1.665\ \Delta\nu_D.$$

In practice the line width is often given in velocity units rather than frequency units so that

$$\Delta v_{\rm FWHM} = \ \frac{c}{\nu_o}\ \Delta\nu_{\rm FWHM}.$$

For spectral lines that are only thermally broadened, the line width depends on the gas temperature. From the measured FWHM line width, we can determine the Doppler broadening parameter and then use Equation 4.9 to determine the gas temperature. Even if additional line broadening mechanisms are important, the measured FWHM line width still provides an upper limit to the gas temperature.

Example 4.2:

We observe the hydrogen line from Example 4.1, and measure the FWHM line width to be 3.3×10^{10} Hz. If the spectral line is only broadened by thermal motions, what is the temperature of the gas?

Answer:
First, based on the measured FWHM line width, we find that the Doppler broadening parameter is $\Delta\nu_D = \Delta\nu_{FWHM}/1.665 = 1.98 \times 10^{10}$ Hz. The Doppler broadening parameter is given by Equation 4.9; solving for gas temperature we find

$$T_K \ = \ \frac{m}{2k}\left(\frac{\Delta\nu_D\ c}{\nu_o}\right)^2.$$

The mass of a hydrogen atom is 1.67×10^{-24} g and the wavelength of the transition from Example 4.1 is 6.56×10^{-5} cm; thus the frequency of the transition is $\nu = c/\lambda = 4.57\times10^{14}$ Hz. Therefore the temperature is

$$T_K \ = \ \frac{1.67 \times 10^{-24}\ \rm g}{(2)(1.38 \times 10^{-16}\ \rm ergs\ K^{-1})}\left(\frac{(1.98 \times 10^{10}{\rm Hz})(3.00 \times 10^{10}\ \rm cm\ s^{-1})}{4.57 \times 10^{14}\ \rm Hz}\right)^2$$

$$= 1.02 \times 10^4\ \rm K.$$

4.1.3 Spectral Line Radiative Transfer

We can use the radiative transfer equation that we derived in Section 2.1.3 to relate observations of the intensity of a spectral line to the physical properties of the medium producing the line. For simplicity, we consider Equation 2.16, where we have already assumed the medium along the line of sight to be uniform, and substitute Equation 2.11 to explicitly include the dependence of the source function on the emission and absorption coefficients

$$I_\nu \;=\; I_\nu^o\, e^{-\tau_\nu} \;+\; \frac{j_\nu}{\kappa_\nu}\,(1 - e^{-\tau_\nu}).$$

Remember that the optical depth for a uniform medium is just $\tau_\nu = \kappa_\nu L$, where L is the line of sight depth of the medium.

We start by ignoring any background radiation and consider the two limits discussed in Chapter 2. When the medium is optically thin ($\tau_\nu \ll 1$), as in Equation 2.13, the spectral line intensity is

$$I_\nu \;\approx\; j_\nu\, L.$$

Substituting for the value of the emission coefficient (Equation 4.4), we find that the intensity of a spectral line as a function of frequency in the optically thin limit is given by

$$I_\nu \;\approx\; \frac{h\nu\, n_u\, A_{ul}}{4\pi}\; \phi_\nu\, L. \tag{4.10}$$

Another useful measure of a spectral line is the intensity of the line integrated over frequency, which is called its *integrated intensity*. For an optically thin line this is given by

$$\int I_\nu\, d\nu \;\approx\; \int \frac{h\nu\, n_u\, A_{ul}}{4\pi}\; \phi_\nu\, L\, d\nu.$$

The line width is usually a very small fraction of the line frequency; therefore ν varies little over the range in which the line profile function is non-zero. We can, therefore, move ν outside the integral along with other terms that are not frequency dependent. We can rewrite this integral and perform the integration to get

$$\int I_\nu\, d\nu \;\approx\; \frac{h\nu\, n_u\, A_{ul}}{4\pi}\; L \int \phi_\nu\, d\nu \;\approx\; \frac{h\nu\, n_u\, A_{ul}}{4\pi}\; L, \tag{4.11}$$

where we used the fact that the integral of the profile function is unity (see Section 4.1.1.1). Thus, when a spectral line is optically thin, the integrated intensity of the line is just proportional to the product of the average density of atoms in the upper state of the transition and the path length through the medium or $n_u L$. As discussed in Section 2.2.2, the product of the number density of atoms and path length is called column density, usually denoted by N, with units cm^{-2}. In this case, only atoms in the upper state of the transition contribute to the emission, so the product is the column density of atoms in the upper state denoted by N_u, where $N_u = n_u L$. Therefore, we can rewrite Equation 4.11 as

$$\int I_\nu\, d\nu \;\approx\; \frac{h\nu}{4\pi}\; A_{ul}\, N_u, \tag{4.12}$$

which shows that when spectral lines are optically thin, the integrated intensity is proportional to the column density of the atoms in the upper state of the transition.

The optical depth in a spectral line varies with frequency and depends on the frequency offset relative to the line center frequency according to the line profile function. As the column density of atoms in the upper state increases, the optical depth at the center of the line (τ_o) also increases and eventually becomes optically thick ($\tau_o \gg 1$). In this case, at line center we find that $(1 - e^{-\tau_o}) \sim 1$, and the line center intensity, ignoring any background radiation, is given by

$$I_\nu(\text{line center}) \;\approx\; \frac{j_\nu}{\kappa_\nu},$$

and is independent of column density. Using Equations 4.4 and 4.5, expressions for the emission and absorption coefficients in terms of the Einstein coefficients, and Equations 4.6 and 4.7, the relationship between the Einstein coefficients, we can rewrite the optically thick line center intensity as

$$I_\nu(\text{line center}) \approx \frac{2h\nu^3}{c^2}\left(\frac{g_u n_l}{g_l n_u} - 1\right)^{-1}. \tag{4.13}$$

We can define a quantity called the *excitation temperature*, T_{ex}, which enters through the Boltzmann factor as follows

$$\frac{n_u}{n_l} = \frac{g_u}{g_l} e^{-\Delta E_{ul}/kT_{ex}}, \tag{4.14}$$

where $\Delta E_{ul} = E_u - E_l$. This is a general expression that relates the ratio of the number density of atoms in the upper state to that in the lower state in terms of the excitation temperature. Note that if $T_{ex} = 0$, then there are no atoms in the upper state and as T_{ex} approaches infinity the fraction of atom in each state is proportional to the degeneracy of the state. If the gas is in thermodynamic equilibrium, then T_{ex} will be the same as the gas kinetic temperature. In Section 4.1.5 we will discuss the properties of a medium that determine the excitation temperature.

If we use this definition of excitation temperature and the fact that $h\nu = \Delta E_{ul}$, we can rewrite the intensity at line center of an optically thick line as

$$I_\nu(\text{line center}) \sim \frac{2h\nu^3}{c^2}\frac{1}{e^{h\nu/kT_{ex}} - 1}. \tag{4.15}$$

You should recognize Equation 4.15 as having the same functional form as the Planck function (see Equation 3.2), which describes the intensity of a blackbody, where the temperature of the blackbody has been replaced by the excitation temperature. Therefore, for optically thick spectral lines, the intensity at line center depends only on the excitation temperature and is given by

$$I_\nu(\text{line center}) \sim B_\nu(T_{ex}). \tag{4.16}$$

If the gas is in thermodynamic equilibrium, the intensity of an optically thick spectral line is given by the Planck function at the gas kinetic temperature. At frequencies offset from the line center the optical depth in the line decreases according to the line profile function. Therefore, even if the spectral line is optically thick at the line center, at frequencies away from line center the optical depth will be smaller and eventually much less than unity, and the intensity will go to zero.

4.1.4 Kirchhoff's Rules for Spectroscopy

The nineteenth century scientist Gustav Kirchhoff provided an empirical explanation for the presence of dark lines (absorption lines) and bright lines (emission lines) in spectra. His three rules of spectroscopy are (1) a hot solid or gas under high pressure produces a continuous spectrum, (2) a hot gas at low pressure produces a bright-line or emission-line spectrum, and (3) a cold gas at low pressure viewed against a hot source of continuous emission produces a dark-line or absorption-line spectrum. The first of Kirchoff's rules concerns blackbody emission discussed in Chapter 3. The other two rules can be derived more quantitatively using the equations of radiative transfer, as we show below.

Independently of whether a spectral line is optically thick or thin, the source function for the line is given by $B_\nu(T_{ex})$. Therefore we can rewrite Equation 2.16 for spectral lines to be

$$I_\nu = I_\nu^o \, e^{-\tau_\nu} \; + \; B_\nu(T_{ex}) \, (1 - e^{-\tau_\nu}), \tag{4.17}$$

where $B_\nu(T_{ex})$ is the Planck function at temperature T_{ex} and the optical depth $\tau_\nu = \kappa_\nu L$ is given by

$$\tau_\nu \;=\; \frac{h\nu}{4\pi} \, [n_l \; B_{lu} - n_u \; B_{ul}] \; L \; \phi_\nu .$$

We can rewrite the optical depth using the relation between the B coefficients (see Equation 4.7) and the definitions of excitation temperature (see Equation 4.14) and column density ($N_l = n_l L$) to yield

$$\tau_\nu \;=\; \frac{h\nu}{4\pi} \; B_{lu} \; N_l \; (1 - e^{-\Delta E_{ul}/kT_{ex}}) \; \phi_\nu . \tag{4.18}$$

In the above equation, the term in parentheses is often called the correction for stimulated emission.

Equation 4.17 provides some insight into Kirchhoff's rules of spectroscopy. For simplicity, in the following discussion we will assume that the background radiation I_ν^o is a continuous spectrum, and that the atoms are in thermodynamic equilibrium so that $T_{ex} = T_K$, the gas kinetic temperature. We apply Equation 4.17 to two cases. First, we consider the case where the temperature of the medium is low so that I_ν^o is ten times larger than $B_\nu(T_{ex})$. In this case, for small but non-zero values of τ_ν, the first term in Equation 4.17 is slightly decreased while the second term is quite small. The observed I_ν, then, is less than I_ν^o and an absorption line is seen. The net effect of the medium, as we should expect, is to absorb the background radiation. As the optical depth of the medium increases, the first term in Equation 4.17 gets smaller and the depth of the absorption line increases; this effect is illustrated in Figure 4.5. As the optical depth gets large, the first term of Equation 4.17 approaches zero and the second term becomes dominant. As $\tau_\nu \to \infty$, we see that $I_\nu \to B_\nu(T_{ex})$. Hence, the intensity never goes to zero. If $B_\nu(T_{ex}) \ll I_\nu^o$, then the intensity in the line can become very small compared with the intensity of the background radiation at surrounding frequencies. Since the optical depth is the greatest at line center, this is where the intensity approaches $B_\nu(T_{ex})$ first. Note that in Figure 4.5 once the optical depth at line center is greater than unity, the shape of the line is altered and is no longer Gaussian.

In the opposite case, we assume the excitation temperature of the medium is large so that $B_\nu(T_{ex})$ is ten times larger than I_ν^o. In this case, the emission from the medium will be much more important than absorption, and the emergent emission will show an emission-line spectrum, as the emergent intensity in the spectral line is greater than the background intensity seen at the surrounding frequencies. We illustrate an emission line in Figure 4.6 and the growth of the line as the optical depth of the medium increases. Note that when the medium is optically thick, the peak intensity of the spectral line is, again, $B_\nu(T_{ex})$, as we discussed in the previous section. We also see, just as with the absorption line, that when the line center optical depth is large, the shape of the line is no longer Gaussian.

4.1.5 Collisional Transition Rates and Excitation Temperature

The excitation temperature defined in Equation 4.14 is a measure of the ratio of the number of atoms in the upper state of a spectral line transition to the number of atoms in the lower state. The ratio of atoms in the upper state to that in the lower state is often referred to as the *population ratio.* The population ratio depends on a number of physical processes. We have already discussed radiative transitions and these can either excite the atom by

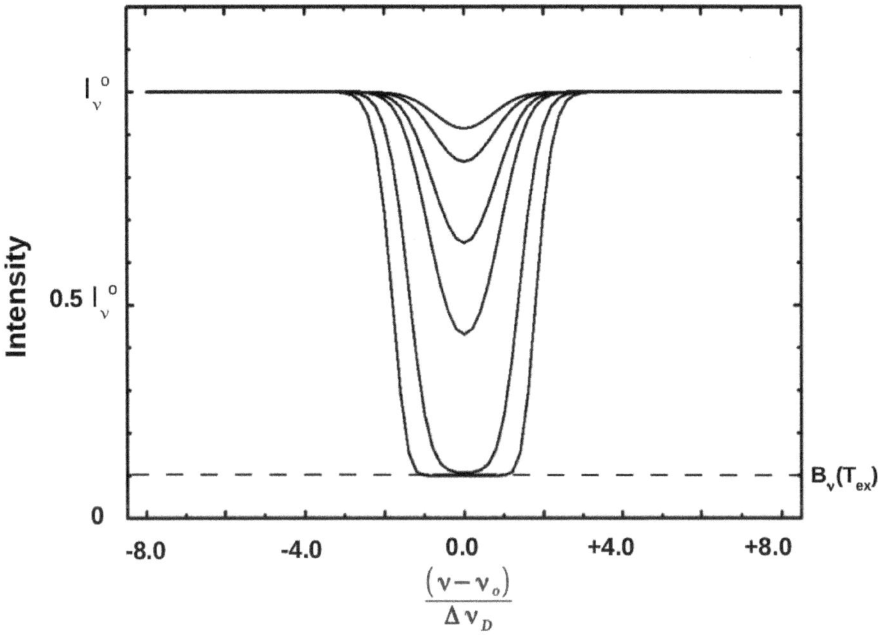

Figure 4.5: Absorption-line spectra, where the background continuum source has intensity, I_ν^o, ten times greater than $B_\nu(T_{ex})$. The absorption lines have increasing optical depth at line center with values of 0.1, 0.2, 0.5, 1, 5 and 20. The x-axis gives the frequency in terms of the Doppler broadening parameter $\Delta\nu_D$ (see Equation 4.9). As the optical depth at line center becomes large, the intensity approaches $B_\nu(T_{ex})$.

absorption or de-excite the atom by spontaneous or stimulated emission. In addition to radiative processes are collisional transitions, in which collisions between atoms transfer energy between thermal energy and the internal excitation energy. Such collisional transfer of energy can lead to either an excitation or de-excitation of the atom. The collisional rates are important and so we need to know how to calculate them.

Consider an atom colliding with some specific type of particle, which we will refer to as the target particle. The target particle can be other atoms, free electrons, or molecules. The atom's collision rate is the inverse of the average time between collisions, $\tau_{\rm coll}$, which is the average distance between collisions divided by the atom's velocity $v_{\rm atom}$. The average distance between collisions is known as the mean free path, which we denote by ℓ. Therefore,

$$\ell = v_{\rm atom}\ \tau_{\rm coll}. \tag{4.19}$$

The collision rate, then, is

$$R_{\rm coll} = \frac{1}{\tau_{\rm coll}} = \frac{v_{\rm atom}}{\ell}. \tag{4.20}$$

To determine the atom's mean free path, imagine the atom moving away from its last interaction, and calculate how far it will travel on average before bumping into the next target particle. The likelihood that an atom will interact with a particular target particle is defined in terms of the interaction *cross-section*, $\sigma_{\rm coll}$, which has units of cm^2 and is the effective cross-sectional area of the target particle as seen by the atom. A larger cross section means the atom is more likely to collide with it. Since the target particle has cross-sectional area $\sigma_{\rm coll}$, in traveling to its next collision, the atom sweeps out a cylinder with end area

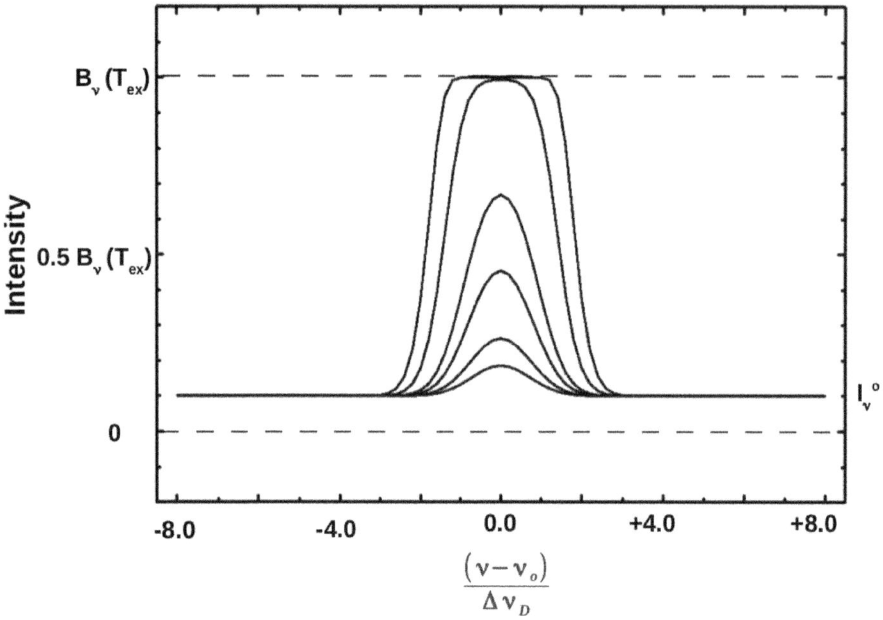

Figure 4.6: Emission-line spectra produced when $B_\nu(T_{ex})$ is ten times larger than I_ν^o. The emission lines have increasing optical depth at line center with values of 0.1, 0.2, 0.5, 1, 5 and 20. As in Figure 4.5, the x-axis gives the frequency in terms of the Doppler broadening parameter $\Delta\nu_D$. As the optical depth at line center becomes large, the intensity of the line approaches $B_\nu(T_{ex})$. Note the intensity well outside of the line is just that of the background radiation I_ν^o.

$\sigma_{\rm coll}$ and length equal to the mean free path, ℓ. The volume of this cylinder, $V_{\rm cyl}$ is the average volume occupied by each target particle, and therefore is the inverse of the density of target particles, $n_{\rm target}$. We have, then,

$$V_{\rm cyl} = \sigma_{\rm coll}\,\ell = \frac{1}{n_{\rm target}}. \tag{4.21}$$

Putting Equations 4.20 and 4.21 together, we have

$$R_{\rm coll} = n_{\rm target}\,\sigma_{\rm coll}\,v_{\rm atom}. \tag{4.22}$$

The rate at which one atom undergoes collisions is then given by the product $n\sigma v$, and has units of a rate ($\rm s^{-1}$). At the microscopic level, thermal motions dominate the kinematics of the atoms and molecules and therefore the relevant velocity for particle collisions is related to the gas kinetic temperature.

To quantify the excitation or de-excitation process, we must know more than just the atom's collision rate. We need the rate at which collisions transform atoms or molecules from one state to another. For this purpose, we define the *collisional rate coefficients*, C_{lu} and C_{ul} such that their product with the density of target particles provides the rate at which collisions excite an atom from a lower state to an upper state (collisional excitation) or from an upper state to a lower state (collisional de-excitation). The collisional rate coefficients can be calculated, although these calculations can be very complex. The collisional rate coefficients play the same role as σv in Equation 4.22 and so have units of $\rm cm^3\ s^{-1}$, but provide the details of how the collisions change the energy state of atoms and are averaged over the velocity distribution of the atoms. For a thermal or Maxwell-Boltzmann velocity distribution, the two collisional rate coefficients are related to each other by

$$\frac{C_{lu}}{C_{ul}} = \frac{g_u}{g_l} e^{-\Delta E_{ul}/kT_K}, \tag{4.23}$$

where ΔE_{ul} is the energy difference between the upper and lower states and T_K is the gas kinetic temperature. With these definitions, the rate per unit volume at which atoms are collisionally excited from a lower state l to an upper state u is given by $n_l\, n\, C_{lu}$, where n is the density of the appropriate target particle. Similarly, the rate per unit volume at which atoms in the upper state are collisionally de-excited is given by $n_u\, n\, C_{ul}$.

We now examine all of the radiative and collisional processes that can excite or de-excite an atom. For multiple energy levels this can be very complicated, so we will illustrate this for an atom with only two levels. The radiative and collisional processes are shown in Figure 4.7, giving the rate per unit volume for each process. In equilibrium, the rate from the lower state to the upper state must equal the rate from the upper state to the lower state. Therefore equilibrium considerations require

$$n_l B_{lu} I_\nu \; + \; n_l n C_{lu} \; = \; n_u A_{ul} \; + \; n_u B_{ul} I_\nu \; + \; n_u n C_{ul}.$$

We can rearrange this equation, solving for the equilibrium population ratio to yield

$$\frac{n_u}{n_l} = \frac{g_u}{g_l} e^{-\Delta E_{ul}/kT_{ex}} = \frac{B_{lu} I_\nu + n C_{lu}}{A_{ul} + B_{ul} I_\nu + n C_{ul}}, \tag{4.24}$$

where I_ν is the intensity of the radiation field that the atoms are exposed to at the frequency of the transition and we have used the definition of excitation temperature (see Equation 4.14).

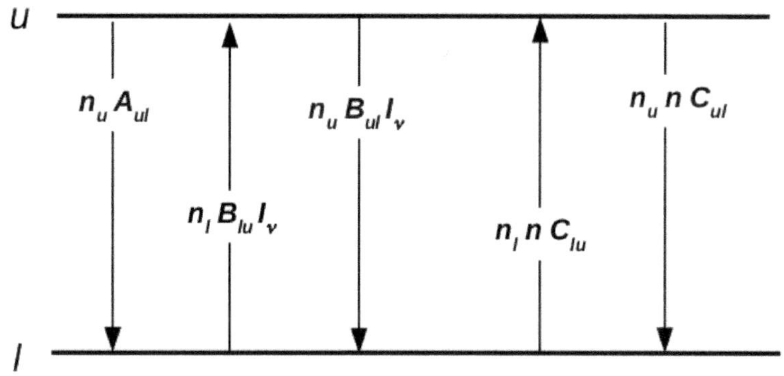

Figure 4.7: A schematic showing two energy states of an atom or molecule. Marked with arrows are the radiative and collisional processes that can change the energy state. Included with each process is its rate per unit volume.

It is common to represent the intensity of the radiation field in Equation 4.24 by a quantity called the *radiation temperature*. The radiation temperature is the temperature that when put in the Planck function yields the same intensity as the radiation field. At longer wavelengths, where the Rayleigh-Jeans approximation can be used, this would be the same as the brightness temperature of the radiation field. Using this convention, we can characterize the intensity of the radiation field by

$$I_\nu \; = \; B_\nu(T_R) \; = \; \frac{2h\nu^3}{c^2} \, [e^{h\nu/kT_R} - 1]^{-1}.$$

This radiation field can be due to a number of different sources. Many of the radio spectral lines arise from the cold interstellar medium and in this environment the dominant source of radio wavelength radiation is often the cosmic microwave background (CMB), which is discussed in more detail in Chapter 10. In addition to the CMB radiation, the atoms themselves are emitting and this emission will also contribute to the radio wavelength radiation. If the CMB is the only source of radio wavelength radiation, and since the CMB is 2.73 K blackbody radiation, then $T_R = 2.73$ K. Using the definition of T_R and the relationship between the Einstein coefficients (Equations 4.6 and 4.7) and the collisional rate coefficients (Equation 4.23), we can rewrite equation 4.24 as

$$e^{-\Delta E_{ul}/kT_{ex}} = \frac{A_{ul}[e^{h\nu/kT_R} - 1]^{-1} + nC_{ul}e^{-\Delta E_{ul}/kT_K}}{A_{ul}[1 - e^{-h\nu/kT_R}]^{-1} + nC_{ul}}. \tag{4.25}$$

We consider two limits, the first in which the collisional rates dominate over the radiative rates, and the second in which the radiative rates dominate. The first case applies when the gas density is very large, and in this case we can ignore the radiative terms in Equation 4.25. In this high density limit we find

$$e^{-\Delta E_{ul}/kT_{ex}} \sim e^{-\Delta E_{ul}/kT_K},$$

and the excitation temperature is approximately equal to the gas kinetic temperature. In this limit we say that the energy level populations are *thermalized* as they follow a Boltzmann distribution at the gas temperature. In the opposite limit, where the density is very small, the collisional rates in Equation 4.25 are unimportant and can be ignored, leading to

$$e^{-\Delta E_{ul}/kT_{ex}} \sim e^{-h\nu/kT_R}.$$

Therefore, since $\Delta E_{ul} = h\nu$, the excitation temperature in the low density limit is equal to the radiation temperature.

The density at which the collisional and radiative de-excitation rates are equal is called the *critical density* (n_{cr}) and is given by

$$n_{cr} = A_{ul}/C_{ul}. \tag{4.26}$$

Because C_{ul} depends on the particle velocity it is temperature dependent, and thus the critical density also depends on the gas kinetic temperature. The temperature of the gas is often larger than the radiation temperature, and therefore, in these cases, the excitation temperature of radio spectral lines is bounded by the gas temperature (at high densities) and by the radiation temperature (at low densities). The behavior of the excitation temperature with density is illustrated in Figure 4.8.

Applying the definition of radiation temperature to Equation 4.17, we can write the line intensity as

$$I_\nu = B_\nu(T_R)\, e^{-\tau_\nu} + B_\nu(T_{ex})\,(1 - e^{-\tau_\nu}).$$

Note that if T_{ex} equals T_R, then the line intensity is just equal to $B_\nu(T_R)$, independent of the optical depth. But this is exactly the same intensity one would measure at frequencies offset from line center where the optical depth is zero. Since in this case the intensity of the line is the same as the background, no emission or absorption line will be produced. Therefore no spectral line results when the excitation temperature equals the radiation temperature.

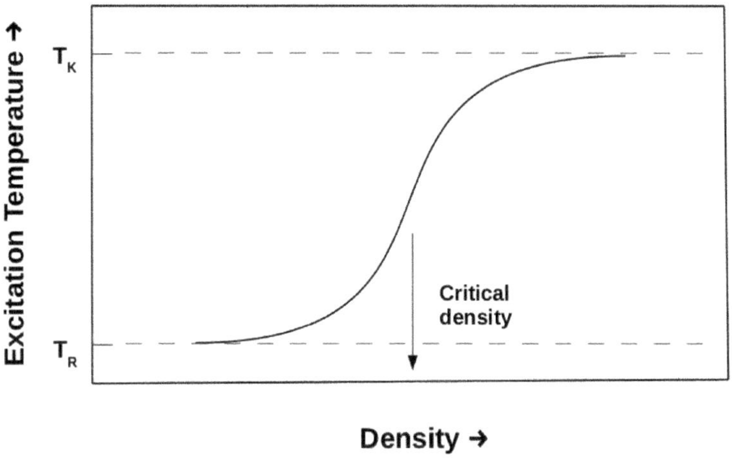

Figure 4.8: A schematic diagram showing how the excitation temperature varies with density. In the low density limit, the excitation temperature is approximately the radiation temperature and in the high density limit the excitation temperature is approximately the gas temperature. At the critical density, there is a rapid change in the excitation temperature.

4.2 RADIO SPECTRAL LINES

In this section we will discuss some specific examples of radio spectral lines. As can be seen by the energy level diagram of hydrogen shown in Figure 4.1, the low-lying electronic energy levels are typically separated by several electron volts of energy. This is generally true for most atoms and molecules. Therefore, transitions between these low-lying electronic levels produce spectral lines that are typically at ultraviolet, optical or infrared wavelengths, and not at radio wavelengths. Hydrogen does have radio frequency spectral lines. These lines arise from transitions between either electronic states with large principle quantum numbers – these are called *radio recombination lines* – or from a hyper-fine splitting in the ground state, producing the *21-cm spectral line*. However, most radio spectral lines arise from molecules which have more complex energy level structures due to the additional quantized motions, specifically rotation. In this section we will discuss all three types of radio spectral lines, starting with the 21-cm line of atomic hydrogen.

4.2.1 21-cm Spectral Line of Atomic Hydrogen

Electrons and protons have an intrinsic angular momentum called *spin*. This property should not be viewed literally as a spinning particle, but as a description of its quantum mechanical angular momentum that has no classical analog. Spin angular momentum is governed by quantum mechanics, and therefore is quantized. This spin quantization restricts the electron in the ground state ($n = 1$) of the hydrogen atom to have only two allowed spin states or orientations, one that is in the same direction as the proton's spin and one that is in the opposite direction. Charged particles with spin, such as the electron and proton, have a magnetic dipole moment, and therefore the electron and proton can interact in a way similar to two bar magnets. Like the orientation of bar magnets, the two spin orientations have slightly different energies due to the interaction of the magnetic moments of the proton and electron. The lower energy state is that in which the spins of the electron and proton are oppositely oriented (anti-parallel) and the higher energy state is that in which the spins have the same orientation (parallel). The splitting of the ground state into these two spin states is called hyperfine structure and is illustrated in Figure 4.9. The energy difference between these two spin states is very small, amounting to only about 1×10^{-17} ergs or about 6×10^{-6} eV. Contrast that with the energy difference between the ground and first excited electronic states of hydrogen which is 10.2 eV!

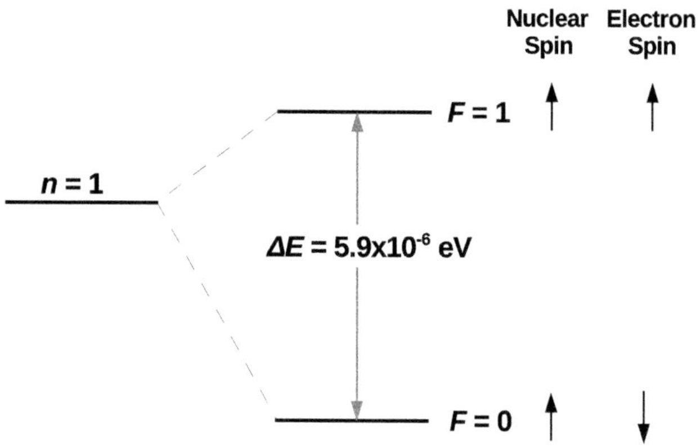

Figure 4.9: An energy level diagram showing the $n = 1$ level of hydrogen split into two hyperfine levels due to the interaction of the magnetic dipole moments of the electron and the nucleus (in this case a proton). The arrows indicate the orientation of the spins of the electron and proton. For the lower state the spins are oppositely oriented and in the upper state the spins have the same orientation.

These two spin states are designated by the total spin quantum number, denoted by F. Both the electron and proton are spin-1/2 particles, so when they are anti-aligned (lower energy state) the total spin is $F = 1/2 - 1/2 = 0$, and when they are aligned (higher energy state) the total spin is $F = 1/2 + 1/2 = 1$. Radiative transitions can occur between these two spin states, but since the energy difference is very small, the resulting spectral line is at a frequency of 1420.406 MHz or at a wavelength of about 21.1 cm. Because the transition between these two states involves a change in the spin orientation of the electron relative to the proton, this transition is often called the *spin-flip* transition of hydrogen. This is a magnetic dipole transition, not an electronic dipole transition, and so it is 'forbidden'. The spontaneous emission rate for this transition is therefore very small and the Einstein A coefficient is only 2.87×10^{-15} s^{-1}, more than 22 orders of magnitude smaller than the A coefficient for the electronic transitions in hydrogen. The average amount of time an individual hydrogen atom in the higher energy spin state waits before undergoing a spontaneous 21-cm emission, which is given by the reciprocal of the A coefficient, is about 3.5×10^{14} s or about 11 million years.

The Dutch astronomer Hendrik van de Hulst in 1944 predicted that hydrogen should have hyperfine structure and could produce detectable emission at radio wavelengths. The first detection was not made until 1951 by Harold Ewen and Edward Purcell at Harvard University. The detection of this transition opened up a new window on the study of the gas in the interstellar medium. Although this transition is intrinsically weak, hydrogen is very abundant in the interstellar medium, so this line is readily detected. This transition represents a major tool for studying the atomic hydrogen in the Milky Way (see Chapter 5) and in other galaxies (see Chapter 8).

Since the 21-cm transition of hydrogen has such a small A coefficient (and therefore correspondingly small B coefficients), the emission we observe in the interstellar medium of the Milky Way and other galaxies is often optically thin. As we showed earlier (see Equation 4.12), if the emission is optically thin, then the integrated intensity of the line is given by

$$\int I_\nu \, d\nu \sim \frac{h\nu \, N_u \, A_{ul}}{4\pi},$$

where N_u is the column density of hydrogen in the upper spin state. Substituting for the constants, the line frequency and the A coefficient give

$$\int I_\nu \, d\nu \sim 2.16 \times 10^{-33} \text{ erg s}^{-1} \text{ cm}^{-2} \text{ sr}^{-1} \left(\frac{N_u}{\text{cm}^{-2}}\right). \tag{4.27}$$

We are of course interested in the total hydrogen column density and not just the hydrogen in the upper spin state of the 21-cm transition. The ratio of the column density or populations in the two spin states can be expressed in terms of the excitation temperature defined in Equation 4.14 by

$$\frac{N_{F=1}}{N_{F=0}} = \frac{g_{F=1}}{g_{F=0}} \, e^{-\Delta E_{ul}/kT_{ex}},$$

where the g's are the statistical weights, and $g = 2F + 1$ for the spin states of hydrogen. Substituting for the constants, the line frequency and the statistical weights, we find

$$\frac{N_{F=1}}{N_{F=0}} = 3 \, e^{-0.068/T_{ex}}. \tag{4.28}$$

Since these states are referred to as spin states of hydrogen, the excitation temperature is often referred to as the *spin temperature*, and denoted by T_s. Remember from Section 4.1.5, the lower bound on the excitation temperature is set by the CMB, and therefore the excitation temperature, or spin temperature, of the hyperfine levels is always equal to or greater than the CMB temperature of 2.73 K. For excitation temperatures $T_{ex} \gg 0.068$ K, the exponential in Equation 4.28 is very close to unity, and so the ratio of column densities is always about 3. Therefore, there are always about 3 times as many hydrogen atoms in the upper spin state as in the lower spin state. Hence, the column density of hydrogen in the upper spin state is always about 3/4 of the total hydrogen column density.

In Section 4.1.5 we showed that the excitation temperature has to be greater than the temperature of the background radiation in order to produce an observable emission line. So what is the excitation temperature of the 21 cm line? Because the energy difference of this transition is so small, it can be readily excited by collisions in gas at any temperature expected to be present in the interstellar medium. For gas temperatures between 100 and 1,000 K, the critical density (see Section 4.1.5) for this transition is between about 0.04 and 0.13 cm^{-3}. Since the density of hydrogen atoms, even in the lower density regions of the interstellar medium, is often larger than this value, the excitation temperature of the transition usually equals the gas temperature. In fact, even if the gas density is less than the critical density, there are effects, such as the Wouthuysen-Field effect, that couple the spin temperature to the Lyman α radiation. This likely drives the excitation temperature to equal the gas kinetic temperature. Since the gas temperature is much higher than the temperature of the cosmic microwave background, the atomic hydrogen in the interstellar medium has sufficient column density to produce observable emission lines.

Including the fact that the column density of hydrogen in the upper spin state is 3/4 of the total hydrogen column density (N_H), we can rewrite Equation 4.27 in terms of the total hydrogen column density as

$$\int I_\nu \, d\nu \sim 1.62 \times 10^{-33} \text{ erg s}^{-1} \text{ cm}^{-2} \text{ sr}^{-1} \left(\frac{N_H}{\text{cm}^{-2}}\right).$$

Solving for the hydrogen column density, we find

$$N_H \sim 6.17 \times 10^{32} \text{ cm}^{-2} \left(\frac{\int I_\nu \, d\nu}{\text{erg s}^{-1} \text{ cm}^{-2} \text{ sr}^{-1}}\right). \qquad (4.29)$$

Therefore, when the emission is optically thin, the total hydrogen column density can be deduced from just the measured integrated intensity of the 21-cm line. A simulated 21-cm emission line is shown in Figure 4.10. In this example the gas is at a temperature of 100 K, the line is assumed to be Doppler broadened only by the thermal motions of the gas, and the line is approximately optically thin at line center. A useful approximation for Gaussian-shaped spectral lines, such as the spectral line in Figure 4.10, is that the integrated intensity of the line is given roughly by the height of the line multiplied by its FWHM linewidth ($\Delta\nu_{FWHM}$).

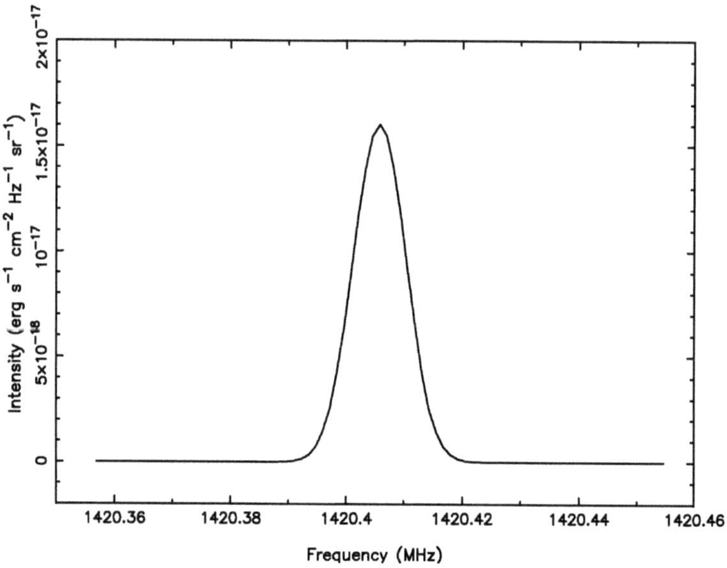

Figure 4.10: simulated 21-cm emission line due to atomic hydrogen. This example was computed for hydrogen gas at a temperature of 100 K and the line is broadened only by thermal motions of the hydrogen atoms. The optical depth at the center of the line is 0.3.

Example 4.3:

Based on the spectral line shown in Figure 4.10, if the emission is optically thin, what column density of atomic hydrogen is implied?

Answer:
First we must estimate the integrated intensity of the spectral line shown in Figure 4.10. We approximate the integrated intensity of a Gaussian-shaped line by its

height multiplied by the full width of the line at half its height. In this case the peak intensity is 1.6×10^{-17} erg s^{-1} cm^{-2} Hz^{-1} sr^{-1} and we multiply by the full width of the line at half intensity which is 1.1×10^4 Hz (or 0.011 MHz) to yield an integrated intensity of 1.8×10^{-13} erg s^{-1} cm^{-2} sr^{-1}. Using Equation 4.29, the column density of atomic hydrogen would then be

$$N_{\rm H} \sim 6.17 \times 10^{32}\ {\rm cm}^{-2} \left(\frac{1.8 \times 10^{-13}\ {\rm erg\ s^{-1}\ cm^{-2}\ sr^{-1}}}{{\rm erg\ s^{-1}\ cm^{-2}\ sr^{-1}}} \right) = 1.1 \times 10^{20}\ {\rm cm}^{-2}.$$

Therefore the column density is 1.1×10^{20} cm^{-2}.

Instead of plotting a spectral line as intensity versus frequency, radio astronomers often express intensity in units of brightness temperature and frequency in units of velocity. If the intensity is expressed in terms of the brightness temperature (see Equation 3.13), then the hydrogen column density is given by

$$N_{\rm H} \sim 3.82 \times 10^{14}\ {\rm cm}^{-2} \left(\frac{\int T_B\, d\nu}{\rm K\ Hz} \right).$$

Likewise, instead of plotting the frequency of the emission we could instead plot the radial velocity of the emitting atoms. The radial velocity is related to the observed frequency and the rest frequency by the Doppler formula given in Equation 1.10. If we plot the spectrum shown in Figure 4.10 in terms of brightness temperature and velocity we get the spectrum shown in Figure 4.11. Astronomers usually express velocity in km s^{-1} as shown in the figure. Note that the spectral line in Figure 4.11 has the line center at a velocity of zero; therefore the emitting gas on average is at rest relative to the observer. If the gas had a radial velocity relative to the observer, then the frequency of the spectral line would be offset from its rest frequency depending on the magnitude and direction of motion. The units of integrated intensity, using the plotting convention in Figure 4.11, are K km s^{-1}. We can rewrite the relation between HI column density and integrated intensity in terms of the intensity expressed in brightness temperature and the velocity in km s^{-1} as

$$N_{\rm H} \sim 1.81 \times 10^{18}\ {\rm cm}^{-2} \left(\frac{\int T_B\, dv}{\rm K\ \ km\ s^{-1}} \right). \tag{4.30}$$

Example 4.4:

Based on the observed spectral line shown in Figure 4.11, if the emission is optically thin, what column density of atomic hydrogen would be implied?

Answer:
We estimate the integrated intensity of the line shown in Figure 4.11 using the same technique as in Example 4.3. The integrated intensity is approximately the peak of the line (26 K) multiplied by the full width at half the peak intensity of the line (2.3 km s^{-1}) resulting in an integrated intensity of roughly 60 K km s^{-1}. Using Equation 4.30, we find

$$N_{\rm H} \sim 1.81 \times 10^{18}\ {\rm cm}^{-2} \left(\frac{60\ {\rm K\ km\ s^{-1}}}{\rm K\ \ km\ s^{-1}} \right) = 1.1 \times 10^{20}\ {\rm cm}^{-2},$$

and therefore the atomic hydrogen column density is 1.1×10^{20} cm^{-2}, the same as we derived in Example 4.3.

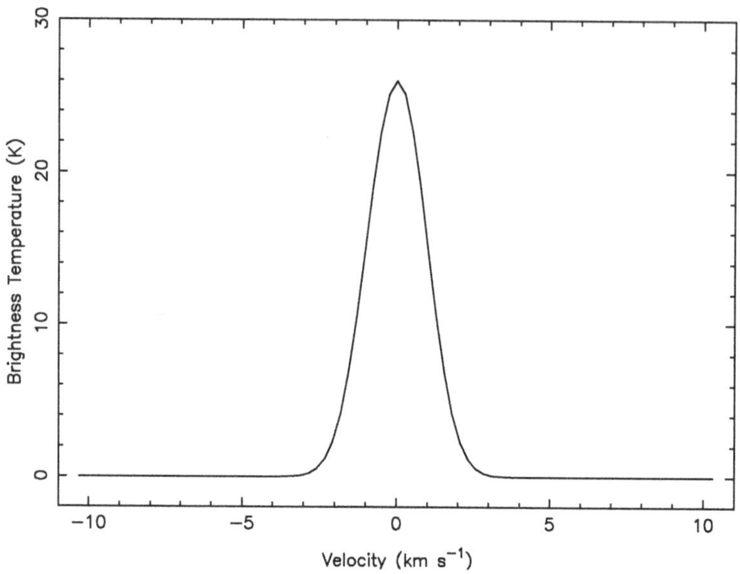

Figure 4.11: The same simulated 21-cm emission line shown in Figure 4.10 plotted as brightness temperature versus velocity.

When the hydrogen atom is placed in an external magnetic field, the interaction between the magnetic field and the magnetic dipole moment of the electron causes further splitting of the F=1 hyperfine energy level of the n=1 electronic state. The external magnetic field breaks the degeneracy of the magnetic sub-levels that make up the F=1 level. The splitting of energy levels by this interaction is called the *Zeeman effect.* The F=1 level is split into three components and the separation of these components in energy is proportional to the strength of the external magnetic field parallel to our line of sight. The 21-cm line is then composed of three separate components that are spaced in frequency by

$$\Delta\nu \;=\; 1.4 \text{ MHz} \left(\frac{B_{\|}}{\text{gauss}}\right),$$

where $B_{\|}$ is the parallel component of the magnetic field. For magnetic field strengths typical of the interstellar medium in our Galaxy, the splitting is much smaller than the Doppler broadened width of the spectral line. However, the emission components at the highest and lowest frequencies are oppositely circularly polarized, while the third component in the middle is unpolarized. By careful observations in which the emission in the two orthogonal circular polarizations is measured and differenced, it is possible to determine the separation of the levels due to the Zeeman effect. This effect is important as it provides a means for measuring the magnetic field in the atomic gas, although only the component of field parallel to the line of sight is measured.

There are other atoms that have spin-flip transitions analogous to hydrogen. Other spin-flip transitions that have been observed at radio wavelengths are from ^{2}H and ^{3}He$^+$. The atom ^{2}H, called deuterium (also called the deuteron) is an isotope of hydrogen that contains one proton and one neutron. The deuteron has spin 1 which interacts with the one electron that has spin 1/2 to produce a splitting of the ground state. Transitions between these two spin states produce a spectral line at a frequency of 327 MHz or a wavelength of 92 cm. The atom ^{3}He$^+$ (also represented as ^{3}He II, see Section 1.1.3) is an ionized isotope of helium in which the nucleus contains two protons and one neutron and the '+' indicates that the atom is singly ionized. The $^3He^+$ nucleus and its remaining electron both have spin 1/2 giving rise to a splitting very similar to hydrogen. The transition between the two spin states in $^3He^+$ is at a frequency of 8665 MHz, or a wavelength of about 3.5 cm.

4.2.2 Radio Recombination Spectral Lines

The absorption of a photon with energy greater than an atom's ionization energy can unbind an electron from an atom resulting in an ionized atom; this process is called *photo-ionization*. The central stars in HII regions and planetary nebulae (see Chapter 6) produce a sufficient number of energetic photons that the hydrogen surrounding these stars can be mostly ionized. The opposite process also occurs, in which a free electron recombines with an ionized atom. In this process the atom emits a photon and the electron ends up in a bound electronic state; this process is called a *radiative recombination*. The photon carries away the energy lost by the electron. This type of radiative process is also called a free-bound transition as the free electron ends up in a bound electronic state of the hydrogen atom. In this radiative recombination process the electron can end up in any of the bound electronic states, including, although rarely, ones with very large principle quantum numbers. If the electron recombines to a state with $n > 1$, then by a sequence of spontaneous emissions the electron will cascade down to the ground state ($n = 1$), producing a series of emission lines called recombination lines. Note in the energy level diagram for hydrogen (see Figure 4.1), as the principle quantum number gets larger and larger, the spacing between energy levels becomes smaller and smaller. Therefore if the electron recombines to a high principle quantum number, subsequent spontaneous transitions give rise to radio wavelength emission lines.

One can readily compute the frequencies of hydrogen recombination lines by calculating the energy difference between levels using Equation 4.1. Replacing ΔE with $h\nu$, this gives

$$\nu = 3.28805 \times 10^{15} \text{ Hz} \left(\frac{1}{n^2} - \frac{1}{(n + \Delta n)^2} \right),$$

for a transition from level $n+\Delta n$ to n. We denote the recombination line of hydrogen between levels $n = 110$ and $n = 109$ as H109α, where the H indicates that this is a recombination line of hydrogen, the 109 is the principal quantum number of the final state and α indicates that $\Delta n = 1$. Likewise we use β to indicate $\Delta n = 2$, γ to indicate $\Delta n = 3$, and so on. For example, to calculate the frequency of the H109α recombination line, we set n=109 and $\Delta n = 1$, and find $\nu = 5.0089 \times 10^9$ Hz (5.0089 GHz), which is a wavelength of about 6 cm.

While Δn can have any value, transitions with $\Delta n = 1$ have the largest A coefficients and therefore are the most likely transitions. Hence α-transitions ($\Delta n = 1$) produce the brightest recombination lines. The recombination lines that arise from large values of n are radio frequency recombination lines; however as the electron cascades to lower n by a series of spontaneous emissions, the atom will eventually produce infrared, optical and ultraviolet recombination lines. For instance, the recombination line H2α (from the $n = 3$ to the $n = 2$ level) is the transition called Balmer α, at a frequency of 4.57×10^{14} Hz or a wavelength of 656 nm, which is in the optical part of the electromagnetic spectrum. Thus, the recombination process produces a series of emission lines ranging from ultraviolet to radio wavelengths.

The value of the Einstein A coefficient is proportional to n^{-5}, so recombination lines from large principal quantum numbers have relatively small A coefficients. For example, the A coefficient for the H109α line is only 0.35 s^{-1}, eight orders of magnitude smaller than that of typical optical recombination lines of hydrogen. For this reason, and because the number of hydrogen atoms in these high principal quantum number states is very small, the recombination lines at radio wavelengths are relatively weak. Nevertheless, these lines are readily detectable with today's radio telescopes and receivers, and provide information on the temperature and column density of ionized hydrogen, and equally important, on the radial velocities of distant HII regions and planetary nebulae (see Chapter 6).

Atoms other than hydrogen also produce radio recombination lines. Atoms with one electron excited to a very high principal quantum number state are called *Rydberg atoms.* For such an atom with atomic number Z, the atom has Z protons and $Z - 1$ electrons in low-lying bound electronic states and one electron in a very high principal quantum number state. This highly excited electron therefore experiences a net electric charge of +e, similar to that experienced by the electron in the hydrogen atom. Imagine a singly-ionized carbon atom, when it undergoes a radiative recombination with an electron ending in a high principal quantum number state, then this carbon atom would be a Rydberg atom. Since the Coulomb potential for this excited electron is similar to that of hydrogen, the high principal quantum number states of carbon will have nearly the same energy levels as those of hydrogen. Therefore the frequency of the radio recombination lines will be approximately the same as hydrogen, the difference arising from the differences in the mass of the nucleus. The general expression for these recombination line frequencies is

$$\nu = R\,c \left(1 + \frac{m_e}{M_n}\right)^{-1} \left(\frac{1}{n^2} - \frac{1}{(n+\Delta n)^2}\right), \tag{4.31}$$

where R is the Rydberg constant ($R = 1.0973732 \times 10^5$ cm^{-1}), m_e is the electron mass and M_n is the nuclear mass. For example, the H109α line is at 5.0089 GHz, while the same recombination line in helium (He109α) is at 5.0110 GHz and that of carbon (C109α) is at 5.0114 GHz. These different recombination lines are all closely spaced in frequency and can often be observed simultaneously, although owing to the lower abundance of helium and carbon relative to hydrogen, these recombination lines are weaker than those of hydrogen.

4.2.3 Molecular Rotational Spectral Lines

Molecules can have internal motions or degrees of freedom that atoms do not have, such as vibration and rotation. Consider a simple diatomic molecule, such as carbon monoxide (CO). There is an equilibrium bond length between the carbon and oxygen nuclei, but they can vibrate about this equilibrium. The molecule can also rotate, tumbling end over end, about the center of mass of the two nuclei. Quantum mechanics requires that both the vibrational and rotational motions be quantized; the molecules are only permitted to vibrate or rotate with discrete frequencies. Therefore, in the study of spectral lines emitted by molecules, in addition to electronic energy levels similar to those of atoms, we need to account for both *vibrational energy levels* and *rotational energy levels.* As with atomic energy levels, the spacing of excited electronic energy levels in molecules is several eV, but the energy spacings of excited vibrational and rotational states of molecules are much smaller. The energy spacing of vibrational states are of order 0.1 eV, and therefore radiative transitions between vibrational levels produce spectral lines that are typically in the infrared. The spacing between rotational energy levels is of order 0.001 eV, and therefore radiative transitions between rotational levels produce spectral lines at radio wavelengths. We will not discuss molecular vibration further, but rather focus on the radiative transitions between the rotational energy levels of molecules.

The first detection of a radio-wavelength spectral line from a molecule was made in 1963. The hydroxyl radical (OH) was detected in the interstellar medium by Sander Weinreb and colleagues at a wavelength of 18 cm. A few years later three new molecules (NH_3, H_2O, and H_2CO) were detected at centimeter wavelengths. In 1970, carbon monoxide (CO) was detected by Robert Wilson, Keith Jefferts, and Arno Penzias at a much shorter wavelength of 2.6 mm. The emission from CO was found to be much more widespread in our Galaxy than the other molecules, revealing that molecular gas is an important component of the interstellar medium. As we will see, CO has some unique properties that make it a good probe of the molecular gas in the Galaxy. The development of spectroscopy at millimeter wavelengths opened the floodgate to the detection of dozens of additional molecules in subsequent years. These radio spectral lines, particularly CO, permit us to probe the molecular gas in the interstellar medium of our Galaxy (see Chapter 5) and of other galaxies (see Chapter 8).

We start our discussion of molecular rotational energy levels with the simplest case, *diatomic molecules* (molecules consisting of just two atoms). Such a molecule can rotate about its center of mass, and classically, its angular momentum, L, is given by the product of its moment of inertia, I, and the angular rotation frequency, ω; thus $L = I\omega$. Figure 4.12 shows a schematic of a diatomic molecule, where m_1 and r_1 represent the mass and distance from the center of mass for one nucleus and m_2 and r_2 represent the mass and distance from the center of mass for the other nucleus. The moment of inertia is given by

$$I = m_1\, r_1^2 + m_2\, r_2^2.$$

The definition of the center of mass also tells us that $m_1 r_1 = m_2 r_2$. Therefore we can rewrite the moment of inertia as

$$I = \frac{m_1\, m_2}{m_1 + m_2}\, R^2,$$

where $R = r_1 + r_2$ is the internuclear separation as shown in Figure 4.12. Classically, the rotational energy is given by $E = \frac{1}{2} I\omega^2$ or in terms of the angular momentum by $E = L^2/(2I)$.

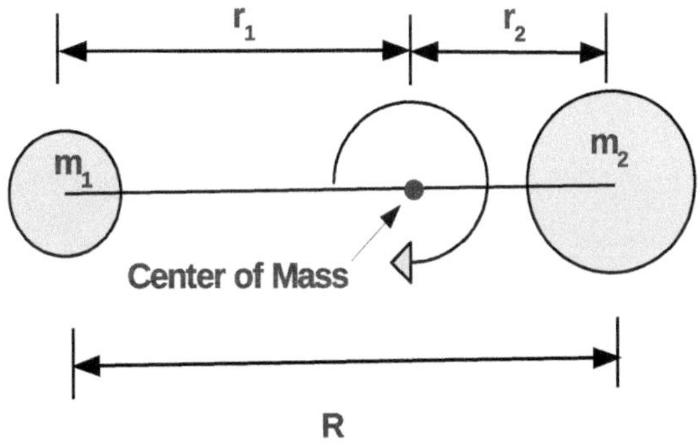

Figure 4.12: A schematic of a diatomic molecule rotating clockwise in the plane of page. The molecule is composed of two atoms, where the nucleus of one has mass m_1 and is located r_1 from the center of mass and the nucleus of the second has mass m_2 and is located r_2 from the center of mass. The separation of the two nuclei is R.

Quantum mechanics dictates that the rotational angular momentum of a molecule is quantized in integer units of $h/2\pi$; thus both the angular momentum and rotational energy can only take on certain discrete values. When the quantization is done properly, solving the appropriate wave equation, we find that the allowed rotational energy states of a diatomic molecule are

$$E = J(J+1)\,\frac{h^2}{8\pi^2\,I}, \tag{4.32}$$

where J is the rotational quantum number, and must have non-negative integer values. It is customary to define a quantity called the *rotation constant* given by

$$B = \frac{h}{8\pi^2\,I}, \tag{4.33}$$

which has units of frequency. With this definition, we can rewrite the rotational energy levels as

$$E = J(J+1)\,hB. \tag{4.34}$$

The rotational energy level diagram for a diatomic molecule is shown in Figure 4.13.

In general, molecules can rotate about any of three orthogonal axes. However, for diatomic molecules, the moment of inertia about the long axis is extremely small, so by Equation 4.32, even the lowest angular momentum state corresponds to an enormous amount of energy, and rotation about this axis can be ignored. In addition, the moment of inertia about the third axis, oriented vertically in the page in Figure 4.12, has the same value as for the axis shown in the figure. Thus the rotation of diatomic molecules can be characterized by a single rotation constant.

Equation 4.34 applies only for an idealized molecule called a rigid rotor, in which the internuclear separation remains fixed. For real molecules the bond length stretches when the molecule rotates faster and this results in a small increase in the moment of inertia and consequently a small decrease in the rotational energy levels. This stretching is called centrifugal distortion and its effects are accounted for with second order terms that are needed to get the precise energy levels and frequencies of the rotational transitions.

For permitted (electric dipole) radiative rotational transitions of diatomic molecules, the change in rotational quantum number is limited to $\Delta J = \pm 1$. Therefore, for spontaneous emission from an upper rotational state J_u to a lower state J_l, the frequency of a rotational transition is

$$\nu = \frac{E_u - E_l}{h} = \frac{[J_u(J_u+1)hB] - [J_l(J_l+1)hB]}{h},$$

Since we require $J_l = J_u - 1$, the frequencies are therefore

$$\nu = \frac{[J_u(J_u+1)hB] - [J_u-1)(J_u)hB]}{h} = 2J_u\,B. \tag{4.35}$$

For example, the rotational transition between the $J = 1$ state to the $J = 0$ state (called the $J = 1 - 0$ transition) is at a frequency of $2B$. The permitted rotational transitions and their frequencies are marked in Figure 4.13. Note that the frequencies of the rotational transition lines increase in steps of $2B$ as J increases.

Although we have focused only on diatomic molecules, the rotation of any *linear molecule* can also be characterized by one rotation constant. Therefore, the equations for the energy levels and frequencies given for diatomic molecules are also valid for linear molecules. Linear molecules are those in which all of the bond angles are exactly 180°. Examples of non-diatomic, but linear molecules, include hydrogen cyanide (HCN) and cyanoacetylene (HCCCN). The bond structure of HCN is illustrated in Figure 4.14.

We can use the carbon monoxide (CO) molecule as an example to examine the energy levels and frequencies of permitted transitions of a diatomic molecule.

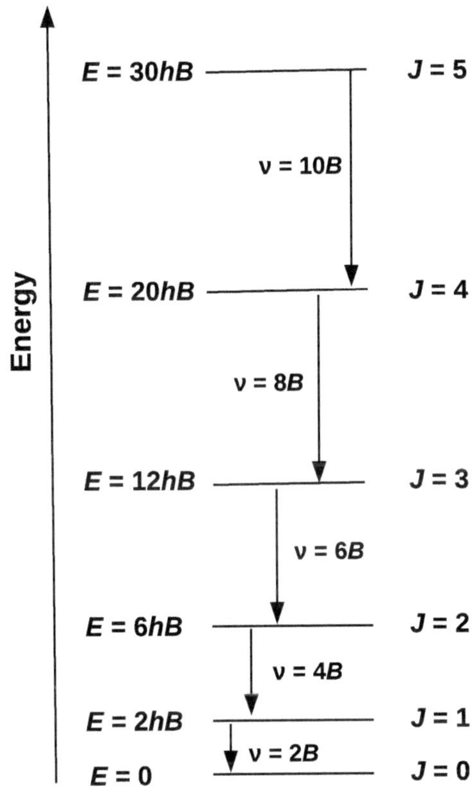

Figure 4.13: The energy level diagram for a simple diatomic molecule. The energy levels are denoted by the rotational quantum number, J, and can take on non-negative integer values. The energies above the ground state are expressed in terms of the rotation constant, B, and Planck's constant. For allowed radiative transitions the rotational quantum number can only change by $\pm$ 1. The allowed spontaneous emission transitions are marked with arrows and have frequencies that are multiples of $2B$.

Example 4.5:

The internuclear separation for CO is $R = 1.13 \times 10^{-8}$ cm. Compute the moment of inertia and the rotation constant of CO.

Answer:
To compute the moment of inertia we need the masses of carbon and oxygen atoms, which are about 2.00×10^{-23} g and 2.67×10^{-23} g, respectively. The moment of inertia is given by

$$I = \frac{m_1 m_2}{m_1 + m_2} R^2 = \frac{(2.0 \times 10^{-23}\ \mathrm{g})(2.7 \times 10^{-23}\ \mathrm{g})}{2.0 \times 10^{-23}\ \mathrm{g} + 2.7 \times 10^{-23}\ \mathrm{g}} (1.13 \times 10^{-8}\ \mathrm{cm})^2$$

$$= 1.46 \times 10^{-39}\ \mathrm{g\ cm^2}.$$

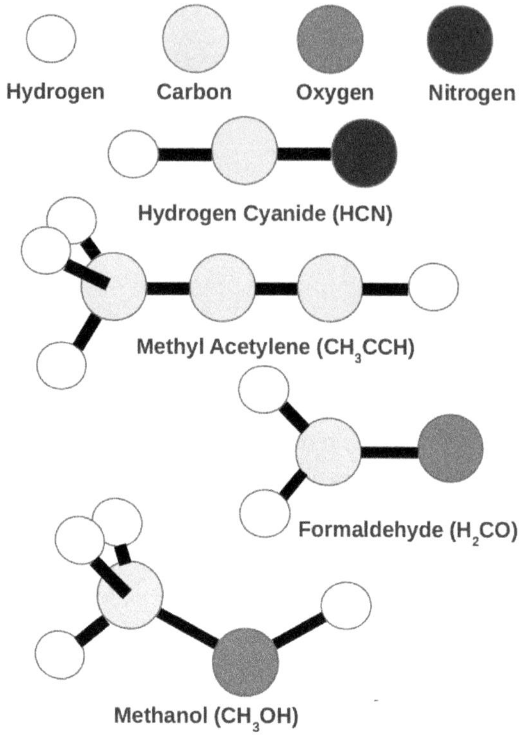

Figure 4.14: The structures of four molecules composed of hydrogen, carbon, oxygen and nitrogen atoms are illustrated. The first molecule shown, hydrogen cyanide (HCN), is a linear molecule. The second molecule from the top, methyl acetylene, also called propyne, is a prolate symmetric-top molecule. The symmetry axis for methyl acetylene is the long axis of the molecule. The moments of inertia around the two axes perpendicular to the symmetry axis have the same value. The last two molecules, formaldehyde and methanol, are asymmetric-top molecules.

Finally, the rotation constant, defined in Equation 4.33, is

$$B = \frac{h}{8\pi^2 I} = \frac{6.63 \times 10^{-27} \text{ erg s}}{(8\,\pi^2)\,(1.46 \times 10^{-39} \text{ g cm}^2)} = 5.76 \times 10^{10} \text{ Hz}.$$

Thus the rotation constant for CO is 57.6 GHz.

Example 4.6:

What are the energies of the first three excited rotational levels of carbon monoxide?

Answer:
Using Equation 4.34 the energy of the $J = 1$ level is

$$E = 2hB = 2\,(6.63 \times 10^{-27} \text{ erg s})\,(5.76 \times 10^{10} \text{ Hz}) = 7.64 \times 10^{-16} \text{ erg}.$$

In a similar manner, we find that the the energies of the the $J = 2$ and $J = 3$ levels are 2.3×10^{-15} erg and 4.6×10^{-15} erg, respectively.

Using the result from Example 4.6, we find that the $J = 1 - 0$, $J = 2 - 1$ and $J = 3 - 2$

rotational transitions are at frequencies ($\nu = (E_u - E_l)/h$) of 115, 230 and 345 GHz respectively, corresponding to wavelengths of 2.6, 1.3 and 0.85 mm respectively. The first excited rotational energy level in CO lies 4.8×10^{-4} eV above the ground rotational state, while the first excited vibrational state lies 0.27 eV above the ground state and the first excited electronic state lies 8.1 eV above the ground state.

The moment of inertia of CO computed in Example 4.5 assumes that the molecule is composed of the most abundant isotopes of carbon (^{12}C) and oxygen (^{16}O). However, there are other stable isotopes of carbon (^{13}C) and oxygen (^{17}O and ^{18}O), and if substituted, because of the mass difference, will cause a change in the moment of inertia. These isotopically substituted molecules will therefore have different frequencies for their rotational lines. For instance, ^{13}C^{16}O is an *isotopologue* of carbon monoxide and its rotation constant is 55.1 GHz, leading to frequencies of 110, 220 and 330 GHz for its $J = 1-0$, $J = 2-1$ and $J = 3-2$ rotational transitions, respectively; these are significantly offset from the rotational lines of ^{12}C^{16}O. Other isotopologues of carbon monoxide include ^{12}C^{17}O, ^{12}C^{18}O, and ^{13}C^{18}O, all of which have been detected in emission in the interstellar medium.

Because the moment of inertia is different for every molecule, each molecule has a unique set of rotational spectral lines. Many of the diatomic and triatomic molecules have their first rotational transitions at wavelengths of a few millimeters. However, for long linear molecules such as cyanoacetylene (HCCCN), the moment of inertia is much larger than for molecules such as carbon monoxide, and according to Equation 4.33 their rotation constant is much smaller. Consequently, the first rotational transitions for these long linear molecules are at much longer wavelengths. For example, the $J = 1-0$ rotational transition for cyanoacetylene is at a wavelength of about 3 cm, about 10 times longer in wavelength than that for carbon monoxide.

For molecules to have permitted (electric dipole) rotational transitions they must have a *permanent dipole moment.* Although the total charge of a molecule is often zero, the electrons can be asymmetrically distributed relative to the center of mass, producing a charge separation within the molecule. The amount of charge separation in units of charge times distance is described by the molecule's dipole moment, symbolized by μ. A commonly used unit of dipole moment is the debye, where 1 debye $= 1\times10^{-18}$ esu cm $= 1\times10^{-18}$ g$^{1/2}$ cm$^{5/2}$ s^{-1}. Note that one has to be careful in converting the units of the dipole moment between cgs and SI units as the definition of charge differs (see Section 1.1.1).

Symmetric linear molecules, such as molecular hydrogen (H_2), carbon dioxide (OCO) and acetylene (HCCH) have their electrons distributed symmetrically, and therefore have no permanent dipole moment. Although these symmetric molecules have no permitted rotational transitions, they can have much weaker forbidden transitions if they have an electric quadrupole moment or magnetic dipole moment.

The value of the electric dipole moment varies from molecule to molecule. One of the smallest dipole moments is found for carbon monoxide, with a value of only about 0.1 debye, while typical dipole moments are of order 2 to 5 debye. Recall that the rate of radiative transitions is given by the Einstein A coefficient, which in turn depends on the molecule's dipole moment. For linear molecules the A coefficient for a transition between level J+1 to level J is given by

$$A = \frac{64\pi^4}{3hc^3}\,\mu^2\,\frac{J+1}{2J+3}\,\nu^3. \tag{4.36}$$

Because A is quadratic in μ, molecules with larger dipole moments have substantially higher rates of radiative transitions.

Most molecules in nature are not linear and the rotational energy levels for these molecules are more complicated. In general, molecules can be classified as either linear, *symmetric-top* or *asymmetric-top* molecules; examples of each are shown in Figure 4.14. The rotation of a molecule can usually be described by rotation about its principle axes and their corresponding moments of inertia. As we mentioned, in the special case of a linear molecule the moment of inertia about one of the axes is very small and the other two moments of inertia are equal. Some non-linear molecules have two equal moments of inertia, and these molecules are classified as a symmetric-top. If all three moments of inertia for a molecule are different, the molecule is classified as an asymmetric-top. If a molecule has an axis of symmetry, as is the case for methyl acetylene and formaldehyde (shown in Figure 4.14), then this is one of the principal axes of the molecule. If the molecule is rotated about the symmetry axis by an angle of $2\pi/n$ radians and remains unchanged, it has n-fold symmetry. In order for a molecule to have two equal moments of inertia and be a symmetric-top molecule it must have at least 3-fold or higher symmetry. For example, the molecule methyl acetylene (CH_3CCH) has 3-fold symmetry so it is a symmetric-top molecule; however the molecule formaldehyde (H_2CO) has only 2-fold symmetry so it is an asymmetric-top molecule.

Linear molecules can be described with just one moment of inertia, and therefore only one quantum number is needed to describe their energy levels. Symmetric-top molecules have two distinct moments of inertia and therefore need two quantum numbers to describe their energy levels. For symmetric-top molecules, the two quantum numbers are the ones that measure the total rotational angular momentum, denoted by J, and the projection of the rotational angular momentum onto the symmetry axis of the molecule, denoted by K. The rotational energy levels for symmetric-top molecules like methyl acetylene are shown in Figure 4.15, where the energy levels for the different K-values are plotted separately; these are called *K-ladders*. Note, because K is the projection of the total rotational angular momentum onto the symmetry axis, K can never be greater than J. Methyl acetylene is a prolate symmetric-top, because the moment of inertia about the symmetry axis is the smaller of the two moments of inertia. For prolate symmetric tops, the same J levels have increasing energies for larger values of K. The opposite is true for oblate symmetric-top molecules. For permitted (electric dipole) transitions, the selection rules are that $\Delta J = \pm 1$ and $\Delta K = 0$. The fact that radiative transitions cannot move a molecule between different K-ladders is important for determining the temperature of the gas, as will be discussed in Section 5.2.3. The more complex asymmetric-top molecules are beyond the scope of this book; the interested reader should consult a book on molecular spectroscopy.

There are two common effects that can alter the rotational energy level structure of molecules, even linear molecules. The first occurs for molecules in which at least one nucleus has a non-zero spin. A nucleus with a spin quantum number, I, greater than or equal to 1 will have an electric quadrupole moment. The interaction between this quadrupole moment and the rotation of the molecule produces a splitting of the rotational energy levels. If J is greater than or equal to I, then the rotational level is split into $2I + 1$ sublevels, and if J is less than I, it is split into $2J + 1$ sublevels. These sublevels are denoted by the quantum number F, where $F = J \pm I$. For example, the nitrogen nucleus in the linear molecule hydrogen cyanide (HCN) has a spin of I=1. Although the J=0 energy level remains the same (since J is less than I, it has only $2J + 1 = 1$ sublevels), all higher rotational energy levels are split into three sublevels (when J is greater or equal to I, each level has $2I + 1$ = 3 sublevels). The effects of this interaction for splitting the HCN energy levels are shown in Figure 4.16. The energy separation of the three sublevels in the J=1 rotational level is very small, only about one part in 10^5 of the energy of the level. Permitted changes in the quantum number F are $\Delta F = 0, \pm 1$, so each of the rotational transitions now consists of multiple, finely-spaced spectral features as illustrated in the insert to Figure

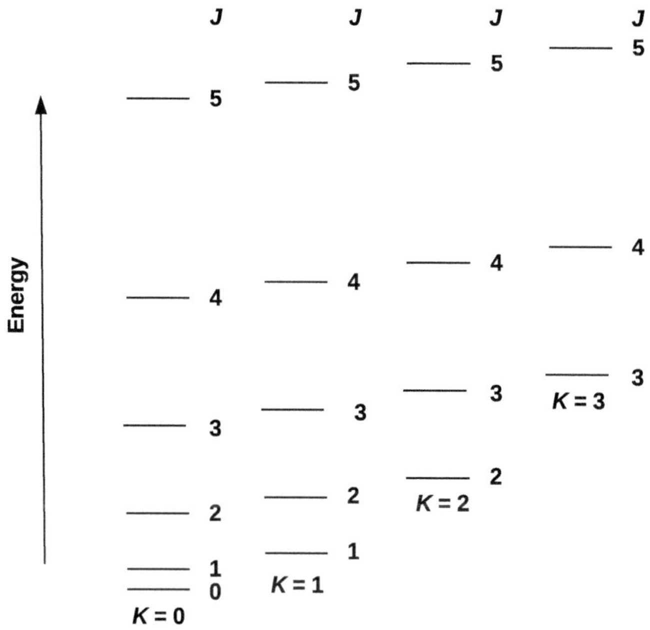

Figure 4.15: The energy level diagram for a prolate symmetric-top molecule such as methyl acetylene, including J-levels up to 5 and K-levels up to 3. As shown, the different values of K are usually plotted separately, and each of these separate stacks is called a K-ladder. For a prolate symmetric-top, for the same value of J, the energy increases for increasing values of K.

4.16. For HCN the J=1-0 transition is at a frequency of about 88.6 GHz and is composed of three components spaced by about 1 MHz in frequency; this splitting is called *quadrupole hyperfine structure.*

The second effect that alters the molecular rotational energy levels occurs when the molecule has a net (non-zero) electronic angular momentum in the ground electronic state. Examples of such molecules include the ethynyl radical (CCH) and the cyano radical (CN). Both of these molecules have an unpaired electron, giving rise to a net electronic spin angular momentum in the ground electronic state of the molecule. The interaction between the net electronic spin and the molecular rotation produces splitting of the rotational energy levels. The details are somewhat complicated; the interested reader should consult a book on molecular spectroscopy. Some molecules, such as CN, have both quadrupole hyperfine structure and structure produced by the non-zero electronic angular momentum.

We now discuss the information that can be obtained from molecular rotational lines. What can be learned depends on which molecular lines are observed. In Section 4.1.5 we discussed the excitation temperature and critical density of a transition. Whether the populations can be described by a thermal distribution depends on the density of the gas relative to the critical density. When the gas density is well below the critical density, collisional excitation is not important, and the rotational levels will be populated due to radiative excitations. In this case, the excitation temperature approaches the radiation temperature and the emission line will be very weak and often unobservable. On the other hand, when the gas density is well above the critical density, then excitations by collisions dominate. In

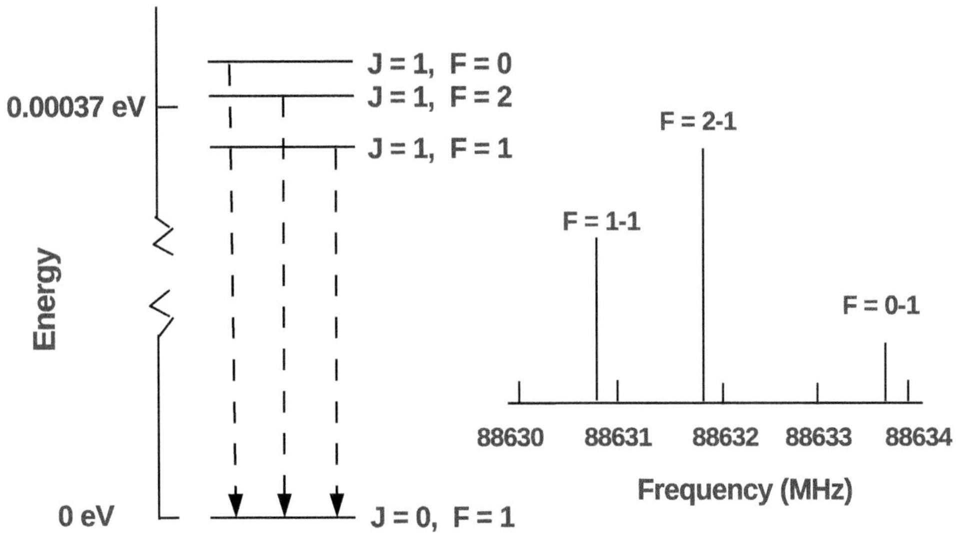

Figure 4.16: The lowest two rotational energy levels of hydrogen cyanide (HCN) are shown on the left, illustrating the splitting of the $J = 1$ level due to the interaction of the quadrupole moment of the nitrogen nucleus with the rotation of the molecule. The $J = 1$ level is split into three closely spaced energy levels (the splitting has been greatly exaggerated in this figure relative to the energy difference between the $J = 0$ and $J = 1$ rotational levels). The allowed transitions between split levels are shown, producing three closely spaced spectral features that constitute the $J = 1 - 0$ transition; these are illustrated to the right. This splitting is called quadrupole hyperfine structure.

this case, the transition's excitation temperature approaches the gas kinetic temperature and an emission line may be observable, depending on the column density of the molecule.

Recall that the critical density is given by $n_{\rm cr} = A_{ul}/C_{ul}$ (see Equation 4.26). The collisional de-excitation rate coefficient, C_{ul}, is similar for all simple linear molecules, and for a temperature of 20 K has a value of order 3×10^{-11} cm^3 s^{-1}. The A coefficient, though, varies significantly between molecules because it depends on the square of the dipole moment, see Equation 4.36. As we noted, the dipole moment for carbon monoxide is exceptionally small, about 0.1 debye and therefore its A coefficients are correspondingly small, about 7×10^{-8} s^{-1} for the $J = 1 - 0$ transition. Therefore, the critical density for the $J = 1 - 0$ transition of CO is of order 10^3 cm^{-3}, while the critical density for the same transition of HCN, which has a dipole moment of order 3 debye, is roughly 10^6 cm^{-3}. To produce significant emission above the background radiation (see Section 4.1.5), the excitation temperature has to be greater than the background radiation temperature, and thus the gas density must at least be of order of the critical density.

Carbon monoxide is unique in having a low critical density and large abundance; therefore we can usually infer the gas kinetic temperature because the line center is likely to be optically thick and its brightness temperature will equal the gas kinetic temperature. However, for many molecular lines, low optical depth can be assumed. For example, since the rarer isotopes of carbon and oxygen are far less abundant, the lines from isotopologues of CO, such as ^{13}CO, are usually optically thin. Emission lines from these isotopologues are useful in determining the column density of that molecule. Starting with Equation 4.12 and converting intensity and frequency to brightness temperature and velocity, we have

$$\int T_B \, dv \;=\; \frac{h\, c^3}{8\pi\, k\, \nu^2}\, N_u\, A_{ul}. \tag{4.37}$$

Equation 4.37, though, yields a measure only of the column density of the molecules in the upper state of that transition. If, for example, one measured the $J = 1 - 0$ transition of ^{13}CO, the integrated intensity would directly give the column density of ^{13}CO in the $J = 1$ rotational level.

The rotational energy levels of ^{13}CO are excited by collisions with other particles. Because the energies of these rotational levels are so small, a number of rotational levels are populated, even if the gas is relatively cold. However, since observations of a single transition of ^{13}CO only provide information on the column density in one of the rotational levels, we would have to observe many transitions to determine the total column density of ^{13}CO. Alternatively, we can determine the total column density of ^{13}CO if we know the fraction of the ^{13}CO molecules in the $J = 1$ level. Because of the relatively small critical density for the transitions of the isotopologues of CO, we can usually assume that the energy states are thermalized and so the relative populations in various rotational levels are given by a Boltzmann distribution. Therefore, the population in the ith rotational level is proportional to

$$n_i \;\propto\; g_i e^{-E_i/kT_K},$$

where g_i is the statistical weight of the ith level, E_i is the energy of the ith level above the ground state, and T_K is gas kinetic temperature. The sum of this function over all rotational levels is called the partition function, Z, and is given by

$$Z \;=\; \sum g_i\; e^{-E_i/kT_K}.$$

When $kT_K > hB$, where B is the rotation constant for the molecule, a useful approximation to the partition function for a linear molecule (in which $g_i = 2J + 1$) is given by

$$Z \;\sim\; \frac{kT_K}{hB} + \frac{1}{3} + \frac{1}{15}\,\frac{hB}{kT_K}. \tag{4.38}$$

Using the partition function, the fraction of the total population in the ith level, f_i, is given by

$$f_i \;=\; \frac{g_i\; e^{-E_i/kT_g}}{Z}. \tag{4.39}$$

Thus, if the molecule is thermalized and the gas temperature is known, then the column density of the molecule in all rotational levels can be calculated from the column density in the ith level by

$$N_{tot} = N_i/f_i. \tag{4.40}$$

Therefore, an observation of a single transition of a thermalized molecule often yields a determination of the column density of that molecule. If the molecule is not thermalized, for example when the gas density is below the critical density, to compute the fraction in any one state requires solving a set of coupled equations, each similar to Equation 4.24, including all relevant rotational levels. Such a calculation requires knowledge of both the density and temperature of the gas, but if these are known the equations can be solved.

When observing an emission line of a molecular transition with a large critical density, such as HCN $J = 1 - 0$, emission will only be detected from the higher density regions. In lower-density molecular gas, those molecules with low $n_{\rm cr}$ will be detected in emission while those with higher n_{cr} will not, even though both types of molecules likely exist in the emitting medium. Molecules with higher n_{cr} (such as HCN, SiO, CS, or HCO^+) are considered high density tracers because their emission reveals the high density gas.

An important phenomenon occurs when the populations of the upper and lower levels of a transition are inverted, which means that the upper level has a population relative to the lower level greater than would be expected from their degeneracies, that is $n_u/n_l > g_u/g_l$. Note that when using Equation 4.14, this implies that the excitation temperature is negative and when using Equation 4.18 that the optical depth is negative. Basically, in these cases the rate of stimulated emission is greater than the rate of absorption; thus the medium acts to amplify a background light source at the frequency of the transition. This is the principle at work in a laser (light amplification by the stimulated emission of radiation). When this inversion occurs for rotational molecular lines, these are called *masers* (microwave amplification by stimulated emission of radiation). The causes for a population inversion can be complicated, but in general involve at least three energy levels in the molecule. The molecule is excited from the lower energy states to the highest state by either collisions or by radiative excitation, and as the molecule de-excites, the intermediate level is preferentially populated over the lowest level, producing a population inversion of the lower two levels. When photons of energy equal to the difference between the two lower states pass through the gas, large amounts of stimulated emission occur. The inversion of the lower two states, in effect, acts as an amplifier of background radiation at the frequency of the transition. Since this amplification involves stimulated emission, the maser light output is coherent. Several molecules have known population inversions producing astronomical masers, and include H_2O and OH, among others.

QUESTIONS AND PROBLEMS

1. An important spectral line in astronomy is the radiative transition between the n=2 and n=1 electronic energy levels of hydrogen. This spectral line is part of the Lyman series of lines in hydrogen and this transition is called the Lyman α line. What are the wavelength and frequency of this transition?

2. The Lyman α transition (see Question 1) has an Einstein A coefficient of 6.26×10^8 s^{-1}. If the column density of hydrogen atoms in the n=2 energy state is 1×10^{12} cm^{-2} (making the line optically thin), what would be the integrated intensity of the Lyman α emission line?

3. (a) If the hydrogen atoms in Question 2 are at a temperature of 10,000 K, what is the value of the Doppler broadening parameter for the Lyman α line in frequency units and in velocity units?
(b) For the column density of the n=2 energy state given in Question 2, what is the peak intensity of the line?

4. In Section 4.2.1 we mentioned that one can approximate the area of the Gaussian shaped spectral line by its peak intensity multiplied by its FWHM line width. What is the fractional error incurred in using this approximation?

5. Derive the expression for the intensity at line center for an optically thick line given in Equation 4.13.

6. Derive the relationship between the Einstein coefficients given in Equations 4.6 and 4.7 by considering the case where the two levels of an atom are in radiative equilibrium.

Hint: start with Equation 4.24 and assume that the density is sufficiently low that you can eliminate the terms involving collisional excitation and de-excitation. The intensity of the radiation field is given by

$$I_\nu \;=\; B_\nu(T_R) \;=\; \frac{2h\nu^3}{c^2}\,[e^{h\nu/kT_R}-1]^{-1},$$

where in radiative equilibrium, $T_R = T_{ex}$.

7. The 21-cm line of atomic hydrogen has an Einstein A coefficient of 2.87×10^{-15} s^{-1}. What are the values of the two Einstein B coefficients? Remember that the statistical weights of the spin states are given by 2F+1.

8. Assume some atomic hydrogen gas is at a temperature of 100 K, and that the two spin states of the 21-cm line are in thermal equilibrium (therefore the excitation temperature is also 100 K). Using the results from Question 7, compute the absorption coefficient for the 21-cm transition. Why do you have to be very careful when computing the absorption coefficient for this transition?

9. A cloud of atomic hydrogen gas is at a temperature of 100 K and the 21-cm transition is in thermal equilibrium. If the 21-cm spectral line from this cloud is Doppler broadened solely by thermal motions and the atomic hydrogen column density is 1×10^{20} cm^{-2}, what is the peak brightness temperature of the 21-cm emission line?

10. Compute the frequencies for the H90α and C90α recombination lines.

11. The molecule silicon monoxide (SiO) has a rotation constant of 21.71 GHz. What are the frequencies of the $J = 1-0$, $J = 2-1$ and $J = 3-2$ rotational transitions?

12. The dipole moment for the molecule SiO is 3.1 debye, and the collisional de-excitation rate coefficient for the $J = 2-1$ transition at a temperature of 20 K is $C_{21} = 1 \times 10^{-10}$ cm^3 s^{-1}. What is the critical density for the $J = 2-1$ transition?

13. Assume the SiO molecule is in thermal equilibrium at a temperature of 20 K. What is the fraction of SiO molecules in the $J = 0$, $J = 1$, $J = 2$, $J = 3$, $J = 4$, and $J = 5$ rotational levels?

14. Derive Equation 4.37.

15. The molecules hydrogen isocyanide (HNC) and hydrogen cyanide (HCN) are isomers — molecules with the same chemical formula, but different molecular structures. In this case, both are linear molecules, but the ordering is different. The $J = 1-0$ transition frequencies are about 90,663 and 88,632 MHz for HNC and HCN, respectively, while their dipole moments are 3.05 and 2.98 debye. Which molecule has a longer lifetime in the $J = 1$ state before undergoing a radiative transition to the ground state?

CHAPTER 5

The Cold Interstellar Medium of the Milky Way

THE space between the stars in our Galaxy is not empty, but filled with gas and dust that we refer to as the interstellar medium. In addition to gas and dust, the *interstellar medium* contains radiation, magnetic fields and high-energy charged particles called cosmic rays. In this chapter, we concentrate on the gas and dust in the interstellar medium, as observations at radio wavelengths have had an enormous impact on our understanding of the physics and spatial distribution of these components in the Milky Way. Much of the gas and dust in the interstellar medium is too cold to produce detectable electromagnetic radiation at optical or infrared wavelengths, and one must observe at radio wavelengths to detect the continuum or spectral line emission from this cold component.

Cold gas and dust can absorb optical light and observations in the early part of the 20th century provided evidence for this cold component in the interstellar medium. The presence of gas was first identified by the optical absorption lines it produced in the spectra of background stars, while the first evidence for dust was through extinction of starlight at optical wavelengths. If the gas and dust are sufficiently close to a hot star, they can be heated to sufficiently high temperatures to produce optical and infrared emission. However, the focus of this chapter is on the cooler atomic and molecular gas and its associated dust, which are best studied at radio wavelengths. This cooler component dominates the mass of the interstellar medium. Before beginning our discussion of radio observations of the gas and dust, we provide a brief overview of the properties of the interstellar medium.

The interstellar medium is a dynamic medium and is affected by a number of impulsive stimuli, such as supernova explosions and expanding HII regions. The densest regions of the interstellar medium are also influenced by gravity, and it is in these dense regions that gravitational collapse of the gas and dust leads to the formation of stars. Because of its dynamical nature, the interstellar medium is composed of a number of distinct phases with different temperatures and densities. Pressure plays an important role in this medium, as regions of overpressure will expand into the lower pressure surroundings and regions of underpressure will be overtaken by the higher pressure surroundings. The interstellar medium will try to achieve pressure equilibrium, and as we will now show, the time scale for regions to achieve this equilibrium is short enough that different regions cannot stay out of equilibrium for very long.

The speed at which a medium adjusts to pressure differences is the speed of sound, c_s, given by

$$c_s = \sqrt{\frac{3kT}{m}},$$

where m is the average mass of the atoms or molecules that make up the medium. For gas at a temperature of $T = 100$ K composed primarily of hydrogen atoms, the speed of sound is 1.6×10^5 cm s^{-1} or 1.6 km s^{-1}. Therefore, the timescale for a parcel of gas of size scale L to achieve pressure equilibrium with its surroundings is of order

$$t \sim \frac{L}{c_s} \sim 6 \times 10^5 \text{ yr} \left(\frac{L}{\text{pc}}\right).$$

For a region 10 pc in diameter this time scale is about 6 million years. This may seem like a long time, but it is relatively short compared to other time scales that affect the disk of our Galaxy.

Although the pressure of the gas throughout the interstellar medium may be similar, it does not have the same temperature and density. Remember that gas pressure is given by $P = nkT$, where n is the gas number density (the number of atoms or molecules per unit volume). Therefore, if in pressure equilibrium, regions of high temperature must have low densities and regions of low temperature must have high densities. The mean pressure of the interstellar medium has a value of order 5×10^{-13} dyne cm^{-2} or 5×10^{-13} g cm^{-1} s^{-2}. Instead of using pressure, astronomers often characterize the interstellar medium by the quantity $P/k = nT$; and the mean value of P/k in the interstellar medium is about 3600 cm^{-3} K.

Regions of different temperatures and densities are referred to as different phases of the interstellar medium, and astronomers recognize at least five distinct phases. A brief summary of each of these is presented below, in order of increasing temperature.

1. The Molecular Medium. This phase of the interstellar medium is primarily composed of molecules. Since hydrogen is the most abundant element in the Universe, it should not be surprising that the most abundant molecule is molecular hydrogen (H_2). The temperature of this molecular gas can be as low as 10 K, and the number density of molecules can be greater than 10^3 cm^{-3}. Unlike all other phases of the interstellar medium, gravity can play an important role in the evolution of the Molecular Medium. This phase of the interstellar medium has a pressure greater than what is generally found in the interstellar medium and it is the self-gravity of this dense gas that keeps this phase from expanding into the lower pressure surroundings. The dense Molecular Medium is the gas phase most directly connected with star formation. This phase is best probed by observations of the spectral lines from rotational transitions of molecules at radio wavelengths which were discussed in Section 4.2.3.

2. The Cold Neutral Medium (CNM). This gas is composed primarily of neutral atomic hydrogen, or HI, at a temperature of about 80 K. This is the colder of the two neutral, atomic gas phases in the interstellar medium. In the early two-phase model of the interstellar medium, which was characterized by denser clouds immersed in a lower density inter-cloud medium, the CNM was the cloud component. The CNM has an average density of about 50 cm^{-3}. This phase of the interstellar medium is most readily probed by the emission of the 21-cm line of neutral atomic hydrogen discussed in Section 4.2.1.

3. The Warm Neutral Medium (WNM). This gas, like the CNM, is primarily composed of neutral atomic hydrogen. Although its temperature is poorly determined, it is much warmer than the CNM, and most likely in the range of 10^3 to 10^4 K. This component is the inter-cloud medium in the early two-phase model of the interstellar medium. Its density is estimated to be of order 1 cm^{-3}, and thus, has a pressure similar to the

CNM. This phase, like the CNM, is best probed by observing the 21-cm transition of neutral atomic hydrogen discussed in Section 4.2.1.

4. The Warm Ionized Medium (WIM). The WIM has temperatures and densities similar to the WNM, but unlike the WNM the hydrogen is fully ionized. Fully ionized gas is found in HII regions (see Chapter 6) associated with newly formed massive stars in which stellar ultraviolet radiation photo-ionizes the surrounding gas. It has been generally assumed that all of the hydrogen ionizing photons (called Lyman continuum photons) emitted by these stars were used in maintaining the ionization in HII regions. However, the presence of a more pervasive ionized gas component in the interstellar medium suggests that some ionizing photons escape the immediate surroundings of the stars, and are responsible for producing the WIM. The free electrons in this phase produce a time delay in radio signals from pulsars which we detect as the dispersion measure (see Section 2.2.2). The distribution of electrons and ionized hydrogen in the WIM can be determined from measurements of the pulsar dispersion measure (see Section 7.4.5) and from the very faint hydrogen recombination line emission at optical wavelengths.

5. The Hot Ionized Medium (HIM). Gas heated by supernova explosions can be very hot, with temperatures exceeding 10^6 K and densities of order 10^{-3} cm^{-3}. This hot gas is sometimes referred to as coronal gas, because it is as hot as the Sun's corona. The hot bubbles of gas produced by supernovae are regions of overpressure and will expand until they reach approximate pressure equilibrium with their surroundings. Since this gas is not continuously heated, it will eventually cool, and merge with the surrounding cooler interstellar medium. Gas at these low densities and very high temperatures cools very slowly, and these hot bubbles last for millions of years. The hot and fully ionized gas of the HIM can be probed by its X-ray emission and the ultraviolet absorption lines it produces in background stars due to the highly ionized metal atoms in this phase.

A very simple, but elegant, model of the gas in the interstellar medium was proposed by George Field, Donald Goldsmith and Harm Habing[1]. They assumed that the gas was in pressure equilibrium and that the rates of heating and cooling of the gas were equal so that the temperature of the gas did not change with time (thermal equilibrium). They found that thermal and pressure equilibrium naturally led to a two-phase model for the interstellar gas, and these two components we now recognize as the CNM and WNM. The fact that these two components could co-exist in pressure and thermal equilibrium explained the emission and absorption measurements of the 21-cm line of HI, observations which will be discussed in Section 5.1.4. Although some of the details of this model, such as the heating rates in the interstellar medium, have been updated, this concept is still useful today. However, the two-phase model was not the complete story of the interstellar medium, as only a few years later a much hotter component of the interstellar medium was identified.

In the early 1970s, both diffuse X-ray emission and absorption lines in the ultraviolet spectrum of stars from highly ionized atoms (such as OVI, in which five of the oxygen atom's eight electrons have been removed) were detected; this required gas with a temperature of order a million degrees. This hot gas is the coronal gas or HIM phase of the interstellar medium. Donald Cox and Barham Smith[2] recognized the role supernova explosions should have on heating of the interstellar gas. Even if there are only 1 or 2 supernovae

[1] 1969, Astrophysical Journal Letters, vol. 155, p. L149.
[2] 1974, Astrophysical Journal Letters, vol. 189, p. L105.

per century in the Milky Way, a network of hot, interconnected gas tunnels can be produced throughout the interstellar medium. Since gas at these high temperatures cools very slowly, they suggested that this hot phase (HIM) can live sufficiently long for it to dominate the volume of the interstellar medium. Following the paper of Cox and Smith, Christopher McKee and Jeremiah Ostriker[3] produced a self-consistent multi-phase model of the interstellar medium. In this model they explained many of the observed traits of the interstellar medium, including the HIM, WIM, WNM and CNM phases.

In this chapter, we will discuss the role radio observations have played in defining the properties of the cold atomic (both the CNM and WNM phases) and molecular gas in the interstellar medium as well as the cold dust. We start with a discussion of the observations of the 21-cm spectral line of HI and summarize how observations of this radio spectral line can be used to deduce the distribution of atomic gas in our Galaxy and the physical properties of the atomic gas. This is followed by a discussion of the observations of the molecular rotational spectral lines and what they tell us about the molecular medium of the Milky Way. Finally, we discuss the observed thermal emission at radio wavelengths from cold interstellar dust grains.

5.1 21-CM SPECTRAL LINE OF ATOMIC HYDROGEN

The physics of the 21-cm spectral line due to neutral, atomic hydrogen (HI) was presented in Section 4.2.1. In this section we discuss how this spectral line has been used to map the distribution of atomic hydrogen in the Milky Way. For observations of the Galaxy, it is most convenient to use the Galactic coordinate system that was discussed in Section 1.3.2. Recall that directions in Galactic coordinates are defined by Galactic longitude and latitude; the definition of these coordinates is illustrated in Figure 1.7.

5.1.1 Observations of the 21-cm Line

In Figure 5.1 we display the observed spectrum of the 21-cm line from the Milky Way in the direction $\ell = 29°$ and $b = 0°$. Note the axis labels of this plot. As discussed in Section 4.2.1, radio spectral lines are usually plotted as brightness temperature (a measure of intensity) versus velocity (a measure of frequency and calculated from the Doppler effect). Since our observations are performed on Earth, which both rotates on its axis and orbits the Sun, the observed Doppler shifts of celestial objects will vary with time. To produce astronomical data that are independent of the observation time and location, astronomers generally correct for the Earth's motions (both rotational and orbital) and quote velocities relative to the Sun, called heliocentric velocities. An additional correction that is often made is to correct for the Sun's motion relative to its local region of the Milky Way. The Sun's orbit about the center of our Galaxy is not perfectly circular, nor is it exactly oriented in the plane of the Galaxy. Astronomers define a velocity reference frame, called the *local standard of rest* (LSR), located at the same distance from the Galactic Center as the Sun and moving in a perfectly circular orbit in the plane of the Galaxy. We can reference velocities relative to the local standard of rest and these velocities are referred to as V_{LSR}. The Sun has a motion of about 15 km s^{-1} relative to the LSR; thus the velocity scale for a spectrum plotted using heliocentric velocity will be slightly different from the same spectrum plotted using LSR velocity.

As we discussed in Chapter 4, the CMB is a pervasive continuum background at radio wavelengths. In Figure 4.6 we illustrate examples of emission-line spectra of varying

[3] 1977, Astrophysical Journal, vol. 218, p. 148.

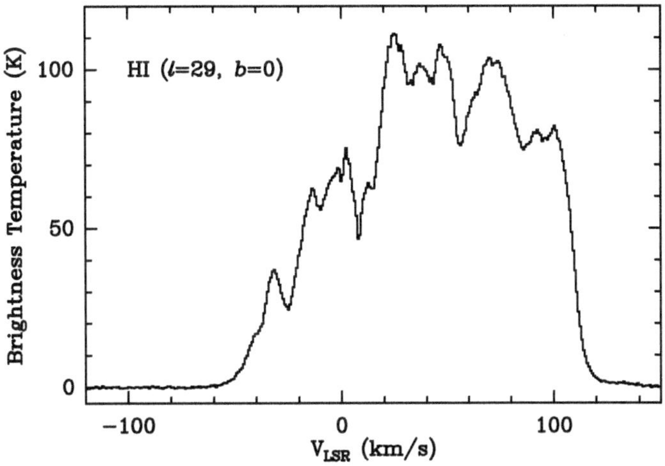

Figure 5.1: This HI spectrum was obtained in the direction $\ell = 29°$ and $b = 0°$. These data were obtained with the Green Bank Telescope and provided by Felix J. Lockman.

optical depths, and we note that outside of the line the intensity is not zero but that of the background radiation. However, observations made with radio telescopes are usually made by a differencing technique (see Chapter 4 in Volume I). An often-used technique is position-switching, where the signal toward blank sky is subtracted from the signal toward the target of interest. Since the CMB is nearly isotropic, its continuum signal is the same in both directions and will be removed by this differencing technique. Therefore, even though the spectrum shown in Figure 5.1 has an intensity of zero outside of the line, this is a consequence of the observing technique. Remember that zero intensity in Figure 5.1, and all spectra shown in this chapter, is really the intensity of the CMB. Thus, the line emission should be on top of the CMB continuum radiation, as illustrated in Figure 4.6, as well as any additional extended Galactic thermal or non-thermal emission.

The spectrum shown in Figure 5.1 is far more complex than the emission from a single cloud of HI that is thermally broadened. Emission is seen over a velocity range of over 170 km s^{-1}, from a V_{LSR} of −50 to +120 km s^{-1}. This broad velocity range is due to large-scale motions of the HI gas relative to the LSR velocity frame, a consequence of Galactic rotation. The Galaxy does not rotate as a rigid body, but each parcel of gas has its own orbital motion about the Galactic center. The line of sight of the observation of Figure 5.1 is depicted in Figure 5.2, which shows a face-on view of the Galactic disk. There is HI gas all along this path; however we will concentrate on gas at six marked locations (A, B, C, D, E, and F). The gas at each location is moving in a nearly circular orbit about the Galactic center and the orbital velocity vector of each parcel is shown and varies in direction and magnitude. The velocity vector shown for the Sun represents the local standard of rest velocity. We now examine what V_{LSR} is observed for gas at each location.

The gas at position A in Figure 5.2 has a similar orbit to the local standard of rest, but because it is turned slightly, it has a small receding radial velocity relative to the LSR velocity. Thus, the gas at position A will be slightly redshifted relative to the local standard of rest velocity. Remember that redshifted gas is defined to have a positive velocity; therefore the gas at position A will have a small positive V_{LSR}. The gas at position B is turned further than gas at A, and so will have a larger recessional velocity relative to the local standard of rest than gas at position A. The gas at position C is at a position we call the *tangent point*, as the line of sight is tangent to a circle centered on the Galactic center that passes through

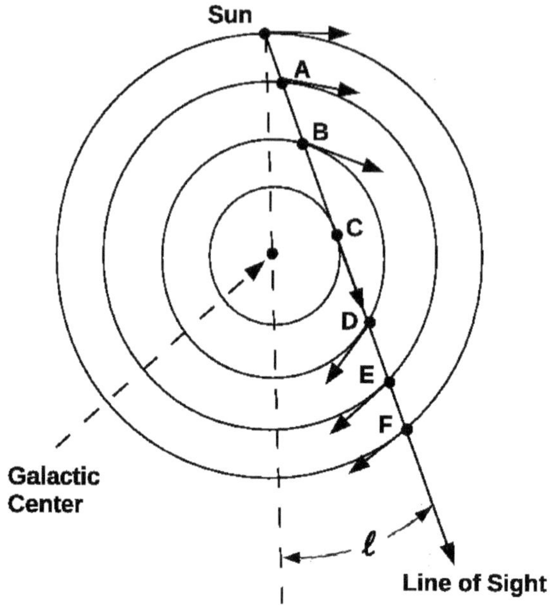

Figure 5.2: This diagram shows a face-on view of our Galaxy. A line of sight through the Galaxy at Galactic longitude ℓ is shown. Along this line of sight there are parcels of gas at locations A, B, C, D, E, and F. The arrows at each of these locations show the velocity vector of the gas parcel in its nearly circular orbit about the Galactic center. The arrow at the position of the Sun is the velocity vector for the local standard of rest.

this point. Position C is the point closest to the Galactic center along this line of sight and this gas at the tangent point produces the largest recessional velocity of any gas along this Galactic longitude. If we look at the spectrum in Figure 5.1, we see that the highest velocity emission is at a $V_{\rm LSR} \sim +105$ km s^{-1}; this is gas near the tangent point at this Galactic longitude.

Gas beyond the tangent point, at position D in Figure 5.2, will have a smaller recessional velocity relative to the the local standard of rest than gas at the tangent point. Thus the gas at position D will have a smaller redshift than gas at position C. Gas at position E will have a yet smaller recessional velocity than either gas at positions C or D. It may not be obvious from the figure, but gas at position F, which lies on the *Solar circle*, a circle that is centered on the Galactic center that passes through the Sun, will have no radial component of motion relative to the local standard of rest velocity frame, and thus a $V_{\rm LSR} = 0$. For gas further away along this same line of sight, and therefore beyond the solar circle, its radial velocity relative to the LSR velocity frame is negative, and therefore this gas will produce blueshifted emission and have a negative $V_{\rm LSR}$. The gas will have progressively larger negative velocities the further the gas lies beyond the solar circle. The HI emission seen in Figure 5.1 with negative velocities is emission from gas that lies beyond the solar circle along this line of sight.

The spectrum shown in Figure 5.1 was taken looking toward the inner part of the Galaxy. However, the appearance of the 21-cm spectrum changes dramatically depending on the direction one observes. For instance, Figure 5.3 shows a spectrum obtained in the direction $\ell = 135°$ and $b = 0°$, thus looking toward the outer part of the Galaxy. Since the Sun is located well-offset from the center of the Galaxy, there is less emission in this direction than toward the inner Galaxy. In addition, the V_{LSR} of the gas emission is almost entirely negative, indicating that the gas is approaching relative to the velocity frame of the LSR. Finally, Figure 5.4 shows a 21-cm spectrum obtained toward $\ell = 29°$, $b = +10°$, thus looking at the same Galactic longitude as the spectrum in Figure 5.1, but now looking in a direction inclined relative to the plane of the Galaxy. There is much less emission in this direction than in the direction shown in Figures 5.1, indicating that atomic hydrogen is concentrated in the Galactic plane.

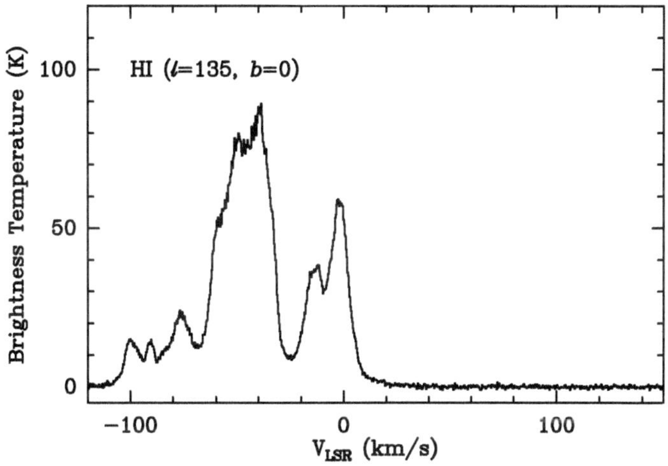

Figure 5.3: This HI spectrum was obtained in the direction $\ell = 135°$ and $b = 0°$. These data were obtained with the Green Bank Telescope and provided by Felix J. Lockman.

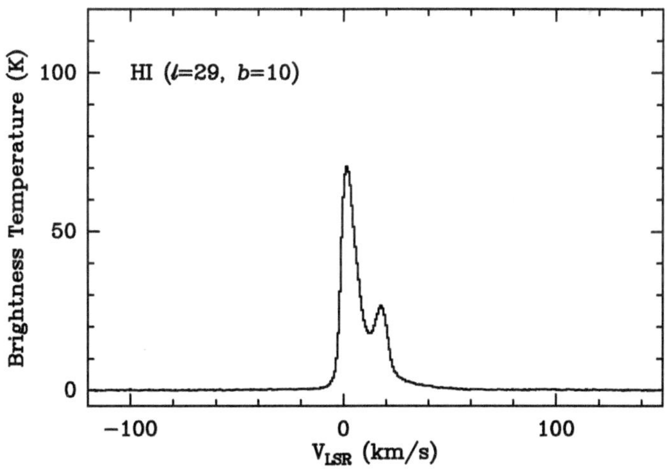

Figure 5.4: This HI spectrum was obtained in the direction $l = 29°$ and $b = +10°$. These data were obtained with the Green Bank Telescope and provided by Felix J. Lockman.

The 21-cm emission from atomic hydrogen has been observed over the entire sky. We can display these results using a method similar to an Aitoff projection used to show the surface of the Earth in a flat projection. The projection uses Galactic coordinates and is illustrated in Figure 5.5. Using this all-sky projection, the distribution of integrated intensity of the 21-cm line over the entire sky is presented in Figure 5.6. This map shows that the 21-cm emission is strongly concentrated to the Galactic plane ($b = 0°$) and is brightest for Galactic longitudes within about 90 degrees of the Galactic center, and thus toward the inner Galaxy. We can readily infer that the hydrogen is distributed in a flattened disk.

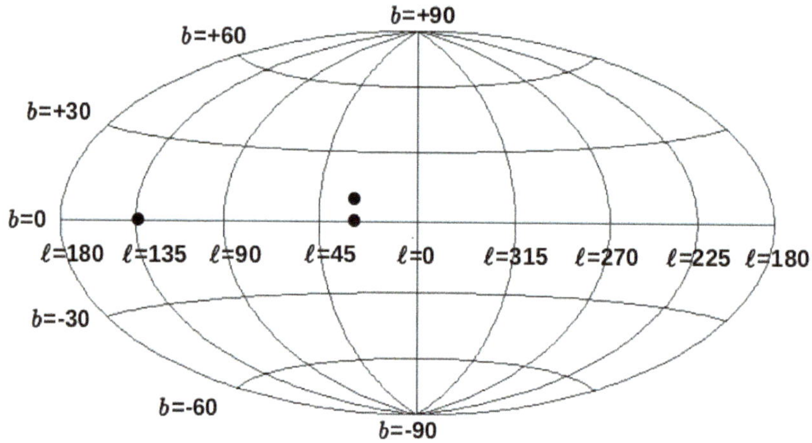

Figure 5.5: An all-sky projection of the sky in Galactic coordinates. This projection is similar to an Aitoff projection used for mapping the surface of the Earth onto a flat plane. Lines of Galactic longitude (ℓ) and latitude (b) are shown. The filled circles indicate the positions of the HI spectra shown in Figures 5.1, 5.3 and 5.4.

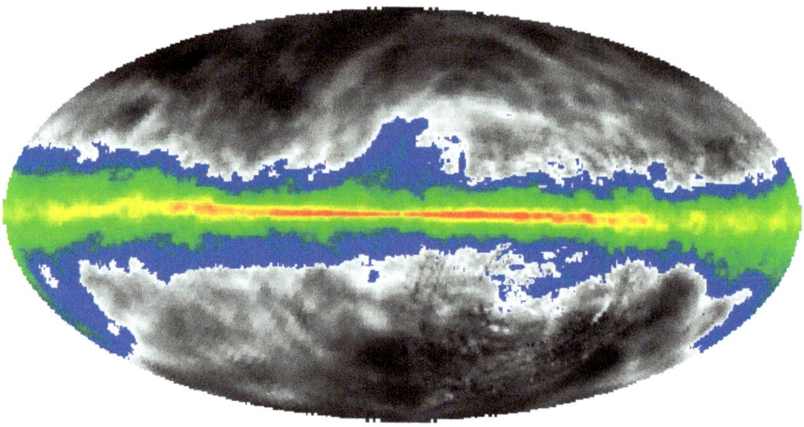

Figure 5.6: An all-sky image showing the distribution of 21-cm hydrogen emission. The colors indicate the integrated intensity; regions with the highest integrated intensity is shown in red, followed by regions of successively decreasing integrated intensity shown in yellow, green, blue, white, grey and black. This image was provided by Felix J. Lockman.

Our location within the Milky Way is not advantageous for determining how the atomic gas is radially distributed in the Galaxy. To understand the radial distribution of HI gas we will need to use the velocity of the emission to determine where along the line of sight the gas is located. However, to determine the location from velocity, we must first determine the orbital speed of the gas as a function of Galactic radius.

5.1.2 Rotation Curve of the Galaxy

The rotation of the Galaxy's disk about the Galactic center can be determined by measuring the radial velocity of the HI gas. Although the gas follows nearly circular orbits, the disk does not rotate as a rigid body. The speed of rotation, V, depends on the distance, R, from the center; the function $V(R)$ describes the *rotation curve* of the Galaxy. Thus, we wish to determine the orbital velocity of the gas at different Galactic radii. As we showed in Figure 5.1, the 21-cm spectra of the Galactic disk have emission over a wide range of velocities. The trick is, in short, to make clever use of the tangent point (point C in Figure 5.2) along a line of sight that passes through radii inside the Solar circle. The tangent point has a readily calculable distance from the Galactic center and has the largest velocity in the observed spectrum. As discussed in Section 5.1.1, gas at the tangent point in the first quadrant of the Galaxy ($\ell = 0°$ to $90°$) has the largest redshifted or positive velocity along the line of sight. Similarly, gas at the tangent point in the fourth quadrant of the Galaxy ($\ell = 270°$ to $360°$) has the largest blueshifted or negative velocity along the line of sight. We can take advantage of this property of the tangent point gas to determine the rotation curve of the Milky Way.

Figure 5.7 illustrates the geometry of the gas at the tangent point (point C) along a line of sight at Galactic longitude ℓ. The triangle formed by lines joining the Sun, the Galactic center and the tangent point C is a right triangle. Therefore the distance between the tangent point gas and the Galactic center (R) is given by

$$R = R_o \sin \ell,$$

where R_o is the Sun's distance from the Galactic center. Gas at the tangent point is moving directly away from the Sun; therefore its entire motion contributes to a Doppler shift. The $V_{\rm LSR}$ we measure, though, is relative to the velocity frame of the LSR, and so includes a component due to orbital motion of the LSR in the tangent point direction. The component of the LSR velocity in the direction of the tangent point is given by $V_o \sin \ell$, where V_o is the circular rotation speed of the LSR. Therefore, the true orbital velocity of gas at the tangent point is

$$V(R) = V_{\rm LSR} + V_o \sin \ell. \tag{5.1}$$

To determine $V(R)$, we need information about R_o and V_o. In 1985, the International Astronomical Union (IAU) adopted values of $R_o = 8500$ pc (or 8.5 kpc) and $V_o = 220$ km s^{-1}. Recent studies suggest that the distance may be somewhat smaller and the velocity somewhat larger than the values adopted by the IAU; however we will continue to use the IAU values.

Example 5.1:

The 21-cm spectrum of HI shown in Figure 5.1 is toward the Galactic longitude of $\ell = 29°$. For this line of sight, what is the distance from the tangent point to the center of the Galaxy?

Answer:
The distance R to the Galactic center is given by $R = R_o \sin \ell$. For $\ell = 29°$ and $R_o = 8.5$ kpc, $R = 8.5 \text{ kpc} \times \sin(29°) = 4.1$ kpc.

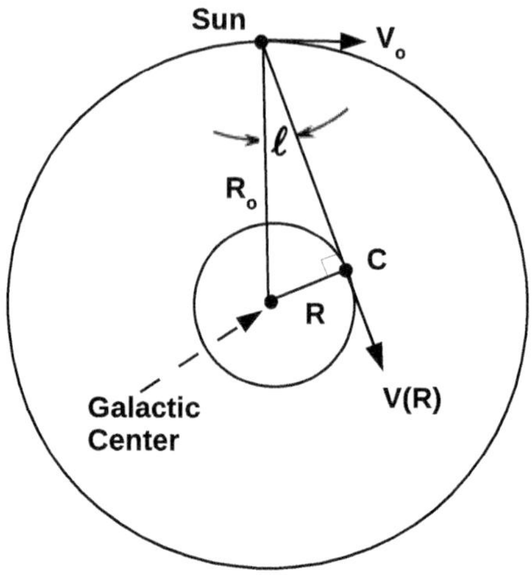

Figure 5.7: A face-on view of the Galaxy showing the geometry of the tangent point. The line of sight is at a Galactic longitude of ℓ and it crosses the tangent point at position C. The triangle formed by the Sun, the Galactic center, and the tangent point position is a right triangle. The gas at position C is at a distance R from the Galactic center and is orbiting with a circular velocity of $V(R)$. The local standard of rest (vector at the Sun) is moving with a circular velocity of V_o and is located a distance R_o from the Galactic center.

Example 5.2:

Based on the 21-cm spectrum shown in Figure 5.1 what is the rotation speed at a galactocentric radius of 4.1 kpc?

Answer:
The highest velocity emission along this line of sight, which is due to gas at the tangent point, is at $V_{\rm LSR} = 105$ km s^{-1}. The rotation speed is given by Equation 5.1

$$V(R) = V_{\rm LSR} + V_o \sin \ell = 105 \text{ km s}^{-1} + (220 \text{ km s}^{-1}) \sin(29°) = 212 \text{ km s}^{-1}.$$

Thus the rotation speed at a distance of 4.1 kpc from the Galactic center is 212 km s^{-1}.

Observing along lines of sight over a range of Galactic longitudes, and therefore at different tangent point distances, permits measurements of the rotation speed of the Galaxy at different radii. The tangent point technique has also been performed using emission from rotational transitions of molecules in the interstellar medium (see Section 5.2.1). However, the tangent point method only works for $R < R_o$ and so other methods are needed beyond the Solar circle. Measures of the rotation curve beyond the Solar circle require information about the distance of the emitting gas, which is often difficult to determine accurately.

One determination of the Galactic rotation curve is shown in Figure 5.8. Measurements within the Solar circle ($R_o = 8.5$ kpc) are based on observations of the 21-cm transition of hydrogen and the J=1-0 rotational transition of carbon monoxide at the tangent point. Measurements beyond the Solar circle are based on HII regions (see Chapter 6) whose distances are estimated from the properties of the exciting stars. Spectra of the molecular gas associated with these HII regions provide the V_{LSR}. If we know the V_{LSR}, distance and Galactic longitude, we can compute the rotational speed and the Galactic radius of the HII region. You will note that the points beyond the solar circle in Figure 5.8 have a much larger scatter and much larger uncertainty, which reflect the difficulty in measuring accurate distances to HII regions and, thus, determining the rotation curve accurately.

The rotation speed in the Milky Way is determined by the enclosed mass and orbital radius following Newton's Laws. Therefore, measuring the rotation curve of the Milky Way allows one to determine the total mass distribution within our Galaxy. The nearly constant rotation speed in the outer part of the Milky Way has an important implication concerning the presence of dark matter. We discuss the analysis of these data and its significance in Section 8.1.2.

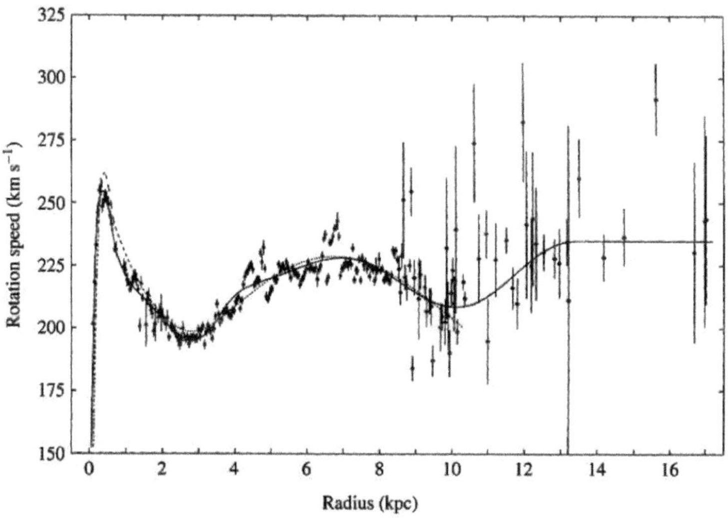

Figure 5.8: The rotation curve for the Milky Way, assuming that $R_o = 8.5$ kpc and $V_o = 220$ km s^{-1}. Within the Solar circle ($R < R_o$), the rotation curve is based on the tangent point analysis using both the transitions of the 21-cm line of hydrogen and the rotational line of carbon monoxide. Beyond the solar circle, the rotation curve was determined from the velocity of molecular gas associated with HII regions of known distance. This figure is from a paper by Dan Clemens (1985, Astrophysical Journal, vol. 295, p. 422).

5.1.3 Distribution of HI in the Milky Way

With the establishment of a rotation curve of the Galaxy, we can use the observed radial velocities to determine where the gas is located along the line of sight. A desirable goal is to use the all-sky surveys of the 21-cm HI emission and the measured rotation curve to determine distribution of atomic hydrogen in the Milky Way. If the 21-cm emission is optically thin, its integrated intensity provides a direct measure of the column density of atomic hydrogen in each parcel of gas at each velocity along each line of sight. However, the emission along lines of sight near the Galactic plane are often not optically thin, and a correction must be applied to determine the column density of atomic hydrogen.

There is a significant complication in determining the distribution of atomic hydrogen. In observations toward the inner Galaxy, there are two positions along a line of sight that have the same $V_{\rm LSR}$. For instance, gas located at positions A and E in Figure 5.2, which are equally distant from the Galactic center, have the same $V_{\rm LSR}$. Thus the emission from gas at these two locations will be detected at the same frequency and coincide in the spectrum. There is no way of distinguishing how much gas is at each location. We call position A the near-distance and position E the far-distance; likewise, the emission from positions B and D will coincide. The confusion of the emission from these two positions is known as the kinematic distance ambiguity. This ambiguity only exists for positions within the Solar circle. For gas beyond the Solar circle, in any direction, the observed radial velocity provides a unique location in the Galaxy. The ambiguity in the inner galaxy distances prevents a thorough detailed mapping of the HI. However, because the two locations with the same $V_{\rm LSR}$ are at the same Galactic radius, R, one can more readily obtain a measure of the HI mass as a function of radius in the Galaxy, as will now be shown.

In Figure 5.9 we show the same line of sight through the Galaxy as in Figure 5.2. Figure 5.9a shows that the points B and D are the same distance from the Galactic center and Figure 5.9b focuses on only point B. Note also the angles ϕ and θ defined in Figure 5.9b. The rotational motion of the gas at point B can be separated into two components, one along the line of sight (V_r) and a second transverse to the line of sight (V_t). The component of the rotational motion along the line of sight is $V(R)\sin\phi$. It is easy to show that $\phi = 180° - \theta$, where θ is given in the law of sines by $R\sin\theta = R_o \sin\ell$. After subtracting the component of the LSR velocity along the line of sight ($V_o \sin\ell$), we find that the $V_{\rm LSR}$ of the gas at position B is

$$V_{\rm LSR} = \left[\frac{V(R)}{R} - \frac{V_o}{R_o}\right] R_o \sin\ell. \quad (5.2)$$

For a given Galactic longitude, ℓ, Equation 5.2 enables us to calculate the Galactic radius, R, for each velocity $V_{\rm LSR}$, in the 21-cm spectrum. Even though emission from gas within the Solar circle has two possible distances from the Sun, there is a unique Galactic radius for each velocity (for each ℓ). Equation 5.3 applies to gas at any Galactic longitude, and anywhere along the line of sight, even beyond the solar circle.

The near/far ambiguity presents some complications for models of the radial distribution of the atomic gas. Despite the fact that the near and far distances are at the same galactocentric distance, we need the distance from the Sun to compute the mass of a gas parcel. The mass of a gas parcel depends on its physical size, and since we only measure the angular size, distance is essential to convert to physical size. Therefore, for gas in the inner Galaxy, we need to make an additional assumption, and the one commonly made is that the distribution of atomic gas in the Galaxy is axially symmetric. The review of the distribution of atomic gas in the Milky Way presented by Peter Kalberla and Jurgen Kerp[4] provides more details of how the mass distribution is modeled.

Emission from the 21-cm line of HI has been detected well beyond a Galactic radius of 30 kpc, which is much larger than the radial extent of the stars. Since the HI gas is concentrated in a flattened disk, it is useful to describe the radial distribution of HI gas in the Galaxy by using a quantity called the *mass surface density*. The mass surface density is the total mass of atomic gas in a cylinder of unit area that passes vertically through the disk. Mass surface density has units of mass per unit area, and is generally reported in $M_\odot$ kpc^{-2}. The mass surface density of HI as a function of Galactic radius derived by Hiroyuki Nakanishi and Yoshiaki Sofue[5] is shown in Figure 5.10. They find that the mass surface density of HI peaks at a Galactic radius of 12 kpc, and declines nearly exponentially

[4] 2009, Annual Reviews of Astronomy and Astrophysics, vol. 47, p. 27.
[5] 2016, Publications of the Astronomical Society of Japan, vol. 68, p. 5.

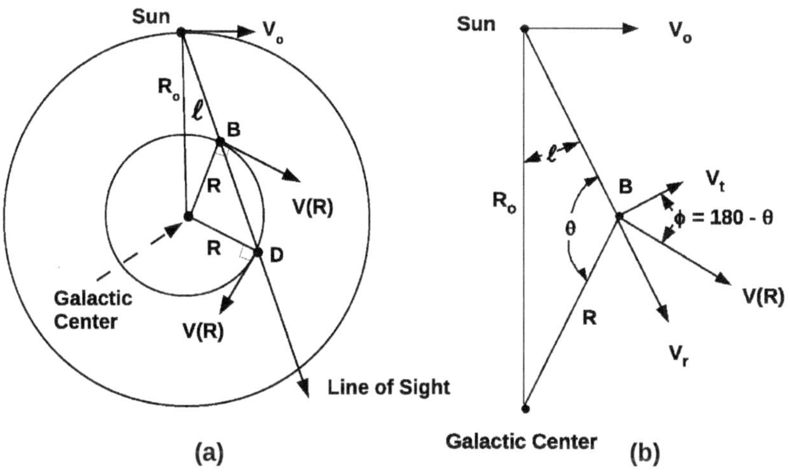

Figure 5.9: (a) A face-on view of the Milky Way showing a line of sight at Galactic longitude ℓ. Gas at positions B and D is equally distant from the Galactic center. The observed radial velocity of the gas emitting at these two locations is the same. (b) An expanded view of the triangle in (a) formed by the Sun, the Galactic center and point B. We define two new angles in this figure, angles θ and ϕ.

at larger radii. Inside the solar circle at 8.5 kpc, the HI mass surface density decreases. There is more uncertainty in the mass surface density of HI in the inner-most region of our Galaxy, and the values shown in Figure 5.10 are somewhat different than those presented in the review article by Kalberla and Kerp.

The studies of both Nakanishi and Sofue and of Kalberla and Kerp estimate that the total mass of HI in the Milky Way is about 8×10^9 $M_\odot$, which is about 10 times smaller than the estimated mass in stars. In the inner galaxy (inside the Solar circle) the HI gas disk is quite thin, with a thickness of only a few hundred parsecs. However beyond the solar circle, the thickness of the HI gas disk increases, and Kalberla and Kerp estimate a disk thickness of about 2 kpc at a Galactic radius of 30 kpc and further the thickness increases to about 5 kpc at a Galactic radius of 40 kpc.

5.1.4 Absorption Lines - Warm and Cold Gas

We showed in Section 4.2.1 that when the HI 21-cm line is optically thin, then the line emission depends only on the column density of atomic hydrogen and not on its temperature. So how might we go about determining the temperature of this gas? A clever technique was developed in the 1960's, in which observations of the 21-cm line in both emission and absorption were combined to determine the excitation, or spin, temperature. An absorption spectrum of HI can be obtained by observing the atomic hydrogen against the continuum emission from a bright background source, such as a quasar. This technique involves measuring the absorption spectrum, and then measuring the emission at positions offset from the background source and using the average of these offset positions to determine what the emission spectrum would be like in the absence of the continuum source. So what does

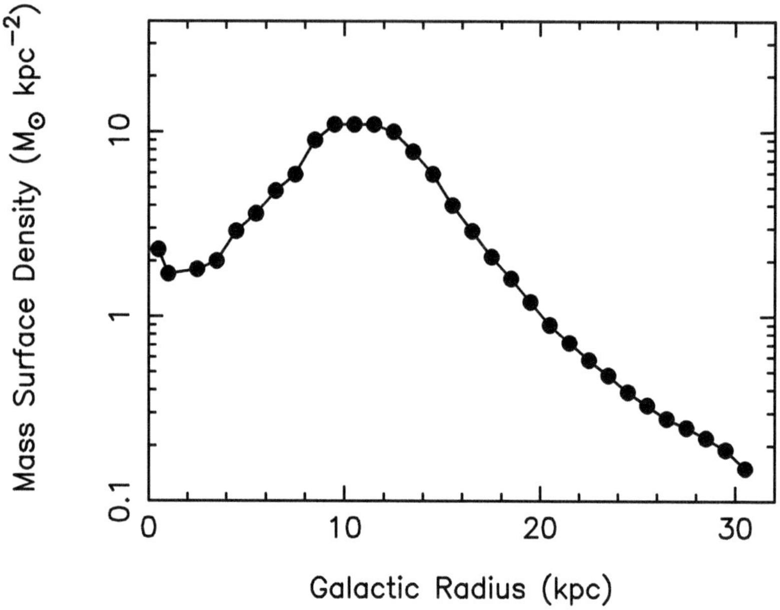

Figure 5.10: The mass surface density of HI as a function of Galactic radius for the Milky Way. The figure is based on data presented by Hiroyuki Nakanishi and Yoshiaki Sofue (2016, Publications of the Astronomical Society of Japan, vol. 68, p. 5).

this give? Before we can answer this question, we first need to delve a little deeper into the absorption of the 21-cm line.

Starting with Equation 4.18, and substituting $h\nu$ for ΔE_{ul} we get

$$\tau_\nu = \frac{h\nu}{4\pi} N_l B_{lu} [1 - e^{-h\nu/kT_{\rm ex}}] \phi_\nu.$$

The quantity in brackets is often called the correction for stimulated emission, as stimulated emission reduces the overall absorption in the line. For the frequency of the 21-cm transition and for reasonable excitation temperatures the exponential term is nearly 1. Therefore the stimulated emission is nearly equal to the absorption for the 21-cm transition, and the optical depth is greatly reduced.

As we discussed in Section 4.2.1, for the 21-cm line we always have $h\nu \ll kT_{ex}$, and therefore we can approximate the exponential term in this equation as

$$e^{-h\nu/kT_{\rm ex}} \sim 1 - \frac{h\nu}{kT_{\rm ex}}.$$

Using this approximation the optical depth is

$$\tau_\nu = \frac{h\nu}{4\pi} N_l B_{lu} \frac{h\nu}{kT_{\rm ex}} \phi_\nu.$$

Also discussed in Section 4.2.1, the total column density of atomic hydrogen is always roughly equal to four times the column density in the lower state of the 21-cm line, thus $N_{tot} \sim 4N_l$. If we use this relation and substitute in for the frequency of the line, the physical constants and the value of the Einstein absorption coefficient for the 21-cm line, we can rewrite the optical depth as

$$\tau_\nu = 2.6 \times 10^{-15} \left(\frac{N_{\rm tot}}{\rm cm^{-2}}\right) \left(\frac{T_{\rm ex}}{\rm K}\right)^{-1} \left(\frac{\phi_\nu}{\rm Hz^{-1}}\right). \tag{5.3}$$

What is important to note in this equation is that the optical depth depends not only on the column density of atomic hydrogen but also on the excitation temperature. In contrast, remember that the HI emission depends only on the column density. Since the absorption depends on the inverse of temperature, a given column density of hot gas produces far less absorption than does an equal column density of cold gas.

We show an example of the emission and absorption in the 21-cm line in Figure 5.11. The absorption spectrum was obtained with an interferometer (the Very Large Array), while the emission spectrum was obtained with a large single-dish telescope (the Parkes radio telescope). The solid angle subtended by the beam of the single-dish telescope is filled with HI emission, but is orders of magnitude larger than the solid angle of the background radio continuum source, which is not the case for the interferometer observation. Therefore, the emission seen by the single dish is almost unaffected by the absorption produced against the very small background source. The differences in the appearance of the emission and absorption spectra are very striking. Not all of the gas detected in emission shows very strong absorption, and this must be related to the temperature of the gas. Although, if optically thin, equal column densities of cold or hot gas produce equal amounts of 21-cm emission, the cold gas produces much deeper absorption features than does the warm gas. We interpret the differences in these spectra to mean that some of the gas along the line of sight is relatively cold, responsible for the relatively narrow and deep absorption features, and the rest of the gas is much warmer and does not produce detectable absorption. The narrow line widths of the absorption features are also consistent with the gas being cold (see the discussion on linewidth in Section 4.1.2). Observations similar to these shown in Figure 5.11 were made in the 1960's, and gave rise to the model of a two-phase interstellar medium — comprising cold clouds immersed in a warmer inter-cloud medium.

As in our discussion in Section 4.1.4, if the background continuum source has a very large intensity, then the emission from the foreground HI is negligible, and we will detect a pure absorption line spectrum. In this case the intensity will be

$$I_\nu = I_\nu^o \, e^{-\tau_\nu}.$$

The optical depth of the absorption line can then be determined from the ratio of the intensity in the absorption line to that of the continuum. The upper panel in Figure 5.11 plots the quantity $e^{-\tau}$ as a function of velocity. Using Equation 4.17, and assuming the intensity of the background is negligible, the emission in a spectral line can be written, using the definition of brightness temperature (see Section 4.2.1) and assuming that $h\nu/kT_{ex} \ll 1$, as

$$T_B = T_{ex} \, (1 - e^{-\tau}). \tag{5.4}$$

Therefore, if we obtain the optical depth from the absorption spectrum, we can use the observed brightness temperature in the surrounding emission spectra to determine the excitation temperature, and thus the gas kinetic temperature. This technique permits a measurement of the gas temperature for any gas that produces absorption.

Carl Heiles and Thomas Troland[6] carried out an extensive absorption line survey using the Arecibo telescope (Millennium Arecibo 21 Centimeter Absorption-Line Survey). Using

[6] 2003, Astrophysical Journal, vol. 586, p. 1067.

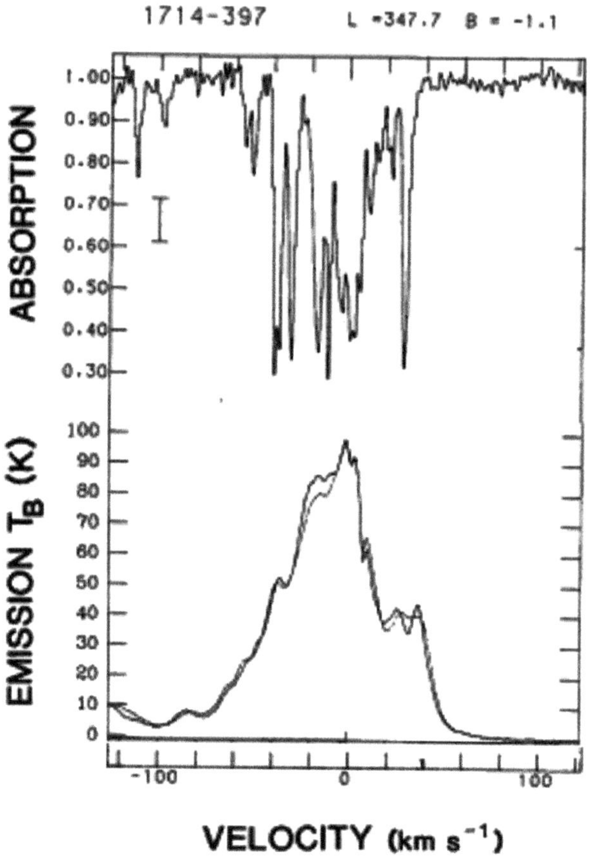

Figure 5.11: The upper panel shows the absorption spectrum of the 21-cm line observed toward the bright radio continuum source 1714−397 obtained with the VLA. The y-axis is labeled absorption and equals the quantity $e^{-\tau}$. The lower panel shows the 21-cm emission spectrum obtained by averaging positions around the continuum source obtained with the Parkes radio telescope. This is a modified version of a figure from the paper by John Dickey and his collaborators (Dickey, Kulkarni, van Gorkom and Heiles, 1983, Astrophysical Journal Supplement, vol. 53, p. 592).

the absorption/emission technique described above and information inferred from the observed 21-cm line widths, they examined the properties of the cold and warm phases of the HI gas. They found that the cold (CNM) phase had a temperature of about 70 K. Because the optical depth of the 21-cm line is inversely proportional to temperature, it is difficult to detect the gas in absorption if it is hotter than a few thousand Kelvin. Thus the absorption measurements only yielded lower limits to the gas temperature of the warm gas (WNM). However, the line widths observed for the emission provide further information about the temperature (see Section 4.1.2). If the lines are broadened by both thermal motions and turbulent motions, then the observed line width sets an upper limit on the thermal motions and therefore on the gas temperature. Heiles and Troland concluded that the temperature of the WNM ranged from 500 to 20,000 K. A more recent and more sensitive absorption survey was carried out by Claire Murray and her colleagues[7]. By combining the absorption features from many different lines of sight, they were able to directly detect the collective

[7]2014, Astrophysical Journal Letters, vol. 781, p. L41.

absorption of the WNM gas and estimated the gas temperature of the WNM to be about 7200 K. The HI gas appears to be well-described by a mix of cold, dense CNM clouds immersed in a warmer, low density WNM gas. Heiles and Troland also estimated that in the vicinity of the Sun about 60% of the HI gas by mass is in the WNM phase and about 40% is in the CNM phase.

5.1.5 Magnetic Field

As discussed in Section 4.2.1, when atomic hydrogen is placed in a magnetic field the interaction between the magnetic field and magnetic dipole moment of the atom causes further splitting of the F=1 hyperfine level, called Zeeman splitting. The amount of splitting of the Zeeman components depends on the strength of the magnetic field parallel to the line of sight. Circular polarization measurements (see Section 4.2.1) permit measurements of the splitting and consequently the magnetic field strength. Since only the line of sight component of the magnetic field is measured, corrections for the other components are needed. Many measurements of the Zeeman effect in HI have been made and a summary of these measurements is presented in the review by Richard Crutcher[8]. For the atomic gas, the total magnetic field is estimated to be about 6 μG (6×10^{-6} gauss). Crutcher also summarizes Zeeman measurements for OH and CN that probe higher density regions of the interstellar medium. These measurements, combined with the HI measurements, suggest that for gas denser than about 300 cm^{-3}, the magnetic field strength increases with increasing density.

The magnetic field produces a magnetic pressure given by $B^2/8\pi$, where B is the magnetic field strength in gauss. For a magnetic field strength of 6 μG, the magnetic pressure is about 10^{-12} dyn cm^{-2}. This pressure is comparable to the thermal pressure in the interstellar medium that we mentioned in the introduction to this chapter. Therefore the magnetic pressure likely plays an important role in the evolution of the gas in the interstellar medium.

5.2 OBSERVATIONS OF THE ROTATIONAL LINES OF MOLECULES

Although by 1970 a number of molecules had been detected in the interstellar medium, it was not believed that molecular gas was a major constituent of the Galaxy. The conventional wisdom was that ultraviolet radiation produced by hot stars throughout the disk of our Galaxy would be sufficient to photo-dissociate most molecules that formed. The importance of the Molecular Medium in the Milky Way was not fully appreciated until wide-spread emission was detected in the fundamental rotational transition of carbon monoxide. The molecule CO was first detected in the interstellar medium in 1970 by Robert Wilson, Keith Jefferts and Arno Penzias[9]. This was the first molecule to be detected at millimeter wavelengths, and the development of millimeter wavelength spectroscopy permitted the study of the molecular interstellar medium to progress rapidly in subsequent years. Today, hundreds of molecular species have been detected in the interstellar medium, creating the new field of astrochemistry.

It is now understood that in the denser regions of the interstellar medium, chemical reactions can rapidly convert the atomic gas into molecular gas, and molecules that form in these regions can be effectively shielded from harmful dissociating radiation. These denser regions of the interstellar gas are almost entirely in molecular form. Observations of permitted molecular rotational lines provide a means to study this dense gas. This dense molecular

[8] 2012, Annual Reviews of Astronomy and Astrophysics, vol. 50, p. 29.
[9] 1970, Astrophysical Journal, vol. 161, p. L43.

gas is an important component of the interstellar medium as it is most closely connected with the formation of stars.

Molecular hydrogen (H_2) is the most abundant molecule in the interstellar medium; however, because of its symmetry, it does not have a permanent dipole moment and thus no permitted rotational transitions. It does have very weak, electric quadrupole transitions, and this emission could be detectable if there were a significant population of molecules in the excited states. However, because H_2 is composed of two light atoms, its moment of inertia is very small and so the rotational energy levels are far apart. Additionally, the selection rules for quadrupole transitions are $\Delta J = \pm 2$. Therefore, the lowest rotational transition is the J = 2-0 transition, whose wavelength of 28 μm is in the infrared. More importantly, the upper state of this transition lies about $\Delta E/k = 500$ K above the ground state. Since the molecular interstellar medium is cold, with temperatures typically of only 10 to 30 K, the upper state of this transition cannot be significantly populated by collisions. Thus, we cannot detect emission from H_2 molecules in the cold molecular gas. The second most abundant molecule is CO which is unique in having an unusually low critical density for excitation (see discussion in Section 4.2.3). These properties make observations of the rotational transitions of CO ideal for probing the molecular component of the interstellar media of the Milky Way and other galaxies.

5.2.1 Molecular Clouds

Although CO emission is found throughout the Galaxy, the emission is not as widespread as the HI emission. The molecular emission is confined to many discrete regions of space with distinct velocities. These discrete emitting regions are called *molecular clouds.* A spectrum of the $^{13}C^{16}O$ emission shown in Figure 5.12 was obtained in the plane of the Galaxy ($\ell = 30°$, $b = 0°$). This spectrum shows spectral features due to eight molecular clouds that lie along this line of sight (the emission from the three highest velocity clouds is blended together). The highest velocity cloud is at V_{LSR} = 102 km s^{-1} and likely lies near the tangent point along this line of sight (see Section 5.1.1). You can compare this spectrum with the HI spectrum shown in Figure 5.1 of the 21-cm line obtained along nearly the same line of sight. The HI spectrum shows continuous emission covering all velocities up to the tangent point velocity, indicating that atomic gas is distributed all along this line of sight. Thus, unlike the atomic gas, the molecular clouds must only fill a small volume of the disk of the Galaxy; the fraction of the disk volume filled with molecular clouds is estimated to be about 1%.

To demonstrate how we obtain information about molecular clouds, we use as an example the molecular cloud associated with the HII region Sharpless 140 (or S140). The S140 molecular cloud is located at Galactic coordinates $\ell = 106.24°$, $b = +4.48°$ at a distance of about 900 pc. Spectra toward the center of this molecular cloud are shown in Figure 5.13, and include the J=1-0 lines of $^{12}C^{16}O$ (denoted as ^{12}CO or just CO, as it is the most abundant isotopologue), $^{13}C^{16}O$ (shorthand notation ^{13}CO), and $^{12}C^{18}O$ (shorthand notation $C^{18}O$). Remember from the discussion in Section 4.2.3, the different isotopologues of CO have slightly different moments of inertia and therefore different rotational line frequencies. For the J=1-0 transition, the frequencies for CO, ^{13}CO, and $C^{18}O$ lines are 115.27 GHz, 110.20 GHz, and 109.78 GHz, respectively.

As was the case for HI observations, molecular line observations are usually made by a differencing technique, such as position-switching, where the signal toward blank sky is subtracted from the signal toward the target of interest. Since the CMB is nearly isotropic, its continuum signal is the same in both directions and will be removed by this differencing

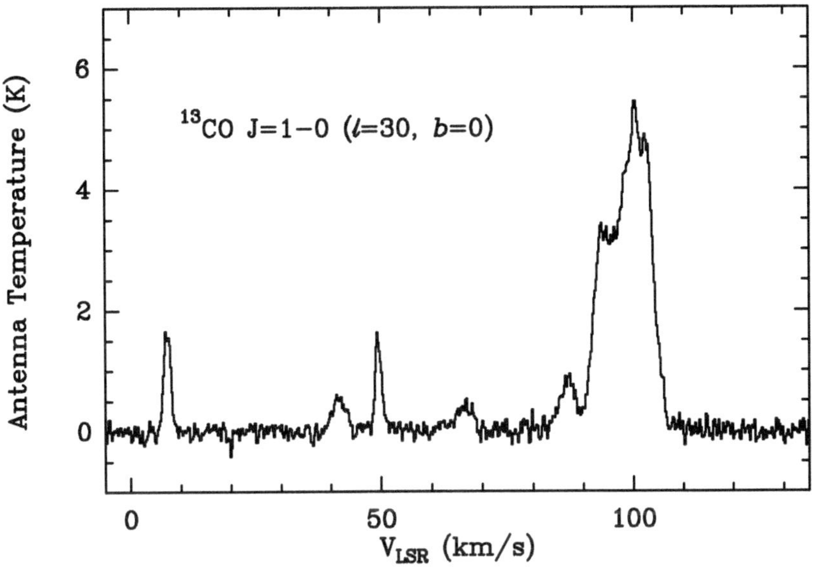

Figure 5.12: A spectrum of the J=1-0 rotational line of ^{13}CO toward Galactic longitude $\ell = 30°$ and Galactic latitude $b = 0°$. This spectrum shows a number of discrete velocity features produced by molecular clouds along this line of sight. These data were obtained with the 14-m telescope of the Five College Radio Observatory and were provided by Mark Heyer.

technique. Therefore, even though the spectrum shown in Figure 5.13 has an intensity of zero outside of the line, this is a consequence of the differencing technique (see further discussion in Section 4.1.4). To determine the true intensity of a molecular spectral line, one must add back in the emission from the CMB expressed in units of brightness temperature (see Section 3.2.3).

The ratios of the isotopic abundances in carbon and oxygen are found to vary with Galactic radius. For instance, the ratio of ^{12}C/^{13}C has a value around 22 near the Galactic center and this ratio rises to about 70 at the Solar circle (see review by Tom Wilson and Robert Rood[10]). Therefore, for the nearby cloud S140 we would expect that the abundance ratio, and consequently the column density ratio, of CO to ^{13}CO to be of order 70. We discussed in Chapter 4 that if the spectral lines are optically thin, then the integrated intensity is proportional to the column density in the upper transition state. If both the CO and ^{13}CO lines were optically thin, then the CO line should have an integrated intensity about 70 times larger than the ^{13}CO line; however from the spectra shown in Figure 5.13 this is clearly not true. We must conclude that the CO line in S140 is optically thick. In fact we find that the CO emission is generally optically thick in molecular clouds.

If the CO emission toward the center of S140 is optically thick, how can we determine its optical depth? We can estimate the optical depth by assuming that the excitation temperature is the same for both CO and ^{13}CO and that the emission satisfies the Rayleigh-Jeans approximation (see Section 3.2.2). Using Equation 5.4, the ratio of brightness temperatures can be expressed as

$$\frac{T_B(\mathrm{CO})}{T_B(^{13}\mathrm{CO})} \approx \frac{(1 - e^{-\tau})}{(1 - e^{-\tau_{13}})}, \tag{5.5}$$

[10]Annual Reviews of Astronomy and Astrophysics, 1994, vol. 32, p. 191.

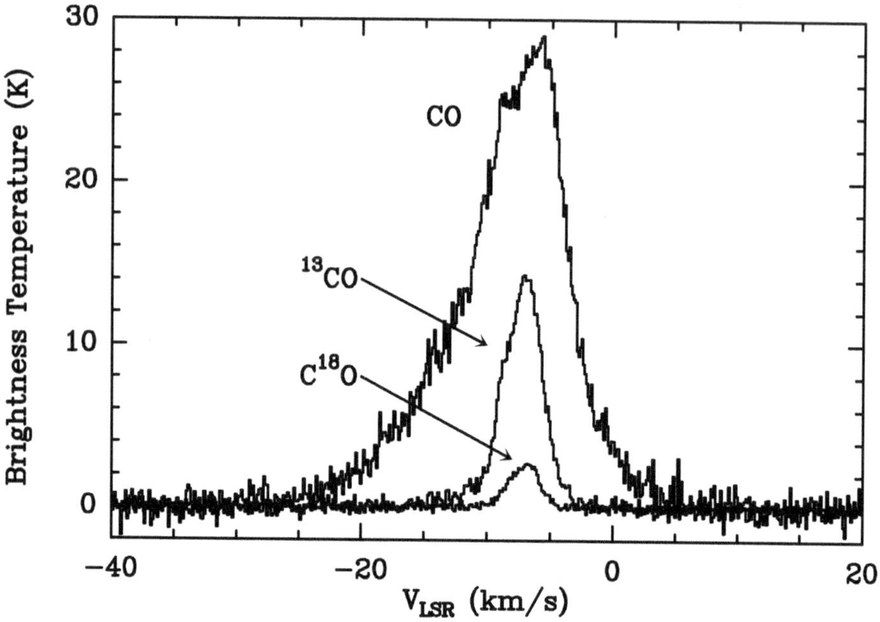

Figure 5.13: Spectra of the J=1-0 rotational lines of CO, ^{13}CO, and C^{18}O toward the molecular cloud S140. Although these lines are at different frequencies, they line up when plotted in $V_{\rm LSR}$. These data were obtained by Ronald Snell with the 14-m telescope of the Five College Radio Observatory.

where τ and τ_{13} are the optical depths of CO and ^{13}CO, respectively. Since the ratio of optical depths is the same as the isotopic ratio, we have $\tau_{13} = \tau/70$. We can then rewrite τ_{13} in terms of τ, and if we know the brightness temperature ratio, we can iteratively solve Equation 5.5 for τ. For the observed ratio of the peak brightness temperatures of 2 toward the center of S140, the line center optical depth of CO can be calculated to be about 50. Therefore the optical depth of ^{13}CO at line center is about 0.7. The observed brightness ratio of the ^{13}CO line to the C^{18}O line is similar to the isotopic abundance ratio for these two species, and therefore consistent with both of these spectral lines being optically thin. As we discussed in Section 4.2.3, since these rarer isotopologues of CO are generally optically thin, they are often used to map the gas column density in molecular clouds.

A map of the integrated intensity of the ^{13}CO J=1-0 line for the S140 molecular cloud is shown in Figure 5.14. The cloud is irregularly shaped, and its longer axis has an angular extent of about 0.7°, which for the distance of S140, corresponds to a linear size of about 11 pc. Since the CO emission is optically thick, Equation 4.16 can be used to relate the intensity of this line to the excitation temperature, which, because we have assumed CO is in thermodynamic equilibrium, equals the gas temperature. For the optically thin ^{13}CO emission, the observed integrated intensity is directly proportional to the column density in the J=1 rotational state of the molecule. To get the total ^{13}CO column density, we need to correct for the population in the other rotational states of the molecule. Because the critical density of CO (see Section 4.2.3) is quite low, it is generally assumed that ^{13}CO is in thermodynamic equilibrium. Therefore, if we obtain the gas temperature from the CO observations, we can use the partition function discussed in Section 4.2.3 to determine the total ^{13}CO column density.

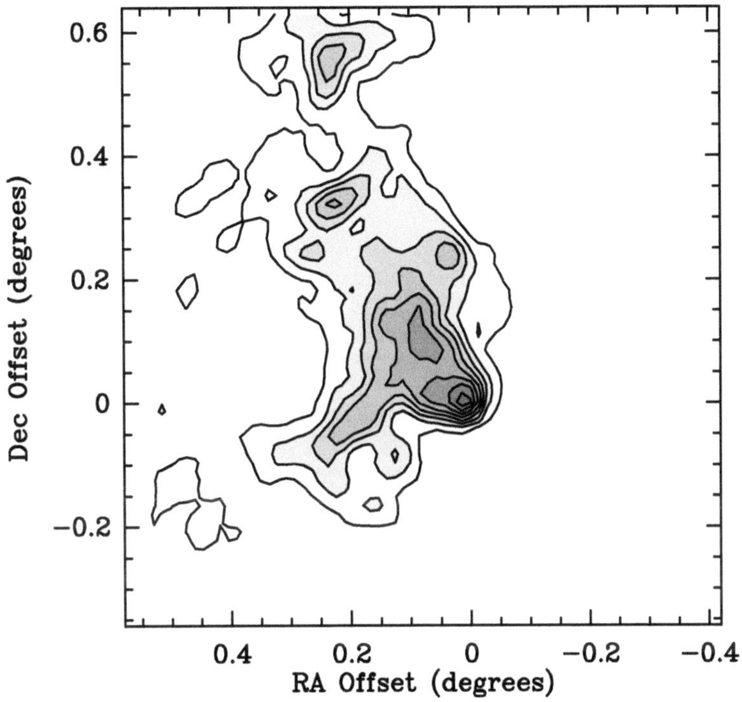

Figure 5.14: A map of the integrated brightness temperature of the ^{13}CO J=1-0 line from a portion of the S140 molecular cloud is shown. The map is plotted in equatorial coordinates, with Right Ascension and Declination offsets given in degrees relative to the reference position RA(J2000) = $22^{h}\ 19^{m}\ 18^{s}$, Dec(J2000) = $+63^{\circ}\ 18'\ 45''$. The contours of integrated intensity start at 4 K km s^{-1} and increase in steps of 4 K km s^{-1}. The data for this figure were provided by Mark Heyer.

Example 5.3:

The CO spectrum shown in Figure 5.13 has a peak brightness temperature of 29.1 K. What gas temperature does this imply?

Answer:

Since the CO line is optically thick, Equation 5.4 implies that the excitation temperature equals the brightness temperature. However, this equation assumed the line intensities can be well-represented by the Rayleigh-Jeans approximation. Using the 115.27 GHz frequency of the CO line and its measured brightness temperature, we find $(h\nu)/(kT_{\rm B}) = 0.19$, while $(1 - {\rm e}^{-(h\nu)/(kT_{\rm B})}) = 0.17$. Therefore using the Rayleigh-Jeans approximation introduces about a 10% error, so the true temperature is 10% higher. Additionally, since this spectrum was obtained by position switching, we must account for the CMB emission that was removed by this differencing process. At the frequency of this line, the brightness temperature of the CMB as defined by Equation 3.13 is 0.9 K. Incorporating these two corrections results in a gas temperature of about 33 K.

Example 5.4:

The ^{13}CO J=1-0 spectrum shown in Figure 5.13 has an integrated brightness temperature of 56.5 K km s^{-1}. The A coefficient for this transition is 6.3×10^{-8} s^{-1}. What is the total column density of ^{13}CO along this line of sight?

Answer:
The column density in the upper state of this transition (J=1) is given by Equation 4.37. Solving for N$_u$ we find

$$N_u = \frac{8\pi\ k\ \nu^2}{h\ c^3\ A_{ul}} \int T_B\ dv.$$

In using this equation, we need a consistent set of units, so to use cgs units, we must convert the integrated intensity to units of K cm s^{-1}. Substituting in the values for the constants, the Einstein A coefficient for this transition and the integrated intensity, we find that N$_u$ = 2.1×10^{16} cm^{-2}. To get the total column density we need to compute the partition function (see Section 4.2.3). This is approximately given by

$$Z \approx \frac{kT_{\rm K}}{hB} + \frac{1}{3} + \frac{1}{15}\frac{hB}{kT_{\rm K}}.$$

Substituting for the temperature of 33 K, found in Example 5.3, and the ^{13}CO rotation constant (B = 5.5×10^{10} Hz), we find that Z ≈12.8. The fraction of the population in the J=1 state (see Equation 4.39) is given by

$$f_1 = \frac{g_1\ e^{-E_1/kT_{\rm K}}}{Z}.$$

The statistical weight $g_1 = (2J+1) = 3$ and the energy above the ground state (E_1) is 7.3×10^{-16} ergs. Therefore, $f_1 = 0.20$; only 20% of the ^{13}CO molecules are in the J=1 state. Therefore, the total ^{13}CO column density is 1×10^{17} cm^{-2}.

Toward the center of the S140 molecular cloud, we find that the gas temperature is about 33 K and the total column density of ^{13}CO is about 1×10^{17} cm^{-2}. However, ^{13}CO is a trace constituent of the molecular gas, and what we really want to know is the column density of the most abundant molecule, H_2. A number of studies have inferred the column density of H_2 from the measured dust extinction at visible or infrared wavelengths, and by comparing this with the measured column density of ^{13}CO, have found that the abundance of ^{13}CO relative to H_2 is about 1×10^{-6}. Therefore, the H_2 column density toward the center of the S140 molecular cloud can be estimated to be 1×10^{23} cm^{-2}.

By applying the techniques used in Examples 5.3 and 5.4, we can determine the column density in all directions toward the S140 cloud. Since the column density gives us the total number of molecules through the cloud per unit area, we can integrate over the surface area of the cloud to determine the total number of molecules in the cloud and consequently the molecular mass of the cloud. For the region shown in Figure 5.14, the cloud mass is estimated to be about 2.2×10^4 M$_\odot$.

In addition to the intensity of the CO lines, the line widths also provide valuable information on the properties of molecular clouds. As explained in Figure 4.6, the widths of lines can be artificially broadened if the lines have large line-center optical depths. Therefore, the optically thin lines provide the truest measure of the gas motions. Toward the center of S140, the ^{13}CO and C^{18}O lines shown in Figure 5.13 have similar line widths given by

$\Delta v_{FWHM} = 3.3$ km s^{-1}. The Doppler broadening parameter (see Equation 4.8) inferred from these lines is 7.3×10^5 Hz. If the broadening were entirely due to thermal motions, then we could apply Equation 4.9 to infer a gas temperature of nearly 7000 K. However, in Example 5.3 we derived a gas temperature of 33 K; why is there a discrepancy? We have long known that the gas motions in molecular clouds are not dominated by thermal motions, but instead are dominated by turbulent motions. On the length-scale of molecular collisions, the gas motions are purely thermal, so the gas temperature applies when we are computing collisional rates. However, on much larger scales the gas motion is turbulent, leading to super-sonic line widths. The source of this turbulent motion in molecular clouds is an ongoing area of investigation.

There are thousands of molecular clouds in the Milky Way, and they have sizes ranging from 1 to 100 pc, and masses ranging from 10 to greater than 10^6 $M_\odot$. Thus, the S140 molecular cloud is average in size and mass. The most massive molecular clouds, with masses greater than 10^5 $M_\odot$, are often called Giant Molecular Clouds or GMCs. The small, low-mass clouds are the most numerous in the Milky Way, while the massive GMCs are quite rare. The review article by Mark Heyer and Tom Dame[11] suggests that the most massive GMCs in the Milky Way have masses of about 6×10^6 $M_\odot$. The number of molecular clouds as a function of mass, N(M), is a power law given by

$$N(M) \propto M^{-1.7}.$$

Therefore, 10^3 $M_\odot$ clouds are 2500 times more numerous than 10^5 $M_\odot$ clouds which are 50 times more numerous than 10^6 $M_\odot$ clouds.

Molecular clouds are highly structured, so observations probing small-scale details are helpful for molecular cloud studies. The most detailed measurements can be obtained by observing nearby molecular clouds. A well-studied molecular cloud is the Taurus Molecular Cloud at a distance of only 140 pc. This cloud was studied by Gopal Narayanan and colleagues[12] who mapped the emission in the J=1-0 transition of CO and ^{13}CO in a region over hundred square degrees in extent. Using the techniques described above, these observations were used to produce an image of the H_2 column density in the Taurus Molecular Cloud that is shown in Figure 5.15. The highest column density regions of the cloud have a clear filamentary structure. The total mass of this cloud is estimated to be 2.4×10^4 $M_\odot$.

In addition to the spatial structure seen in Figure 5.15, the Taurus Molecular Cloud also has a complex velocity structure. Figure 5.16 shows the CO emission color-coded by velocity and reveals the complex motions of the gas in this cloud. The observed kinematics of the cloud provide valuable information about the large-scale turbulent motions in this cloud that dominate the velocity structure. Information about the spatial and velocity structure of molecular clouds is very important in understanding their dynamics, including how clouds collapse under the influence of gravity to form stars. The highest column density regions in the Taurus Molecular Cloud either have already formed new stars or are currently in the process.

[11] Annual Reviews of Astronomy and Astrophysics, 2015, vol. 53, p. 583.
[12] 2008, Astrophysical Journal Supplement, vol. 177, pg. 341.

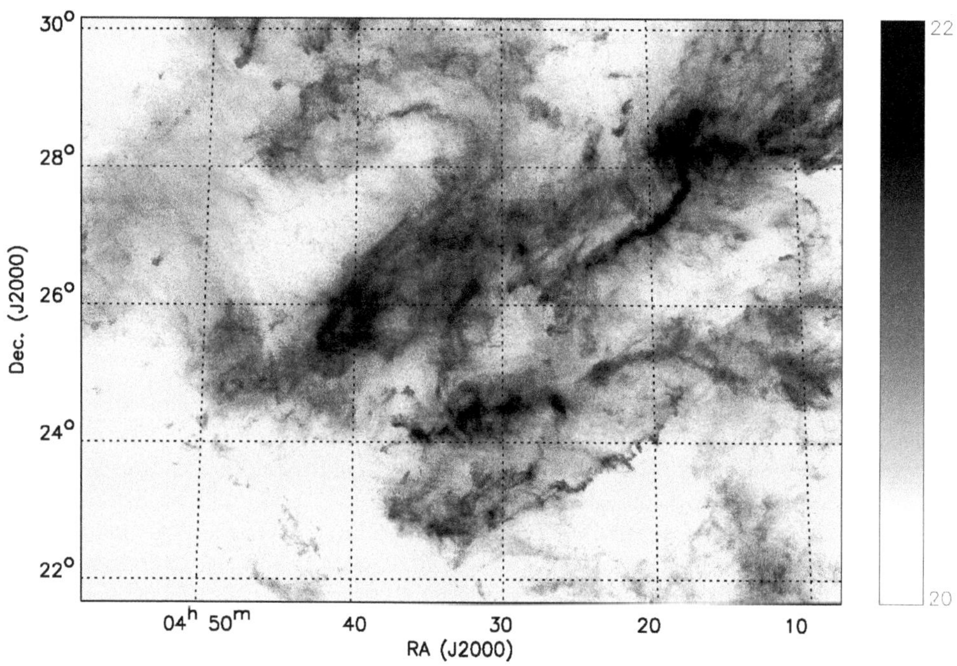

Figure 5.15: A map of the H_2 column density of the Taurus Molecular Cloud inferred from observations of CO and ^{13}CO. This map covers about 100 square degrees of the sky or a region about 25 pc by 20 pc. The wedge on the right-hand side of the figure shows the values of log $N(H_2)$ that correspond to the displayed grey-tones, ranging from less than 10^{20} cm^{-2}, shown in white, to greater than 10^{22} cm^{-2}, shown in black. The data were obtained using the Five College Radio Astronomy Observatory 14-m diameter telescope. This figure is from the paper by Paul Goldsmith and collaborators (2008, Astrophysical Journal, vol. 680, p. 428).

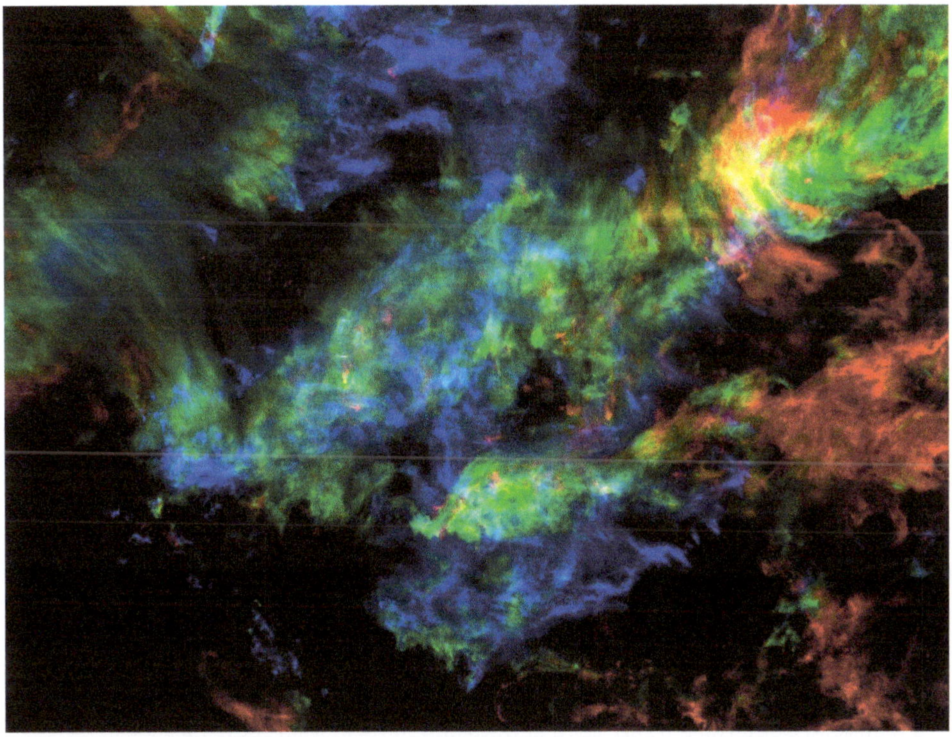

Figure 5.16: A map of the CO emission and velocities in the Taurus Molecular Cloud. This is the same region shown in Figure 5.15. The map is color-coded, so that the emission in the velocity interval of $V_{LSR} = 0$ to 5 km s^{-1} is in blue, the emission in the velocity interval $V_{LSR} = 5$ to 7.5 km s^{-1} is in green and the emission in the velocity interval $V_{LSR} = 7.5$ to 12 km s^{-1} is in red. The emission in green represents the average velocity of the cloud, the emission in blue is gas slightly blueshifted (moving towards us) relative to the average cloud velocity and emission in red is gas slightly redshifted (moving away from us) relative to the average. This is a modified version of a figure that appeared in a paper by Paul Goldsmith and colleagues (2008, Astrophysical Journal, vol. 680, p. 428). This modified image was provided by Mark Heyer.

5.2.2 Distribution of Molecular Clouds in the Galaxy

Numerous large surveys of emission in the J=1-0 transition of CO have been made of the Milky Way and summarized in the review article mentioned earlier by Mark Heyer and Tom Dame. An image showing the emission, in Galactic coordinates, within ±50° of the Galactic plane is shown in Figure 5.17. Unlike the HI emission that is detected all over the sky (see Figure 5.6), the CO emission is more concentrated to both the plane of the Galaxy and to the Galactic center. In fact, for Galactic latitudes greater than about 10° above or below the plane, CO emission is unlikely to be detected.

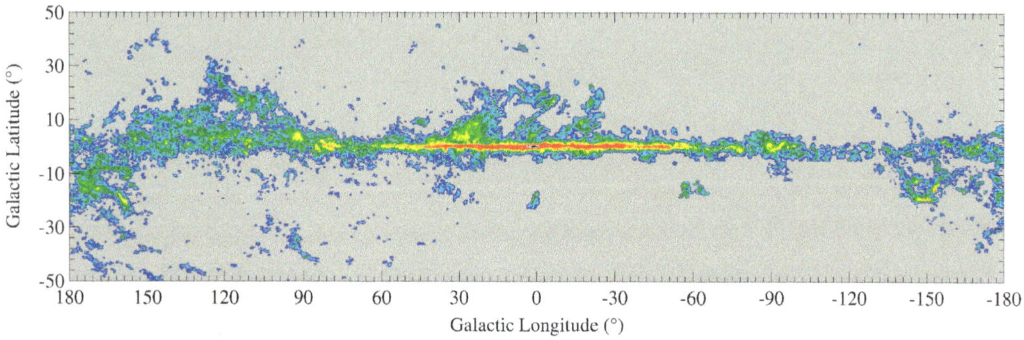

Figure 5.17: A map of the integrated intensity of the J=1-0 transition of CO over a portion of the sky. This map is in Galactic coordinates and shows the region within ±50° of the Galactic plane. The largest CO integrated intensity is displayed in red, with yellow, green and blue colors representing progressively decreasing integrated intensity. Some of the blank areas have not been mapped in CO. This image was provided by Tom Dame.

As we have discussed, the CO emission is usually optically thick, so one might wonder how much we can learn about the distribution of molecular gas from just its CO emission. Only a small portion of the sky has been mapped in the isotopologues of CO, as these lines are much weaker than the CO line. Therefore, the approach we discussed in Section 5.2.1 for individual clouds will be of no use if we only have observations of the CO emission. Despite being optically thick, there is an empirically derived relation between the integrated intensity of the CO J=1-0 line and the column density of H_2 (see review mentioned earlier by Heyer and Dame). This is given by the relation

$$N(H_2) \;=\; \left(\frac{X_{CO}}{\mathrm{cm}^{-2}\ (\mathrm{K\ km\ s}^{-1})^{-1}}\right)\left(\frac{\int T_B\, dv}{\mathrm{K\ km\ s}^{-1}}\right)\ \mathrm{cm}^{-2}, \tag{5.6}$$

where the X-factor, as it is called, has a value of $X_{\rm CO} = 2{\times}10^{20}$ H_2 molecules cm^{-2} (K km s^{-1})$^{-1}$. The X-factor is a global average and will not necessarily be applicable to any one line of sight. The underlying physics of this empirical relation is much debated, but likely arises because molecular clouds are close to being stable, self-gravitating systems. For such systems, a statement of energy balance called the virial theorem requires that twice the internal kinetic energy (thermal and turbulent motions) equals the magnitude of the gravitational potential energy. Why this leads to the relation given above is discussed in the review article by Alberto Bolatto, Mark Wolfire and Adam Leroy[13].

The observed spatial and kinematic information of the CO emission, coupled with the empirical relation between CO integrated intensity and H_2 column density, can be used to

[13]2013, Annual Reviews of Astronomy and Astrophysics, vol. 51, p. 207.

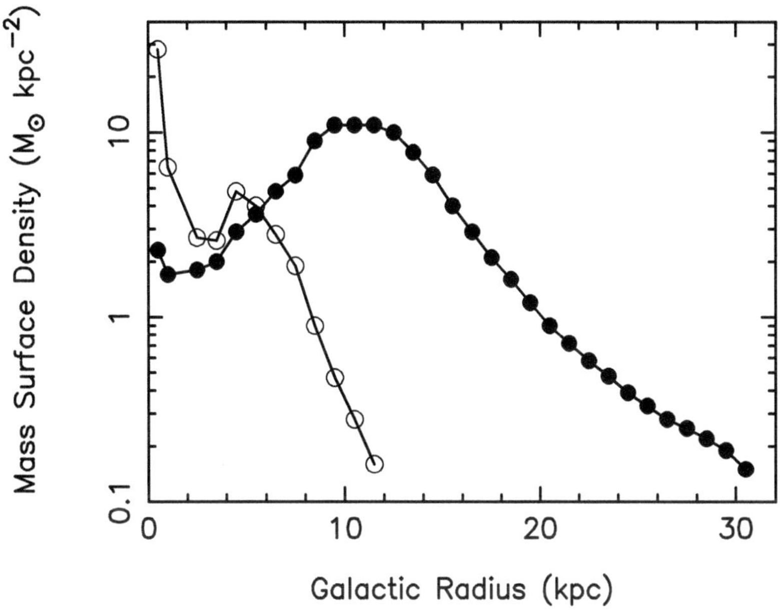

Figure 5.18: The mass surface density of molecular gas (open circles) as a function of Galactic radius is plotted. Also shown on this plot is the mass surface density of atomic hydrogen, repeated from Figure 5.10 (filled circles). This figure is based on the data presented by Hiroyuki Nakanishi and Yoshiaki Sofue (2016, Publications of the Astronomical Society of Japan, vol. 68, p. 5).

determine the distribution of molecular gas in the Milky Way in much the same manner as the atomic hydrogen distribution was derived from 21-cm line surveys, as described in Section 5.1.3. The mass surface density of molecular gas as a function of Galactic radius is shown in Figure 5.18, along with the mass surface density distribution of the atomic gas. Note that inside a radius of about 6 kpc, the mass surface density of molecular gas exceeds that of the atomic hydrogen; thus in the inner part of the Galaxy the interstellar medium by mass is dominated by molecular gas. However, the mass surface density of molecular gas decreases much faster with Galactic radius than does that of the atomic gas, and beyond a radius of about 10 kpc, the interstellar medium is almost entirely composed of atomic gas. Also evident in Figure 5.18, at radii less than a few kpc, there is a large concentration of molecular gas. The total mass of molecular gas in the Milky Way is estimated to be about 1×10^9 $M_\odot$ and, although the molecular gas dominates in the inner galaxy, the total mass of atomic gas is about eight times larger than that of the molecular gas.

5.2.3 Molecular Cloud Cores

Molecular clouds typically have average number densities of about 10^3 molecules cm^{-3}. There are, though, small regions inside molecular clouds that can have much higher densities; these regions are called *molecular cloud cores.* These cores have densities greater than 10^4 molecules cm^{-3} and often as high as 10^6 cm^{-3}. The formation of dense structures in clouds is complex. These structures may be formed by compression of the gas produced in converging turbulent flows, although at some stage gravity likely plays an important role. Further gravitational collapse of the dense gas in these cores ultimately leads to the formation of stars. These dense cores hence are birth sites for stars. We discussed in Section 4.2.3 that the

CO molecule is unique in having a low critical density, and thus can be used to probe the full extent of molecular clouds. Most other molecules have a much higher critical density, and emission in these molecules is generally only detectable in these dense cores. For instance, the critical density to populate the J=1 rotational state of the molecule hydrogen cyanide (HCN), producing emission in the J=1-0 transition, is of order 10^6 cm^{-3} (see Section 4.2.3).

A spectrum of the HCN J=1-0 emission toward the center of the S140 molecular cloud (at the position in Figure 5.14 with zero offsets in Right Ascension and Declination) is shown in Figure 5.19. Due to the nuclear spin of the nitrogen atom, the J=1 level of HCN is split into three finely spaced energy levels (see Figure 4.16) giving rise to three distinct features in the HCN J=1-0 spectrum. A map of the integrated intensity of the HCN emission in S140 is shown in the left panel of Figure 5.20. The angular extent of the HCN emission is much smaller than the ^{13}CO emission shown in Figure 5.14, indicating the high density gas region is quite small. The S140 molecular cloud core has a diameter of about 0.8 pc and a mass of about 10^3 M$_\odot$.

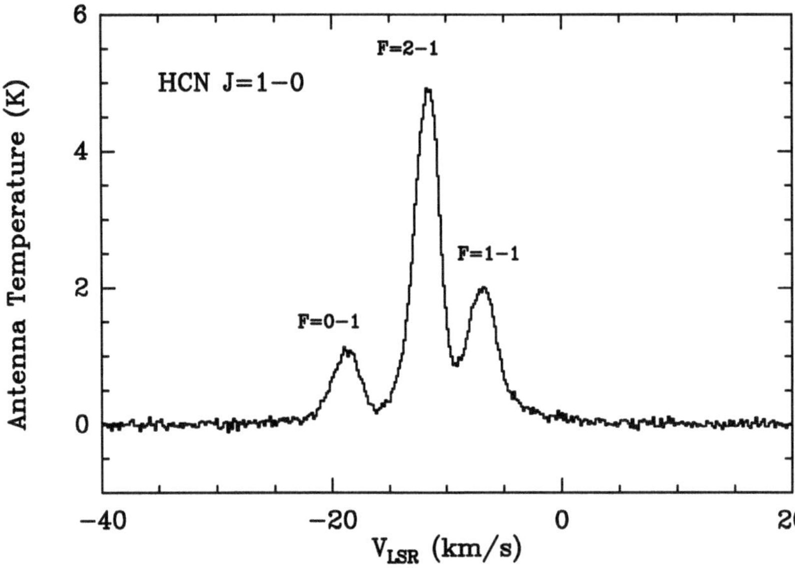

Figure 5.19: The spectrum of the HCN J=1-0 transition toward the center of the S140 molecular cloud at position RA(J2000) = 22^h 19^m 18^s, Dec(J2000) = +63° 18′ 45″. The J=1-0 transition has three distinct hyperfine features (see Figure 4.14). These data were obtained by Ronald Snell using the Five College Radio Astronomy Observatory 14-m telescope.

As we previously mentioned, molecular cloud cores are the sites of star formation, and the dense core in S140 is no exception. An image of the S140 cloud core in the near-infrared, at a wavelength of 2 μm, is shown in the right panel of Figure 5.20 and reveals a recently formed star cluster. Due to the dust associated with this cloud core, these stars are invisible at optical wavelengths; however at longer infrared wavelengths, where the dust extinction is much smaller, the newly forming stars are readily detectable (see more about dust extinction in Section 5.3.1). These newly formed stars in the S140 core are luminous and in these early stages of development can produce very powerful winds that are often highly collimated. This collimated wind impacts the surrounding molecular gas pushing it outward. We can see the effect of this wind on the molecular gas in the CO spectrum shown in Figure 5.13; the CO line does not have a Gaussian shape, but instead has low level emission extending to higher (redshifted) and lower (blueshifted) velocities. This high velocity emission is referred to as line wings. When the emission in just the wings of the line is mapped, the emission shows a small spatial extent with the blueshifted emission offset to the south and the redshifted emission offset to the north (see the paper by Ronald Snell and colleagues[14]). These asymmetric molecular outflows are called *bipolar outflows*, which are an important phenomenon in the star formation process and are discussed further in Section 7.3.3.

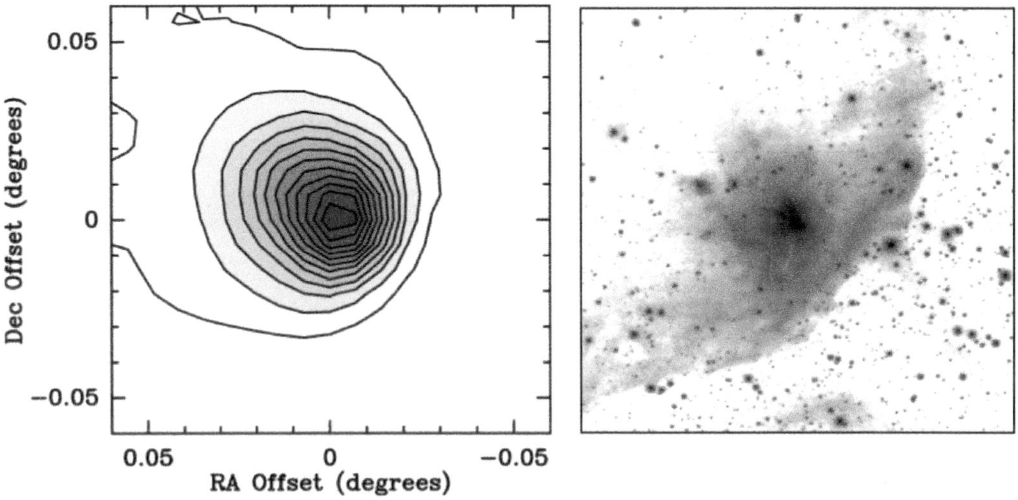

Figure 5.20: On the left is a map of the integrated brightness temperature of the HCN J=1-0 transition toward the S140 molecular cloud. The offsets are relative to the position RA(J2000) = $22^h\ 19^m\ 18^s$, Dec(J2000) = $+63^\circ\ 18'\ 45''$, the same as in Figure 5.14. The contours of integrated brightness temperature start at 4 K km s^{-1} and increase in steps of 4 K km s^{-1}. These data were obtained with the Five College Radio Astronomy 14-m telescope. On the right is a near-infrared image at a wavelength of 2 μm of the same region. The near-infrared image reveals newly formed stars in the S140 core, and was provided by Rob Gutermuth.

The emission from molecules can be an effective diagnostic of the physical conditions of the gas in cloud cores. In addition to the gas column density, the molecular emission can be used to determine the temperature and density. As we discussed in Section 5.2.1, the optically thick CO emission can be used to estimate the gas temperature. Since the CO spectral line is optically thick, one might conclude that the detected emission arises only from the front surface of the cloud and therefore the intensity only gives the temperature at the cloud surface. However, due to motions within the cloud, the gas emission from deep within the cloud can be Doppler shifted, and can therefore escape without being further absorbed. Thus, the CO emission likely arises from various depths into the molecular cloud, providing a temperature measure much deeper into the cloud core than might be expected. The location of the CO emission is a complicated issue and one that only a detailed kinematic model of the cloud can resolve.

Since the interpretation of the temperatures measured with the CO lines is uncertain, independent measures of the gas temperature are helpful. In particular, the symmetric-top molecules discussed in Section 4.2.3 are often used. The emission from these molecules is usually optically thin yet still provides a method for inferring temperature. And, as we will explain, emission from these molecules emanates only from the denser regions and so

[14] 1984, Astrophysical Journal, vol. 284, p. 176.

the inferred temperatures reflect conditions in the molecular cloud core. The rotational energy level diagram for a prolate symmetric-top, shown in Figure 4.15, is defined by two quantum numbers, J and K. There are no permitted K-changing radiative transitions in symmetric-top molecules, and so changes in the K quantum number can only occur by collisions. For this reason, the population of molecules in each K-ladder (see Figure 4.15) is in thermodynamic equilibrium and thus determined only by the temperature of the gas. For example, the relative intensities of the emission in the J = 4-3, K = 0, 1, and 2 transitions are dependent on temperature, not density. Thus, the relative line intensities between different K-ladders directly determine the gas kinetic temperature. Details of this method are described in a paper by Edwin Bergin and colleagues[15].

The prolate symmetric-top molecules detected in the interstellar medium are ammonia (NH_3), methyl cyanide (CH_3CN) and methyl acetylene (CH_3CCH) and the one oblate symmetric-top molecule detected is protonated molecular hydrogen (H_3^+). Molecules such as ammonia or methyl acetylene have relatively large dipole moments and thus need reasonably high densities to produce detectable emission. For this reason, these molecules can be used to determine the gas temperature in dense cloud cores.

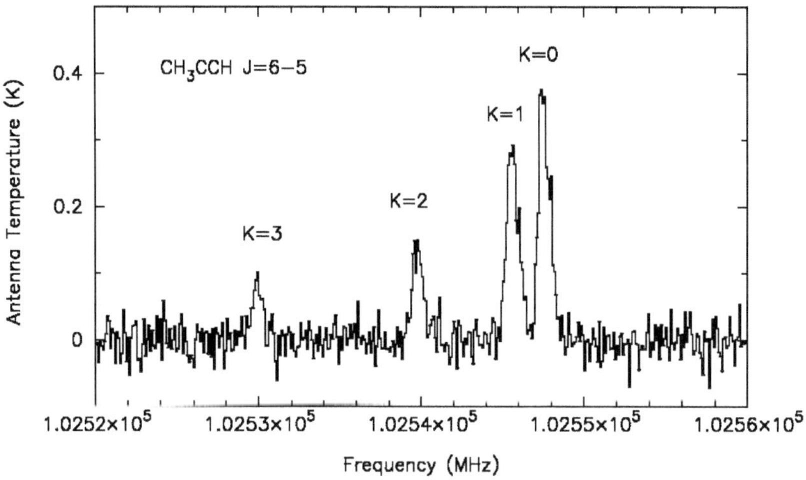

Figure 5.21: A spectrum of the J=6-5, K=0, 1, 2 and 3 transitions of methyl acetylene (CH_3CCH) at a frequency of about 102,540 MHz toward the core of S140. Note that the J=6-5 rotational transitions from the different K-ladders have very similar frequencies and so can be observed in a single spectrum. These data were obtained by Ronald Snell using the Five College Radio Astronomy Observatory 14-m telescope.

The spectrum of the J = 6-5 rotational transition of methyl acetylene obtained toward the core of S140 is shown in Figure 5.21. The J = 6-5 rotational lines from the different K-ladders have similar frequencies, so that the emission from several K-ladders can be obtained simultaneously. Since these lines are usually optically thin, the integrated intensity of one line is given by Equation 4.37 and thus related to the column density of methyl acetylene in the upper state of the transition. The A coefficient for methyl acetylene transitions is given by

$$A_{ul} = 6.6 \times 10^{-12}\ \mathrm{s}^{-1} \left(\frac{\nu}{\mathrm{GHz}}\right)^3 \left(\frac{J_u^2 - K^2}{J_u(2J_u+1)}\right).$$

[15] 1994, Astrophysical Journal, vol. 431, p. 674.

The relative column densities in the K-ladders are dependent only on temperature and therefore the relative fraction in the upper states of each of the K-ladder lines is given by Equation 4.14. The statistical weights for methyl acetylene depend on the degeneracies in the J, K, and I quantum numbers (the I quantum number is the net nuclear spin and is due to the hydrogen atoms in methyl acetylene). The total statistical weight of a level is given by

$$g_u = g_K \; g_I \; (2J_u + 1),$$

where $g_K = 1$ for $K = 0$ and $g_K = 2$ for $K \neq 0$, and $g_I = 1/2$ for $K = 0, 3, 6, 9, \ldots$, and otherwise $g_I = 1/4$. Since we are only interested in the ratio of the K-ladder intensities, any constants will cancel, and we write the integrated intensities of the different K-ladder lines as

$$\int T_B \, dv \propto \frac{N}{Z} \, \nu \, g_K \, g_I \left(\frac{J_u^2 - K^2}{J_u} \right) \, e^{-E_u/kT}, \tag{5.7}$$

where N is the total column density of methyl acetylene, Z is the partition function (see Section 4.2.3) and E_u is the energy of the upper state of the transition. With the J=6-5 transition, for $K = 0$, 1, 2 and 3, the value of $g_K \; g_I \; (J_u^2 - K^2)$ is 3.0, 2.92, 2.67, and 4.5, and similarly the value of E_u/k is 17.2 , 24.4 , 46.0 , and 82.0 K.

Example 5.5:

In the methyl acetylene spectrum shown in Figure 5.21, the integrated intensity of the J=6-5, K=0 transition in S140 is 3.0 times larger than the J=6-5, K=2 transition. What gas temperature does this imply?

Answer:
We can solve for the temperature by setting a ratio using Equation 5.7 for both lines. Since the K-ladder lines all have about the same frequency and the same values of N and Z, we can use the values of E/k given above to find the ratio of the two K-ladder lines. We get

$$\frac{\int T_B(K=0) \, dv}{\int T_B(K=2) \, dv} \sim 3.0 \sim \frac{3.0 \;\; e^{-(17.2 \text{ K}/T_K)}}{2.67 \;\; e^{-(46.0 \text{ K}/T_K)}} = 1.12 \; e^{+(28.8 \text{ K}/T_K)}.$$

We find, then, that the gas temperature is 29 K. The dense core of S140 has a gas temperature similar to the temperature deduced from the optically thick CO emission.

We discussed in Section 4.2.3 the critical density of molecules such as HCN or CS, and found these to have much larger critical densities than CO. However, the critical density also depends on the specific rotational transition. The higher J transitions have a much larger critical density than the lower J transitions of the same molecule, due to the larger Einstein A coefficient for these transitions (see Equation 4.36). Therefore, it takes a much higher density to produce detectable emission in the J = 3-2 transition of HCN than in the J = 1-0 transition. The amount of excitation for a particular molecule, described by the excitation temperature, depends on a competition between radiative and collisional transitions as discussed in Section 4.1.5. Since the collisional rates depend on the temperature and density of the gas, so will the excitation temperature. Thus, the excitation temperature provides insight into the physical conditions of the emitting gas.

Since the intensity of optically thin transitions depends on the population in the upper state, the ratio of intensities of different transitions of the same molecule can be used to compute the population ratio and thus the excitation temperature. If we have measured the gas temperature, then the excitation temperature can be used to infer the density of the gas. Although two transitions can be used to find the excitation temperature, it is even better to observe many rotational transitions from a single molecule and use all of the observed lines together to constrain the physical conditions. The left panel of Figure 5.22 shows the spectrum of four different CS transitions all obtained toward the center of the S140 cloud core. Knowing the radiative rates and collisional rates for CS, and assuming a temperature, we can determine the density that best fits all the observed line intensities. This technique is illustrated in the right panel of Figure 5.22, where the intensity of each of the CS transitions is plotted along with the model with the best fit density. Based on such modeling, we find the gas in the S140 cloud core has a density of $7{\times}10^5$ cm^{-3}. In conclusion, we find that the S140 cloud core has a temperature of about 30 K and a density of nearly 10^6 cm^{-3}.

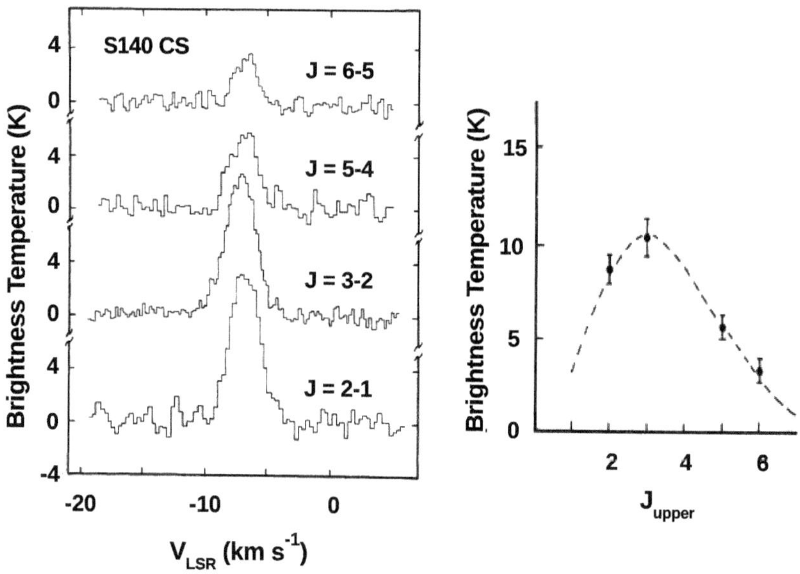

Figure 5.22: The left hand plot shows observed spectra of the CS J = 2-1, 3-2, 5-4 and 6-5 transitions toward the center of the S140 molecular cloud core. The right hand plot shows the intensity of these same CS transitions as a function of J of the upper state of each transition and the curve shows a model with density $7{\times}10^5$ cm^{-3} that fits these data well. Both plots are reproduced from a paper by Snell and collaborators (1984, Astrophysical Journal, vol. 276, p. 625).

5.2.4 Astrochemistry

Nearly 200 different molecular species have been detected in molecular clouds in the Milky Way. Observations are often made of molecular cloud cores where the higher density is sufficient to collisionally excite most molecules and thus produce detectable emission. Over a dozen molecules containing over 10 atoms have been detected so far. The largest molecule detected so far, containing 13 atoms, is cyanopentaactelene (HCCCCCCCCCCCN or $HC_{11}N$), a very long linear molecule. Most molecules are asymmetric-tops and have complex spectra with many spectral lines. An example of the chemical complexity in molecular cloud cores is illustrated in Figure 5.23, which shows the spectrum between 485 and 488 GHz toward the Orion KL molecular cloud core.

The most abundant molecule in molecular clouds is molecular hydrogen (H_2), followed by carbon monoxide, which is about 10,000 times less abundant. The next most abundant molecules (such as HCN, CS, HCO^+, and NH_3) are about another factor of 10,000 times less abundant than CO. The water (H_2O) abundance in molecular clouds is highly variable, and under some conditions can be only slightly less abundant than CO, but in general its abundance is similar to molecules such as HCN or CS. However, after H_2, all other molecules are only trace constituents of the molecular gas.

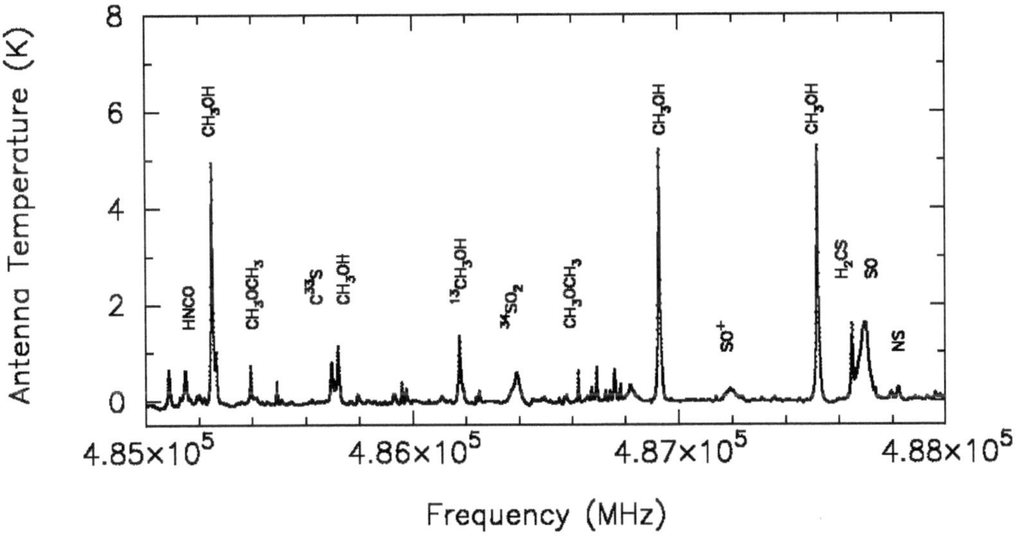

Figure 5.23: The spectrum of the Orion KL molecular cloud core between frequencies of 485,000 to 488,000 MHz obtained with the Herschel Space Observatory. There are over 50 spectral lines detected in this narrow frequency range. The more prominent lines are labeled with the molecular species.

The chemistry in molecular clouds is a field of study in its own right. By laboratory standards, even dense molecular clouds are a very rarefied environment. The chemical reactions in molecular clouds are quite different than those common in terrestrial laboratories. In fact, some molecules, such as protonated carbon monoxide (HCO^+) were first detected in molecular clouds before their spectra were measured in the laboratory. Over the past several decades there has been much work in the area of astrochemistry, and we now have a good understanding of the network of chemical reactions that can reproduce the measured abundances of most molecular species observed in molecular clouds.

5.3 OBSERVATIONS OF THE THERMAL EMISSION FROM DUST

Mixed in with the interstellar gas is small particulate matter astronomers call dust. A review of the properties of dust grains is provided by Bruce Draine[16]. These dust particles

[16] 2003, Annual Reviews of Astronomy and Astrophysics, vol. 41, p. 241.

have a range of sizes, but a characteristic size is of order 0.1 μm or 10^{-5} cm. The sizes of these dust grains are significantly smaller that the wavelength of radio light; this will be important in our discussion. The dust consists of both silicate-rich and carbon-rich dust grains. The silicate dust grains are composed of minerals such as $FeSiO_4$ or $MgSiO_4$ and the carbon grains are nearly pure carbon largely in the form of graphite. By mass, the dust only represents about 1% of the interstellar medium; the other 99% is gas.

Dust grains are believed to form in the winds from evolved stars (such as Red Giant and Red Supergiant stars) and to a lesser extent in planetary nebulae and supernovae explosions. The silicate-rich dust grains can form in the oxygen-rich Red Giant stars, and the carbon-rich dust grains can form in carbon-rich Red Giant stars. Dust grains are destroyed as gas and dust in the interstellar medium are incorporated into stars and by the fast shocks produced by supernovae explosions. The balance between formation and destruction determines the abundance of dust grains in the interstellar medium. The estimated destruction rate, particularly by supernova explosions, is so large that additional sources of grain production are likely needed to explain the observed abundance of dust.

The dust grain surfaces are also sites for the formation of molecules. Although most molecules detected in molecular clouds are formed by gas-phase chemical reactions, there are some reactions that proceed faster on dust surfaces. The dust surface acts as a third body, and permits three-body chemical reactions. In fact, the most abundant molecule in molecular clouds, H_2, is primarily formed on the surfaces of dust grains. Hydrogen atoms can collide and stick to dust grains, but are mobile on the surface. When two hydrogen atoms meet, they can form molecular hydrogen. The heat of formation of the molecular hydrogen is sufficient to eject the molecule off the grain. In addition to H_2, many complex organic molecules are also likely to form on dust grains.

5.3.1 Dust Extinction

Although dust represents only a small fraction of the mass of the interstellar medium, it has a huge impact on astronomy at ultraviolet and optical wavelengths. At these wavelengths, foreground dust blocks the light from astronomical sources and this effect is what astronomers call extinction. For spherical dust grains of radius a, the geometric cross-section is πa^2. The optical depth (see Section 2.1.1) of the dust would be $\tau = \pi a^2 N_d$, where N_{d} is the column density of dust grains. However, dust grains do not simply block the light, but instead have a more complicated interaction with electromagnetic radiation. The details of this interaction depend on the optical properties of the dust grain (for instance the complex index of refraction), the size of the dust grain, and the wavelength of light. The consequence is that the effectiveness of a dust grain in blocking light is wavelength dependent. We often refer to an effective cross-section for the dust, given by $\pi a^2 Q$, where Q is the extinction efficiency and is strongly wavelength dependent.

Dust grains can both absorb light and scatter light. In either case there is a reduction in the intensity of light for a background source. Since extinction is made up of absorption and scattering, the extinction efficiency is the sum of the efficiency for these two processes, therefore $Q = Q_{\mathrm{abs}} + Q_{\mathrm{scat}}$. For wavelengths of light that are similar to the size of the dust grains, both absorption and scattering are important, however at wavelengths much larger than the dust grain size, as is the case at radio wavelengths, scattering is unimportant and can be ignored.

In general the optical depth due to dust is given by:

$$\tau_\nu = \pi a^2 Q N_d. \tag{5.8}$$

The value of Q at ultraviolet and visible wavelengths is of order unity, while at radio wavelengths the dust grains are very inefficient at producing absorption so that $Q \ll 1$ and decreases with increasing wavelength. Dust extinction at ultraviolet and visible wavelengths can be an enormous problem and one that must be corrected for to get the true intensity of a distant astronomical source. At radio wavelengths, because the extinction efficiency of dust is so small, the optical depth of the dust is very small and consequently dust has almost no impact on the propagation of radio light through the interstellar medium.

5.3.2 Dust Emission

Dust grains, being so effective at absorbing the ultraviolet and visible light emitted by stars in our Galaxy, get warmed by the starlight and then, in turn, emit thermal radiation and cool. The dust temperature is determined by the equilibrium between the heating and cooling rates. Since the radiation field in the interstellar medium, far from any star, is relatively small, the equilibrium temperature of dust grains is typically only around 20-30 K. Because the emission and absorption processes depend on the size of the dust grain, different sized grains have slightly different equilibrium temperatures and generally, small dust grains are warmer than large dust grains. Of course location also is important, as a dust grain close to a luminous star will be significantly warmer than a dust grain in the interstellar medium. Because dust is relatively cold, its thermal emission will be predominantly at far-infrared and radio wavelengths and although the emitted intensity of this radiation at this low temperature is small, the pervasiveness of dust throughout the interstellar medium leads to a significant luminosity. The thermal continuum emission from dust, therefore, can be observed at radio wavelengths; Figure 5.24 shows the dust emission at a wavelength of 0.5 mm from the S140 molecular cloud. The region shown in Figure 5.24 is approximately the same as that in Figure 5.14 that shows the molecular line emission. Although the dust image has higher angular resolution than the image of the molecular line emission, the basic features of the molecular cloud are similar in both, as expected since the dust and gas are well-mixed in the interstellar medium.

The dust emission is thermal radiation, and its intensity at frequency ν (see Equation 3.10) is

$$I_\nu = B_\nu(T_d)\,(1 - e^{-\tau_\nu}),$$

where T_d is the dust temperature and τ_ν is the optical depth of the dust at frequency ν. The dust emission at radio wavelengths is almost always optically thin, and so the intensity of the emission can be approximated by

$$I_\nu \approx B_\nu(T_d)\,\tau_\nu.$$

Since a perfectly reflecting (or scattering) surface does not produce thermal radiation, only the absorption efficiency is important in determining the dust optical depth that goes into the above equation. Therefore, for dust with radius a, and with column density N_d, the optical depth is given by

$$\tau_\nu = \pi\, a^2\, Q_{\rm abs}\, N_d,$$

where $Q_{\rm abs}$ is strongly dependent on wavelength or frequency.

A parameter commonly used to quantify the dust content is the mass column density, denoted by σ_d. The dust mass column density is the mass of dust per unit area along a line of sight and is related to the dust column density by

$$\sigma_d = m_{\rm grain} N_d,$$

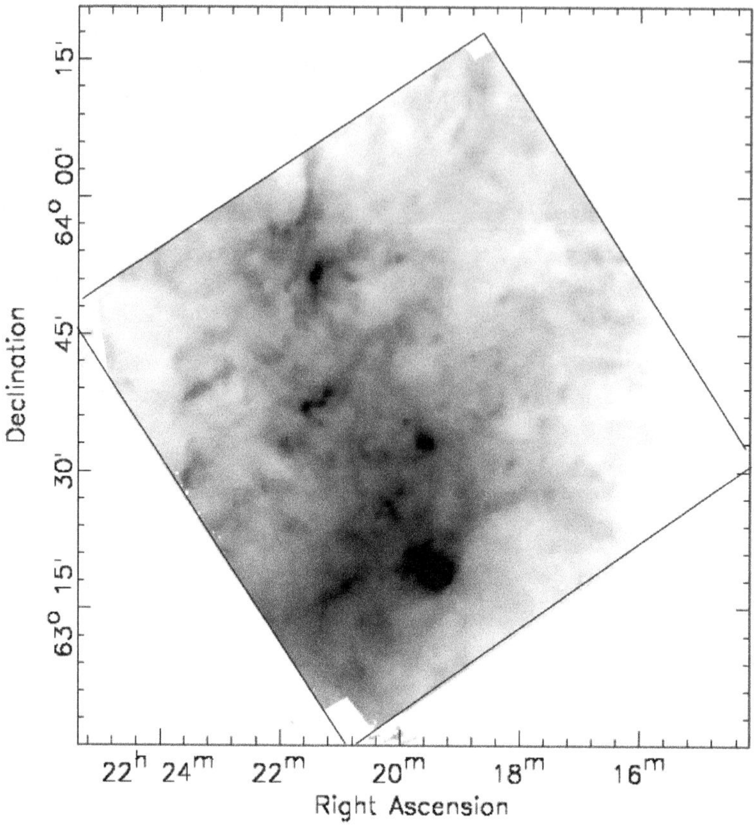

Figure 5.24: An image of the thermal continuum emission of the dust at a wavelength of 0.5 mm from the S140 molecular cloud obtained with the SPIRE instrument on the Herschel Space Observatory. This figure shows roughly the same region as shown in Figure 5.14, but at somewhat higher angular resolution. Herschel is an ESA Space Observatory with science instruments provided by European-led Principal Investigator Consortia and with important participation by NASA.

where $m_{\rm grain}$ is the mass of a typical dust grain. We can rewrite the mass of a dust grain in terms of its volume (assuming spherical dust grains) as

$$m_{\rm grain} = \frac{4}{3}\,\pi\,a^3\,\rho_{\rm grain},$$

where $\rho_{\rm grain}$ is the mass density of the grain material. For both silicate and carbon dust grains, the average density is approximately 2 g cm^{-3}.

We can then rewrite the optical depth as

$$\tau_\nu = \frac{3}{4}\,\sigma_d\,\frac{Q_{\rm abs}}{a\,\rho_{\rm grain}}.$$

The optical depth is strongly frequency or wavelength dependent because Q_{abs} changes dramatically with frequency or wavelength. The wavelength dependence of Q_{abs} is given by

$$Q_{abs} \propto \lambda^{-\beta},$$

where β is the dust emissivity index found empirically to lie between 1 and 2. $Q_{\rm abs}$ also depends linearly on the dust grain size, so at a particular frequency or wavelength the quantity $Q_{\rm abs}/a$ is approximately a constant. Therefore the quantity

$$\kappa_\nu^m = \frac{3}{4}\frac{Q_{\rm abs}}{a\,\rho_{\rm grain}},$$

called the mass absorption coefficient or just *dust opacity* (see Section 2.1.1), is also approximately independent of the composition or the size of the dust grain. The dust opacity has a value of approximately 1 cm^2 g^{-1} at a wavelength of 0.1 cm or 1000 μm.

The observed intensity of the optically thin dust emission can then be written as

$$I_\nu = B_\nu(T_d)\,\kappa_\nu^m\,\sigma_d\,. \tag{5.9}$$

This is often called a modified blackbody spectrum, since the blackbody spectrum is multiplied by a frequency dependent absorption coefficient. Figure 5.25 shows a spectrum of the continuum emission from the dust observed toward the core of S140 at far-infrared and radio wavelengths. The dust emission has a peak at a wavelength of about 90 μm and steadily declines at longer wavelengths. To determine the wavelength dependence of the dust emission at radio wavelengths, we can approximate the Planck function using the Rayleigh-Jeans approximation (see Section 3.2.2) to give

$$B_\nu(T_D) \sim \frac{2kT_d}{\lambda^2}.$$

Including the wavelength dependence of the absorption efficiency, we find that the intensity of the dust emission at long wavelengths has a power-law relation in which the intensity is proportional to $\lambda^{-(2+\beta)}$. Since β is between 1 and 2, the spectral slope at these long wavelengths should be between λ^{-3} and λ^{-4}. The observations shown in Figure 5.25 suggest a spectral slope of about $\lambda^{-3.5}$.

Because the dust emission is a modified blackbody, Wien's law derived in Section 3.2.1 is not applicable in relating the peak wavelength or frequency of the dust emission to the dust temperature. We need to modify this equation to take into account the wavelength dependence of the optical depth. For dust with a dust emissivity index of β, the relation between the dust temperature and the frequency of the peak intensity when expressed as I_ν (not I_λ) is approximately given by

$$\nu_{\rm peak} \approx 2.0\times 10^{10}\ {\rm Hz}\ (\beta+3)\left(\frac{T_d}{\rm K}\right). \tag{5.10}$$

The above approximation is reasonably accurate for β between 0 and 2.

Example 5.6:

The dust spectrum shown in Figure 5.25 peaks at a wavelength of about 90 μm. Assuming that the dust has an emissivity index of $\beta = 1.5$, what is the temperature of the dust?

Answer:
A wavelength of 90 microns or 9×10^{-3} cm corresponds to a frequency of 3.3×10^{12} Hz. Rewriting Equation 5.10 for temperature gives

$$T_d = \frac{5\times10^{-11}\ {\rm K}}{\beta+3}\left(\frac{\nu_{\rm peak}}{\rm Hz}\right).$$

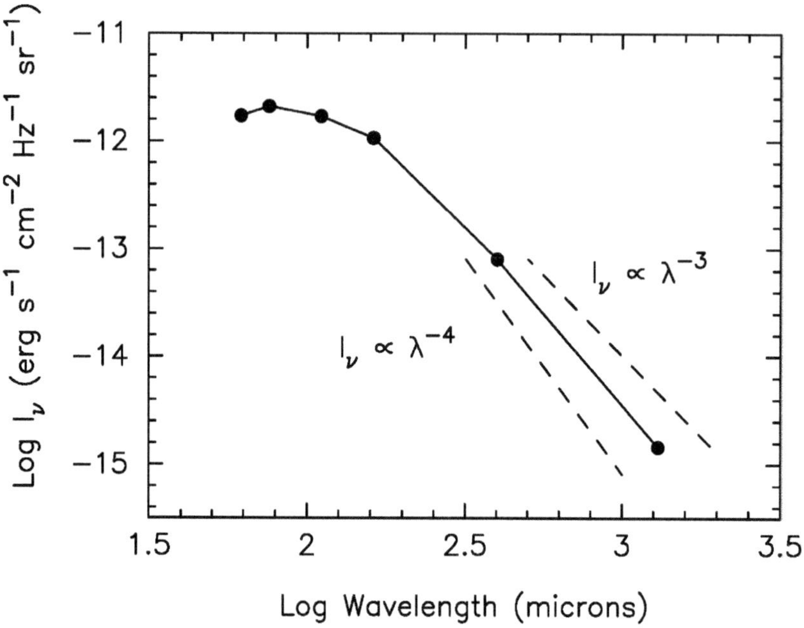

Figure 5.25: A spectrum of the continuum dust emission from the core of S140. Also shown are the spectral slopes for dust emission with a dust emissivity index of $\beta = 1$ and 2 and therefore dust spectral slopes of λ^{-3} and λ^{-4}, respectively. The data suggest a value of β of 1.5, midway between the two slopes plotted. The data here are from observations by Schwartz et al. (1983, Astrophysical Journal, vol. 271, p. 625) and Schwartz, Snell and Schloerb (1989, Astrophysical Journal, vol. 336, p. 519).

Therefore:

$$T_d = \frac{5 \times 10^{-11}\text{K}}{4.5} \left(\frac{3.3 \times 10^{12}\text{ Hz}}{\text{Hz}} \right) = 36.7\text{ K}.$$

In Example 5.6 we find that the temperature of the dust is similar to the temperature of the gas found in Examples 5.3 and 5.5. We can also determine the dust mass column density from the observations shown in Figure 5.25.

Example 5.7:

As shown in Figure 5.25, the intensity of the dust emission toward the core of S140 is $I_\nu = 1.45 \times 10^{-15}$ erg s^{-1} cm^{-2} Hz^{-1} sr^{-1} at a wavelength of 1300 μm or 0.13 cm. If we assume that the opacity of the dust at a wavelength of 0.13 cm is 0.67 cm^2 g^{-1} and the dust temperature is 36.7 K, what is the mass column density of dust along this line of sight?

Answer:
First we need to compute the Planck function for the dust, which is given by

$$B_\nu(T_d) \sim \frac{2kT_d}{\lambda^2} = \frac{2(1.38 \times 10^{-16}\text{ erg K}^{-1})}{(0.13\text{ cm})^2}(36.7\text{ K})$$

$$= 6.0 \times 10^{-13}\text{ erg s}^{-1}\text{ cm}^{-2}\text{ Hz}^{-1}\text{ sr}^{-1},$$

where we have used the Rayleigh-Jeans approximation for the Planck function. Note that the optical depth of the dust is $I_\nu / B_\nu(T_d) = 0.0024$; thus the dust is very optically thin. Using Equation 5.9, the dust mass column density is

$$\sigma_d = \frac{I_\nu}{B_\nu(T_d)} \frac{1}{\kappa_\nu^m} = 3.6 \times 10^{-3} \text{ g cm}^{-2}.$$

In Section 5.2.1 we estimated the total molecular hydrogen column density toward the center of S140 to be 1×10^{23} cm^{-2}. We can convert this to a mass column density of molecular hydrogen by multiplying by the mass of molecular hydrogen (3.34×10^{-24} g) resulting in a value of 0.33 g cm^{-2}. Including the other atoms and molecules in the gas, assuming typical abundances by mass, results in a total gas mass column density of 0.45 g cm^{-1}. Therefore, the gas-mass to dust-mass ratio in S140 is about 125 which is consistent with other values found for the Milky Way. Once we have the gas-mass to dust-mass ratio, the dust emission provides an independent method of determining the mass of interstellar clouds.

5.3.3 Global Distribution of Dust

The Planck Space Observatory has recently provided all-sky images at far-infrared and radio wavelengths. Although its primary mission was to map the Cosmic Microwave Background (CMB), it also provided important information on the foreground dust emission from the Milky Way. Figure 5.26 shows an all-sky image at a wavelength of 1.4 mm. The emission in this image is dominated by thermal dust emission and resembles both the 21-cm HI emission (see Figure 5.6) and the CO emission (see Figure 5.17), as it should, since the dust is mixed with both the atomic and molecular gas in the Galaxy.

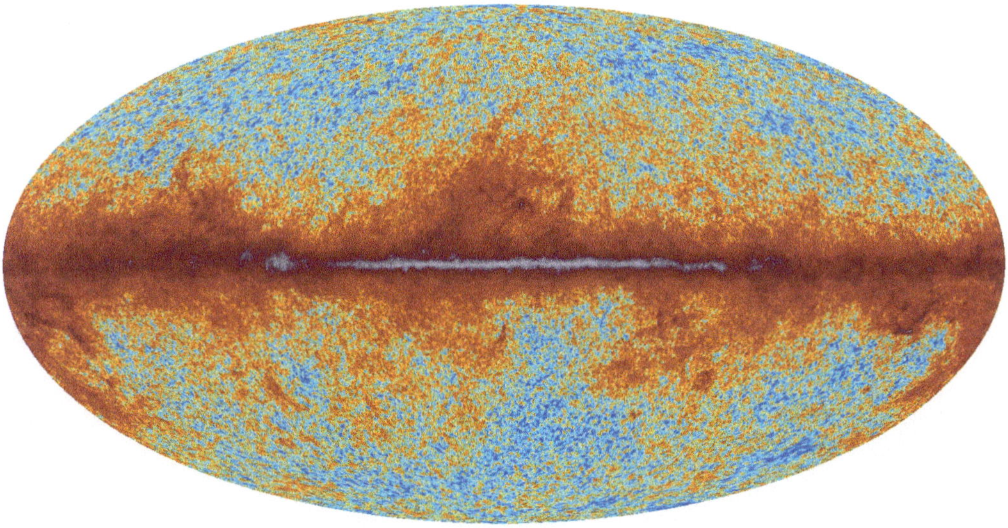

Figure 5.26: An all-sky image of the thermal continuum emission of the dust at a wavelength of 1.4 mm. These data were obtained with the Planck Space Observatory and provided by the ESA/Planck Collaboration.

The dust emission observed by Planck can be used to calculate the dust mass column density in all directions, using the the techniques described in Examples 5.6 and 5.7. However, unlike the HI and CO spectral line emission where we have kinematic information about the emission to assign a distance, we do not have any means to determine the distance of the dust emission. Therefore it is not possible to determine the distribution of dust mass with Galactic radius. The best we can do is to infer the distance by associating the dust emission with features we see in the atomic or molecular line images.

QUESTIONS AND PROBLEMS

1. The HI spectrum toward $\ell = 29°$ and $b = +10°$ is shown in Figure 5.4. Estimate the integrated intensity of the emission. If the emission is optically thin, what is the column density of HI in this direction (see Section 4.2.1)?

2. In Section 5.1.1 we mentioned the Doppler velocity corrections for the Earth's rotation about its axis and its revolution about the Sun. Calculate the range of corrections needed for each of these motions.

3. Assume that the distance from the Sun to the center of the Galaxy is 8.5 kpc and that the circular rotation speed at 8.5 kpc is 220 km s^{-1}. How long does it take the Sun to orbit the Milky Way once? The Sun is 4.7 billion years old. How many times has it orbited the Galaxy since our star was formed?

4. We obtain spectra of the 21-cm line of HI toward three Galactic longitudes ($\ell = 20°$, $\ell = 30°$, and $\ell = 50°$) all at $b = 0°$, and in each direction we observe the emission from HI located at the tangent point. How far is the tangent point from the Galactic center along each of these lines of sight?

5. Along the lines of sight described in Question 4, we measure the maximum radial velocities of the emission from HI to be $V_{LSR} = +133$, $+107$, and $+57$ km s^{-1} for the lines of sight at Galactic longitudes of $\ell = 20°, 30°$, and $50°$, respectively. Calculate the circular rotation speeds at their tangent points.

6. We measure the absorption of HI at 21-cm in front of a very bright radio continuum source. Assume the foreground HI gas has a temperature of 100 K and is in thermal equilibrium. If this gas is Doppler broadened only by thermal motions, what column density of HI is needed to produce an optical depth at line center of 0.1?

7. If the gas in Question 6 is at 1,000 K, what is the required HI column density to produce a line center optical depth of 0.1? What would be the column density if the temperature of the gas was 10,000 K?

8. Use the results from Questions 6 and 7 to explain why it is difficult to detect the warm HI gas in absorption in the 21-cm line.

9. The ^{12}CO and ^{13}CO J=1-0 spectra for a position in the S140 molecular cloud offset from the center is shown in Figure 5.27. Following the approach used in Examples 5.3 and 5.4, estimate the gas temperature and the ^{13}CO column density at this position. Use your answer to then derive the total molecular column density at this position.

10. Using the data presented in Figure 5.27, estimate the optical depth at the center of the ^{12}CO line. Assume that the abundance ratio of CO to ^{13}CO is 70.

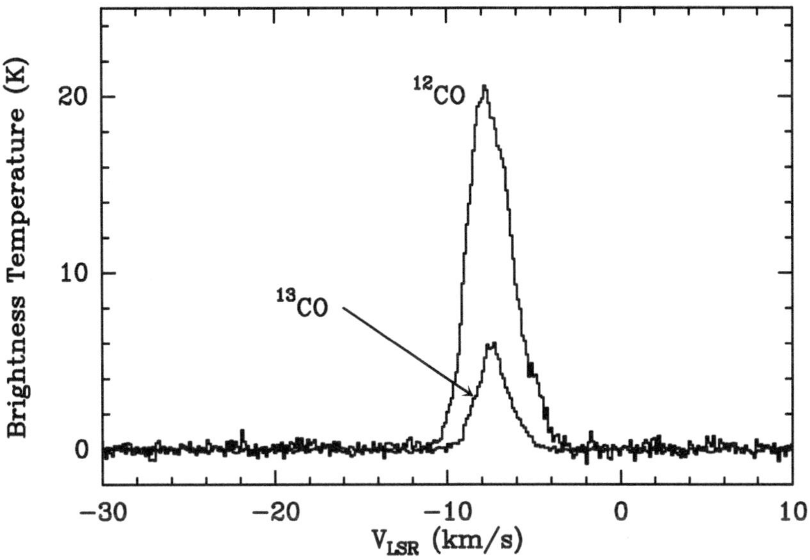

Figure 5.27: Spectra of the J=1-0 rotational lines of CO and ^{13}CO, and C^{18}O toward a position offset from the center of S140.

11. In Example 5.5 we used the ratio of the K=0 and K=2 K-ladder lines from the J=6-5 transition to estimate the gas temperature in the core of S140. The ratio of integrated brightness temperature of the K=0 line to K=1 line is 1.3 and the ratio of integrated brightness temperature of the K=0 line to K=3 line is 6.1. What gas temperature is inferred from each of these two ratios? Do they agree? How do they compare with that derived in Example 5.5?

12. In Example 5.7 the mass column density of dust toward the core of S140 was determined. If the dust grains have a radius of $a = 10^{-5}$ cm and a density of 2 g cm^{-3}, what is the column density of dust grains toward the core of S140?

CHAPTER 6

HII Regions and Planetary Nebulae at Radio Wavelengths

In Chapter 5 our primary focus was on the cold, neutral components of the interstellar medium of the Milky Way. In this chapter we consider a warmer — and ionized — component of the interstellar medium. Although often considered a somewhat exotic state of matter, plasmas (ionized gases) are quite common in astronomical contexts. Ionized regions can range from very extended volumes of space, at very low density, to extremely compact, high-density regions. Their temperatures can range from a few thousand to tens of millions of kelvins. The cause/origin of the ionization can be any one of several physical processes, with photo-ionization and collisional ionization being the principle two. In this chapter we consider two examples of photo-ionized regions: HII regions and planetary nebulae.

6.1 HII REGIONS

HII regions are one of the more common manifestations of thermal bremsstrahlung (or free-free) emission at radio wavelengths (see Section 3.2.4). They are regions of ionized gas that form around massive stars, where a substantial number of stellar photons are in the ultraviolet regime, with energies greater than 13.6 eV, and hence able to ionize hydrogen. Recall from Section 1.1.3 that 'HII' denotes ionized hydrogen.

The traditional model for an HII region is known as a *Strömgren sphere*, after Bengt Strömgren, the Danish astronomer who first described these regions analytically in 1939. The essence of the model is that a star producing photons more energetic than 13.6 eV, called *Lyman continuum photons*, is embedded within an atomic or molecular cloud. The Lyman continuum photons ionize the cloud material, producing a plasma bubble. A Strömgren sphere is the ionized bubble surrounding the star in this simple model.

To a reasonable approximation, most stars emit a continuous spectrum that is well-represented by the Planck function at the temperature of the stellar surface or photosphere. For solar-type stars, with surface temperatures of about 5500 K, Equation 3.5 shows that the average photon energy is about $\langle E_{ph} \rangle = 2.70\ (8.62 \times 10^{-5}\,\mathrm{eV\ K^{-1}})\,(5500\mathrm{K}) = 1.3$ eV, which is in the near-infrared regime. Main sequence stars significantly more massive than the Sun, such as O and early-B spectral types, are much hotter — typically in the 30,000 to 50,000 K range — and so their average photon energies are much higher. At 40,000 K, for example, the average photon energy is 9.3 eV, which is in the near-ultraviolet regime. Thus, for these massive stars, a substantial number of the emitted photons have energies greater than 13.6 eV, and can ionize hydrogen.

Massive stars form fairly quickly and are still surrounded by collapsing molecular cloud material when the star is already emitting copious amounts of UV photons. The gas in these clouds is mostly in the form of hydrogen molecules, which can be dissociated by photons less energetic than 13.6 eV. The resulting atoms are then ionized by photons with energies greater than 13.6 eV. Thus, the gas close to the star will be ionized, producing an HII region, while most of the molecular cloud will remain neutral. What results is a cloud of ionized hydrogen, where scattering between the free protons and electrons results in bremsstrahlung radiation (see Section 3.2.4), and recombination of ions with electrons results in spectral line emission (see Section 4.2.2).

6.1.1 Ionization Structure of HII Regions

There are two fundamental processes that determine the degree of ionization within an HII region: photo-ionization and radiative recombination.

The former process occurs when a neutral hydrogen atom absorbs a photon of sufficient energy to remove the electron, thus ionizing the atom. Any excess energy of the photon, beyond that needed to liberate the electron, will be converted into kinetic energy, primarily of the electron. The latter process (recombination) occurs when a free electron is captured into a bound energy state by a proton, thus forming a neutral atom. In the process, a photon is emitted with energy equal to the energy lost by the electron as it goes into the bound state, plus any excess kinetic energy the electron had before the recombination. The two processes compete with one another, one ionizing the gas, the other neutralizing the gas, until they establish an equilibrium that determines the overall ionization state of the nebula. We examine each of these processes in more detail, and then address the equilibrium state.

Photo-ionization is an interaction between an atom and a photon, and may be described in terms of an interaction cross-section. We introduced this concept in Section 2.1.1 where we denoted the cross-sectional area for interaction as σ_ν. When the interaction is the photo-ionization of hydrogen, the cross-section is commonly denoted by a_ν. It is zero below the threshold frequency of $\nu_0 = 3.29 \times 10^{15}$ Hz (corresponding to 13.6 eV) and decreases roughly as ν^{-3} with increasing frequency. The peak value of the cross-section is at the threshold frequency, and is about $a_\nu \approx 6.3 \times 10^{-18}$ cm^2.

In addition to the cross section, the photo-ionization rate depends on the number of absorbing particles (hydrogen atoms) present and the number of ionizing photons available. The former is given by the density of neutral hydrogen atoms, n_H, while the latter depends on the intensity of the local radiation field. Because intensity is direction-dependent while photo-ionizations occur regardless of the direction of photon propagation, we must use the intensity averaged over all solid angles. This is called the *mean intensity*, denoted by J_ν, and given by

$$J_\nu = \frac{1}{4\pi} \int I_\nu \, d\Omega. \tag{6.1}$$

Then the number of ionizing photons per unit area per unit time is given by

$$\int_{\nu_0}^{\infty} 4\pi \frac{J_\nu}{h\nu} \, d\nu, \tag{6.2}$$

regardless of the direction they are coming from. The lower limit for the integral is the threshold frequency ν_0 corresponding to a photon energy of 13.6 eV; the division by $h\nu$ converts from energy of radiation to number of photons. We can write the rate of photo-ionizations per unit volume as

$$\text{Ionization rate per unit volume} = n_H \int_{\nu_0}^{\infty} 4\pi \, \frac{J_\nu}{h\nu} \, a_\nu \, d\nu. \tag{6.3}$$

The integral is the ionization rate per atom; we multiply this by the volume density of neutral hydrogen atoms, n_H, to obtain the number of ionizations per unit time per unit volume, in units of $\text{cm}^{-3}\ \text{s}^{-1}$. Note that the photo-ionization cross section must be left inside the integral because of its frequency dependence. A consequence of this frequency dependence is that higher energy photons will travel further through the nebula before they cause an ionization. The radiation field varies with distance from the star such that the average photon energy increases with increasing distance. This is what astronomers call "hardening" of the radiation field.

Radiative recombination is also an interaction between two particles, this time a proton and an electron, and it can also be characterized by an interaction cross-section, which we considered in Section 4.1.5. That discussion culminated in Equation 4.22, giving the rate at which an individual particle undergoes collisions as

$$R_{\text{coll}} = n_{\text{target}} \, \sigma_{\text{coll}} \, v_{\text{atom}},$$

which has units of s^{-1}. In the present case, we consider the target particles to be protons, and the colliding particle at velocity v to be an electron.

The cross section and the velocity can be combined into a single term, called the *recombination coefficient*, denoted by α, which has units of $\text{cm}^3\ \text{s}^{-1}$. The collision cross section is roughly proportional to v^{-2} (faster moving particles are less likely to recombine), so the recombination coefficient depends inversely on the average electron velocity and hence is a decreasing function of the electron kinetic temperature. In the study of HII regions it is customary to refer to the electron's kinetic temperature as the "electron temperature", and denote it by T_e. In addition to temperature, the recombination coefficient is also a function of the quantum level to which the electron recombines. The sum of the cross sections over all quantum levels, i.e., the total cross section for recombination to any level, is denoted by the subscript A. To get the *recombination rate* per unit volume, we multiply by the number density of electrons, n_e, which gives us

$$\text{Recombination rate per unit volume} = n_p \, n_e \, \alpha_A(T_e) \tag{6.4}$$

which has units of $\text{cm}^{-3}\ \text{s}^{-1}$, i.e. the same units as the ionization rate in Equation 6.3.

To model the ionization structure of an HII region, the key is to recognize the equilibrium between these processes and equate the photo-ionization rate of the atoms with the radiative recombination rate of the electrons and protons. We consider here the idealized case of a pure atomic hydrogen cloud, without dust, isothermal and uniform in density, at rest with respect to a single star producing ionizing radiation. Setting these two processes equal at all points in the nebula, Equations 6.3 and 6.4 give us

$$n_H \int_{\nu_0}^{\infty} 4\pi \, \frac{J_\nu}{h\nu} \, a_\nu \, d\nu = n_p \, n_e \, \alpha_A(T_e). \tag{6.5}$$

Equation 6.5 is a general expression describing ionization equilibrium; it isn't yet in a form that we can solve to determine the ionization structure of the HII region. To obtain a solvable form of the equation, we need to consider a bit more physics, related to both the photo-ionization and the radiative recombination.

The mean intensity, J_ν, has two main contributions, the direct stellar radiation coming from the central star, and the diffuse radiation produced by the radiative recombinations occurring within the HII region. The former is directional in nature, and can be treated by the radiative transport equation. The latter is isotropic, and we will treat it by considering the recombination process.

First, we consider the stellar radiation. In traveling from the star to some point in the nebula, the stellar radiation will suffer both geometric dilution as it spreads over an ever-larger area, and also absorption as it photo-ionizes neutral hydrogen atoms. The flux density emitted from the surface of the star is given by πI_ν (see Question 8), and so the luminosity per frequency interval of the star is

$$L_\nu = 4\pi R^2(\pi I_\nu), \tag{6.6}$$

where R is the stellar radius. From Equation 6.1, the mean intensity of the star, $J_\nu(\text{star})$, is related to the flux density by

$$4\pi\, J_\nu(\text{star}) \;=\; \int I_\nu(\text{star})\, d\Omega_{\text{star}} \;=\; F_\nu(\text{star}). \tag{6.7}$$

Using the relation between luminosity and flux (Equation 1.4), we find that

$$F_\nu = \frac{L_\nu}{4\pi r^2},$$

where r is the distance from the star. Including the attenuation due to absorption, $e^{-\tau_\nu}$, we have

$$4\pi\, J_\nu(\text{star}) \;=\; \frac{L_\nu}{4\pi r^2}\, e^{-\tau_\nu}, \tag{6.8}$$

where the optical depth, τ_ν, in this case, is given by

$$\tau_\nu = \int_0^r n_H\, a_\nu\, dr'. \tag{6.9}$$

We will substitute the expression for J_ν given by Equation 6.8 into the left side of Equation 6.5.

We can account for the diffuse radiation by considering the recombination process in detail. When an electron and proton recombine, the electron can go directly to the ground state or to some excited state. Every atomic state has a particular probability of occurring in the recombination. The sum of the recombination probabilities over all atomic states gives the total recombination coefficient to any level, α_A. However, recombinations to the ground state result in photons with energies of at least 13.6 eV and these photons can then ionize another hydrogen atom. When this recombination and subsequent reionization occurs, the net number of ionizing photons does not change. These photons are the diffuse radiation component. If we exclude these recombinations to the ground state, then we can also exclude the diffuse radiation field in Equation 6.5 because these two parts contribute equally on both sides of the ionization-recombination equilibrium equation. This is called the *on-the-spot approximation* and assumes that the ionizing photons emitted by recombinations to the ground state immediately ionize other hydrogen atoms; this is a good approximation in most cases. The recombination coefficient excluding recombinations to the ground state is denoted by α_B. For hydrogen gas at 10,000 K, $\alpha_A = 4.18 \times 10^{-13}$ cm^3 s^{-1} while $\alpha_B = 2.59 \times 10^{-13}$ cm^3 s^{-1}. The fraction of recombinations going directly to the ground state is given by $1 - \alpha_B/\alpha_A$, so we see that 38% of recombinations go directly to the ground state, and thus produce photons that can ionize hydrogen.

Combining these considerations, we can rewrite the ionization equilibrium condition (Equation 6.5) in the form

$$n_H \int_{\nu_0}^{\infty} \frac{e^{-\tau_\nu}}{4\pi r^2} \frac{L_\nu}{h\nu} a_\nu \, d\nu = n_e \, n_p \, \alpha_B(T_e). \tag{6.10}$$

Equation 6.10 does not have an analytical solution and must be solved numerically. An example of such a solution, which gives the densities of neutral hydrogen and ionized hydrogen as a function of distance from the star (recall that τ_ν is a function of r), is shown in Figure 6.1. An important result of this solution is that the gas is almost completely ionized out to a particular radius, at which point the ionization fraction rapidly falls to near zero, forming a boundary between the ionized and neutral gas. The distance from the star to this ionization boundary is called the *Strömgren radius*, R_S.

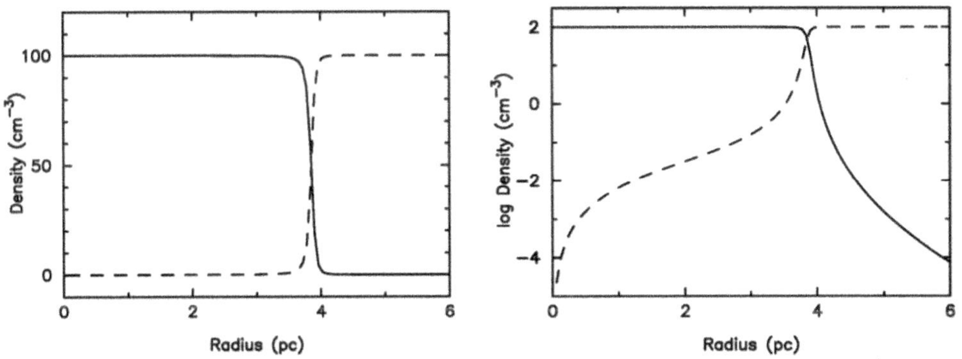

Figure 6.1: The ionized (solid line) and neutral (dashed line) gas densities plotted as a function of distance from the star for a cloud of density 100 cm^{-3} with an ionizing star of temperature 42,000 K. On the left we use a linear density scale while on the right we use a logarithmic scale; the latter shows the variations more clearly. The gas is almost completely ionized out to a radius of nearly 4 pc.

Within the Strömgren sphere, electrons and protons are continually recombining to form atoms, which are promptly photo-ionized by the stellar Lyman continuum photons. Put another way, for a static Strömgren sphere, the Lyman continuum photons that are continually produced by the star simply maintain the ionization of the HII region. This fact provides a very convenient means to calculate the Strömgren radius. We simply equate the rate at which ionizing photons are emitted by the star, which we denote by Q, to the rate of recombinations to excited states occurring throughout the Strömgren sphere:

$$Q = \int_{\nu_0}^{\infty} \frac{L_\nu}{h\nu} \, d\nu = \frac{4\pi}{3} R_S^3 \, n_e n_p \alpha_B(T_e). \tag{6.11}$$

To evaluate the integral we must know L_ν, which largely depends on the surface temperature of the star, and is obtained from stellar atmosphere models. In practice, we often use tabulated values for Q, given as a function of the spectral type of the exciting star. Examples of this are given in Table 6.1[1], which lists, for main-sequence stars, the spectral type of the star, its approximate surface temperature, the stellar luminosity, and the logarithm of the number of Lyman continuum photons (Q; i.e., the number of photons emitted per second

[1]Table values are from Sternberg et al. 2003, Astrophysical Journal, vol. 599, p. 1333.

with energies greater than or equal to 13.6 eV, which can ionize hydrogen). Once Q is known, Equation 6.11 is easily solved for R_S, giving

$$R_S = \left[\frac{3}{4\pi}\frac{Q}{n_e\,n_p\,\alpha_B(T_e)}\right]^{1/3} \tag{6.12}$$

where $\alpha_B(T_e)$ is the sum of the recombination coefficients to all excited levels ($n \geq 2$).

Table 6.1: Massive Star Temperatures, Luminosities and Ionizing Photon Rates

Spectral Type	Temperature (K)	Luminosity $\log(L/L_\odot)$	H Ionizing Photon Rate $\log(Q/s^{-1})$
O3	51,200	6.04	49.87
O4	48,700	5.88	49.68
O5	46,100	5.73	49.49
O6	43,600	5.57	49.29
O7	41,000	5.40	49.06
O8	38,500	5.24	48.75
O9	35,900	5.06	48.47
B0	33,300	4.88	48.02

For smaller clouds, the number of ionizing photons emitted by the star may be sufficient to ionize the entire cloud, in which case we say the HII region is *density-bounded.* For larger clouds, if Q is too low to ionize all the gas, so that the Strömgren sphere is wholly contained within the cloud, then we say the HII region is *ionization-bounded.*

Example 6.1:

An HII region forms in pure hydrogen gas around an O6 star. The density and temperature of the ionized gas are $n_e = n_p = 10^4$ cm^{-3} and $T_e = 10^4$ K.

(a) Calculate the size of the resulting Strömgren sphere.

(b) If the density were 100 cm^{-3}, what would the Strömgren radius be?

Answer:
(a) From Table 6.1, we see that an O6 star produces about $10^{49.29} = 1.95 \times 10^{49}$ Lyman continuum photons per second. Then, using Equation 6.12, we find that the Strömgren radius will be

$$R_s = \left[\frac{3}{4\pi}\frac{1.95 \times 10^{49}\,\mathrm{s}^{-1}}{(10^4\,\mathrm{cm}^{-3})\,(10^4\,\mathrm{cm}^{-3})\,(2.6\times 10^{-13}\,\mathrm{cm}^3\,\mathrm{s}^{-1})}\right]^{1/3}$$

$$= 5.64 \times 10^{17}\ \mathrm{cm} = 0.183\ \mathrm{pc}.$$

(b) Inspection of Equation 6.12 shows that $R_S \propto n^{-2/3}$. So, for a density 100 times lower, R_S will be $0.01^{-2/3} = 21.5$ times its original value. Instead of 0.183 pc, R_S would be $0.183 \times 21.5 = 3.93$ pc. This agrees with the radius shown in Figure 6.1, as expected since a similar stellar temperature and cloud density were used in that case.

Finally, assuming the HII region is ionization-bounded, it is interesting to ask what happens at the boundary of the ionized and neutral regions. Physically, the supply of ionizing photons runs out over short distances. As the distance from the star increases, so does the geometrical dilution. This decreases J_ν, allowing n_H to increase. The increase in n_H leads to more absorption, further decreasing J_ν, and leading to even higher values of n_H. This process rapidly escalates until all of the gas is neutral. A detailed analysis[2] shows that the process becomes complete over a distance of about 10 mean free paths of the ionizing photons, or $10\,(a_\nu\, n_H)^{-1}$, which is inversely proportional to n_H. Using the threshold value of a_ν, this is a distance of $1.6 \times 10^{18}\ n_H^{-1}$ cm for n_H in cm^{-3}. For the densities of 10^4 and 10^2 cm^{-3} of Example 6.1, this gives transition zone widths of $52\ \times 10^{-6}$ pc and 5.2 $\times 10^{-3}$ pc, respectively. Compared to their Strömgren radii of 0.183 and 3.93 pc, these zones are extremely narrow. We see that HII regions are very sharp-edged.

The preceding calculations were for pure hydrogen nebulae; however, real nebulae contain helium, heavy elements (metals) and dust. The heavy elements are trace constituents and have little effect on the ionization structure of the nebulae, although they do produce readily observable emission lines. Helium has a number density about one-tenth that of hydrogen; however the ionization potential for neutral helium is 24.6 eV, considerably higher than for hydrogen. Because the ionization cross-section for hydrogen is maximum for 13.6 eV photons and decreases rapidly for higher energies, the stellar flux of hydrogen-ionizing photons is not affected by the presence of helium. Dust, on the other hand, can have a much greater effect on the ionization structure of a nebula. Although dust is depleted in HII regions relative to the general interstellar medium, it can absorb a substantial fraction of the hydrogen-ionizing stellar radiation. Because dust absorption of stellar radiation reduces the number of ionizing photons in a nebula, dusty HII regions will be smaller than predicted by Equation 6.12 for a given spectral type. In Section 6.2.1 we infer the ionization rate, Q, of a nebula based on its bremsstralung emission. When dust absorption is important, the value of Q, and hence the spectral type of the ionizing star, will be underestimated.

6.1.2 The Temperature of HII Regions

The temperature of an astronomical plasma is determined by the competition between a multitude of heating and cooling processes. The efficacy of these processes depends on a variety of physical parameters, including the gas temperature and density, the energy spectrum of the radiation field, and the metallicity of the plasma (i.e., the abundance of heavy elements), among other factors. Here, we consider only the most important mechanisms; advanced textbooks on the interstellar medium may be consulted for a more-detailed analysis.

The principal heating mechanism for HII regions is the photo-ionization of hydrogen. Many of the photons causing ionization have energies greater than 13.6 eV, and the excess energy appears as kinetic energy of the freed electrons. These electrons scatter off other particles in the plasma, sharing their excess kinetic energy and thereby heating the gas. Thus, the ionizing radiation from the star is the primary source of HII region heating. Hotter stars produce more energetic photons, so the freed electrons will have more kinetic energy and heat the gas more; all other things being equal, hotter stars produce hotter HII regions.

Somewhat paradoxically, the photo-ionization heating rate is determined not only by the energy of the ionizing photons but also by the *recombination* rate. Physically, this can

[2]See, for example, John Dyson & David Williams, 1997 *The Physics of the Interstellar Medium.* Institute of Physics Publishing: Bristol.

be understood based on the nearly 100% ionization fraction within the nebula. In order to inject the excess photon energy, $h(\nu - \nu_0)$, and thus heat the gas, there must be neutral atoms to absorb the photons. Within the HII region the rate-limiting step for heating is how fast the electrons recombine to form atoms — which can then be photo-ionized with the excess photon energy heating the plasma. For lower density HII regions, for example, the recombination rate per unit volume will be smaller and so the heating rate per unit volume by photo-ionization will also be smaller.

The primary cooling mechanism for HII regions is forbidden line emission from metal ions, particularly OII, OIII, NII, and SIII. The outer electrons of these ions have excited states only a few tenths to a few electron volts above their ground states and so they are easily excited to these levels by collisions. The excited atoms can then de-excite either by collisions or by photon emission. Even though these are forbidden transitions (see Section 4.1.1) and so have small radiative transition rates, at the densities of many HII regions the collision rates are also quite low. The photons emitted will not be absorbed by other metal ions but rather will escape the nebula. The small radiative transition probabilities also preclude excitations to these levels by absorbing radiation from the central star and so the excitations occur only via collisions. The net effect, then, is that some kinetic energy of the atoms is converted to excited energy states and then to radiation, which escapes the nebula, thus cooling the ionized gas. For HII regions of higher densities, however, the collisional de-excitation rates are greater, leaving fewer ions to radiatively de-excite, and thus inhibiting the cooling. At densities greater than 10^5 cm^{-3}, the collision rates become fast enough that these forbidden emission lines are less effective for cooling. Thus, higher density HII regions tend to have higher temperatures.

A secondary cooling mechanism occurs through the radiative recombination of hydrogen to excited states. Because the lifetimes of the excited states of hydrogen are so short, when an electron recombines to an excited state, the photons subsequently emitted are unlikely to find excited atoms that can absorb them, and so they readily escape the nebula. The kinetic energy that the electron had before recombining, then, has been carried away. This cooling process occurs even in the absence of metal atoms. Although HII regions are readily detected by their bremsstrahlung emission, this free-free radiation carries so little energy that it is only a minor contributor to the cooling, even in the artificial case of a pure hydrogen nebula.

Based on the competition between the various heating and cooling processes we expect that HII region temperatures will be in the range of about 6,000 to 12,000 K, which is confirmed by observations. The fact that the electron temperature of HII regions is of this order is relevant for several of the arguments we make in this section. For example, the kinetic energies of electrons ($3/2\ kT$) at the temperatures of HII regions are of order 1 eV — far too low to ionize hydrogen by collisions. Therefore, it is clear that UV photons from the star must be responsible for the ionization. A similar argument would not hold, for example, if the plasma had a temperature of 10^6 K. At this temperature the particle kinetic energies are of order 10^6 K / 11,600 K eV^{-1} $\sim$ 86 eV, so collisions would be a likely source of the ionization. As another example, because the electrons of a plasma at 10^4 K will have typical kinetic energies a bit less than 1 eV, if they recombine to the ground atomic state, the emitted photon will have only slightly more than 13.6 eV of energy. As a result, the photo-ionization cross-section of the photon will still be near its maximum value, so the photons won't travel very far before being absorbed to cause another photo-ionization. This makes the on-the-spot approximation a reasonable one.

6.1.3 Time Scales of HII Regions

To obtain an intuitive understanding of the properties and behavior of HII regions it is helpful to consider the timescales for various physical processes occurring within the nebula.

One interesting timescale that we can calculate is how long, on average, a hydrogen atom must wait before being ionized by a stellar UV photon. We can estimate this time in the following way. First, we obtain the number of ionizations per atom per second, ignoring the attenuation of radiation due to absorptions, by taking the number of Lyman continuum photons emitted by the star per second (see Equation 6.11), spreading them over the surface of an imaginary sphere inside the HII region, and multiplying by the photo-ionization cross-section. The reciprocal of this is the average length of time an atom waits to be ionized. Thus, the ionization time scale, τ_i, for atoms a distance r from the star, and ignoring absorptions, is given by

$$\tau_i = \left[\int_{\nu_0}^{\infty} \frac{L_\nu}{h\nu} \frac{1}{4\pi r^2} a_\nu \, d\nu \right]^{-1} . \tag{6.13}$$

This is a lower limit to the ionization time scale because we have neglected the absorption term of $e^{-\tau_\nu}$, which reduces the number of photons available to cause ionizations, and hence increases the average time before atoms are ionized.

Similarly, the timescale for the inverse process, radiative recombination, is useful to know: how long, on average, will it take for a proton to recombine with an electron in an HII region? The essential physical quantities here are the recombination coefficient to all levels, α_A, and the free electron density, n_e. In particular, the recombination time is of order $\tau_r \sim 1/(n_e\, \alpha_A)$, or

$$\tau_r \sim \frac{1}{(n_e)\,(4.18 \times 10^{-13}\ \mathrm{cm^3\ s^{-1}})} \sim (10^5\ \mathrm{yr}) \left(\frac{n_e}{\mathrm{cm^{-3}}} \right)^{-1} . \tag{6.14}$$

The numerical values of Equation 6.14 assume an electron temperature of 10^4 K. The order of magnitude estimates given in Equations 6.13 and 6.14 are useful for back-of-the-envelope calculations, as we demonstrate in Examples 6.2 and 6.3.

Example 6.2:

Consider the HII region of Example 6.1 with a density of 100 $\mathrm{cm^{-3}}$ and a Strömgren radius of 3.93 pc. At an intermediate position in the HII region, of r =1.0 pc, what is the mean lifetime of an atom before being ionized?

Answer:
A proper solution to this problem requires that we numerically calculate the integral of Equation 6.10 to account for the frequency dependence of a_ν and the optical depth of the gas between the stellar surface and the position r = 1.0 pc. We can obtain a crude lower limit to the mean lifetime if we ignore the $e^{-\tau_\nu}$ term and assume that all of these photons have the maximum photo-ionization cross section of $a_\nu = 6.3 \times 10^{-18}$ $\mathrm{cm^2}$. Making these two simplifications, then, apart from the factor of $4\pi\, r^2$, the integral of Equation 6.13 reduces to that of Equation 6.11, which is just the number of ionizing photons emitted by the star, Q. In Example 6.1 we saw that an O6 star produces about $Q = 10^{49.29}\ \mathrm{s^{-1}} = 1.95 \times 10^{49}$ Lyman continuum photons per second. So the mean time for a photo-ionization to occur

at $r = 1.0$ pc $= 3.08 \times 10^{18}$ cm will be

$$\frac{4\pi r^2}{Qa_\nu} = \frac{(4\pi)\,(3.08 \times 10^{18}\,\mathrm{cm})^2}{(1.95 \times 10^{49}\,\mathrm{s}^{-1})\,(6.3 \times 10^{-18}\,\mathrm{cm}^2)} = 9.70 \times 10^5\ \mathrm{s} = 11.2\ \mathrm{days}.$$

Remember that this time represents a lower limit, because we over-estimated the photo-ionization cross section by using the threshold value and over-estimated the number of ionizing photons by neglecting the optical depth of the gas interior to $r = 1.0$ pc. Nevertheless, the salient point is that the photo-ionization process is very rapid.

Example 6.3:

If the exciting star of Example 6.1 were to suddenly turn off, how long would it take the HII region, with a density of 10^2 cm^{-3}, to recombine into neutral gas?

Answer:
Using the approximate expression of Equation 6.14 we have

$$\tau_r \sim (10^5\ \mathrm{yr}) \left(\frac{10^2\ \mathrm{cm}^{-3}}{\mathrm{cm}^{-3}}\right)^{-1} \sim \frac{10^5\ \mathrm{yr}}{10^2} = 1000\ \mathrm{yr}.$$

Had we used the higher density of Example 6.1, of 10^4 cm^{-3}, the recombination time would have been of order 10 years. Either way, the conclusion is that recombination is a much slower process — by orders of magnitude — than photo-ionization.

These two characteristic times, for photo-ionization and for radiative recombination, provide a heuristic means to calculate the ionization fraction within the nebula. In equilibrium, $n_H/\tau_i = n_e/\tau_r$; thus the ratio of neutral atoms to ions, n_H/n_e, is given by the ratio τ_i/τ_r. Using the values from the examples above, in which $\tau_i = 9.70 \times 10^5$ s and $\tau_r = 10^3$ yr $= 3.2 \times 10^{10}$ s, we obtain a ratio of 3.0×10^{-5}. This implies that about 99.997% of the atoms within the HII region are ionized, as confirmed by the numerical solution plotted in Figure 6.1.

6.2 RADIO EMISSION FROM HII REGIONS

6.2.1 Bremsstrahlung Emission from HII Regions

In Section 3.2.4 we discussed the thermal bremsstrahlung emission process and derived several equations relating the radiation intensity to the physical parameters of the emission region. Here we extend that discussion, with an emphasis on the physical properties of HII regions. Using Equation 3.17 for the free-free intensity,

$$I_\nu^{ff} = B_\nu(T_e)\,(1 - e^{-\tau_\nu^{ff}}),$$

using the Rayleigh-Jeans approximation for B_ν and integrating over the source solid angle, Ω_S, we obtain the free-free flux density we would measure at frequency ν:

$$F_\nu = \frac{2k\nu^2}{c^2} T_e \int_{\Omega_S} (1 - e^{-\tau_\nu^{ff}})\, d\Omega. \tag{6.15}$$

For simplicity we have assumed an isothermal HII region, so that the electron temperature is constant over the solid angle of the region and may be taken outside the integral. As discussed in Section 3.2.4, at high frequencies the source is optically thin, with $\tau_\nu \ll 1$, and the flux density may be approximated as

$$F_\nu \approx \frac{2k\nu^2}{c^2} T_e \int_{\Omega_S} \tau_\nu^{ff} \, d\Omega \propto \nu^{-0.1}, \tag{6.16}$$

which is nearly independent of frequency. At low frequencies, the source is optically thick with $\tau_\nu^{ff} \gg 1$, and the flux density may be approximated as

$$F_\nu = \frac{2k\nu^2}{c^2} T_e \Omega_S \propto \nu^2 \tag{6.17}$$

which shows the rising spectrum of the low-frequency side of the Planck function. Both the optically thin and thick behavior are shown in Figure 6.2. The turnover frequency, given by Equation 3.20, and defined by $\tau_\nu^{ff} = 1$, divides the optically thick and thin regimes.

The brightness temperature, in contrast to the flux density, is constant in the optically thick regime, and decreases with increasing frequency in the optically thin regime as $T_B \propto \nu^{-2.1}$ (see Figure 6.2). This is a natural consequence of the relationship between brightness temperature and flux density, given by

$$T_B = F_\nu \frac{c^2}{2k\,\nu^2} \frac{1}{\Omega}. \tag{6.18}$$

In the optically thick regime, the factor of ν^{-2} cancels with the ν^2 dependence of F_ν, leaving T_B independent of frequency, as expected for optically thick gas. In the optically thin regime, where F_ν is nearly independent of frequency, the factor of ν^{-2} gives a rapidly falling brightness temperature.

In general, the optical depth depends on both the gas density and the path length through the source. Because the size of an HII region depends on the gas density (see Equation 6.12), we can relate the optical depth to density alone. Equation 3.19 gives the bremsstrahlung optical depth in terms of the emission measure, which is the integral of the product of densities, $n_e\, n_p$, over the path length through the HII region. The size of an HII region scales as $(n_e\, n_p)^{-1/3}$ (see Equation 6.12); therefore, the optical depth scales as $(n_e\, n_p)^{2/3}$. Hence, at any given frequency, higher density HII regions will be more optically thick than lower density HII regions, and so will have higher turnover frequencies.

The turnover frequency is important from an observational standpoint for several reasons. First, it roughly indicates the frequency at which the HII region presents its highest flux density and hence is easiest to detect. Additionally, the optimal choice of observing frequency depends partly on what property one wishes to measure. To calculate the mass of ionized gas, for example, an optically thin observing frequency must be used, so that all the emission is detected, not just the emission from the outer layers. Conversely, by observing at optically thick frequencies we can infer the electron temperature since it equals the optically thick brightness temperature. Moreover, the choice of observing frequency can introduce an observational bias. For the same exciting star and plasma temperature, all HII regions will have the same optically thick brightness temperature, but higher-density HII regions will be smaller and so will have lower flux densities than less dense HII regions. Hence,

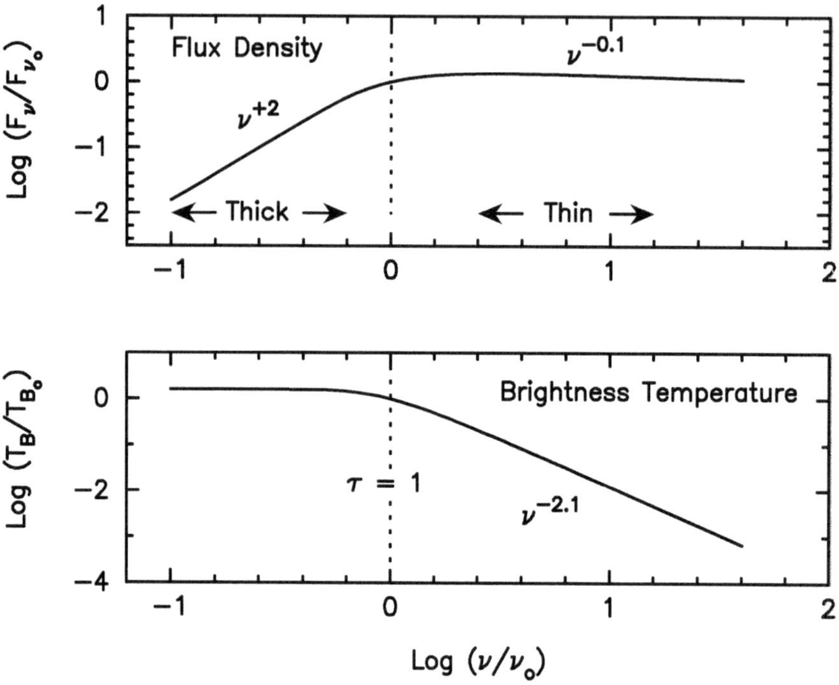

Figure 6.2: The normalized flux density, above, and the normalized brightness temperature, below, plotted as a function of normalized frequency, where ν_0 is the turnover frequency, given by Equation 3.20.

low-frequency observations will have a bias against detecting the dense and highly-compact HII regions.

Example 6.4:

A uniform, isothermal, spherical HII region containing only hydrogen, with Strömgren radius 0.05 pc, at a distance of 5 kpc, has an electron density of 10^4 cm^{-3} and a temperature of 8,000 K. What brightness temperature is measured toward the center of the HII region at a frequency of 15 GHz?

Answer:
We first use the emission measure to calculate the optical depth of the region, which in turn will give us its brightness temperature. The emission measure for this region is

$$n_e^2 L = (10^4\ \mathrm{cm}^{-3})^2\ (0.1\ \mathrm{pc}) = 10^7\ \mathrm{cm}^{-6}\ \mathrm{pc}.$$

Using Equation 3.19 we find that τ_ν is given by

$$\tau_{15\ \mathrm{GHz}} \approx 0.0824 \left(\frac{8000\ \mathrm{K}}{\mathrm{K}}\right)^{-1.35} \left(\frac{15\ \mathrm{GHz}}{\mathrm{GHz}}\right)^{-2.1} \left(\frac{10^7\ \mathrm{cm}^{-6}\ \mathrm{pc}}{\mathrm{cm}^{-6}\ \mathrm{pc}}\right) = 0.015.$$

Thus, at this frequency the region is optically thin and we may use the approximation $T_B \approx \tau_\nu\, T_e$, giving

$$T_B \approx (0.015)\,(8000\ \mathrm{K}) = 120\ \mathrm{K}.$$

Note that if we observe positions toward the edge of the HII region, then the path length through the nebula, and hence the optical depth, is less, leading to a lower brightness temperature.

The calculation demonstrated in Example 6.4 is useful for modeling the radio emission for a given set of HII region parameters. Observationally, the problem is exactly the opposite: we measure the emission and calculate the nebular properties. Precisely how we derive physical parameters of an HII region from observations depends on several factors, including whether the region is resolved by the observations, if it is optically thin, what, if anything, is known about the source geometry, and so forth. A fundamental problem is that the optical depth, and hence the intensity, are functions of position in the nebula. The optical depth depends on the emission measure — the integral of $n_e^2\,dl$ — so even in the idealized case of a uniform electron density, differing path lengths through the HII region give rise to differing optical depths.

If the observations fully resolve the HII region, (which occurs when the beam size of the telescope is much smaller than the solid angle subtended by the HII region) then a simple approach is to use Equation 6.18 with Ω equal to the solid angle of the telescope resolution. Along with the measured flux density, this provides a means to calculate the brightness temperature. For an adopted electron temperature, T_e, and converting intensity to brightness temperature, we use

$$T_B = T_e\,(1 - e^{-\tau_\nu}) \tag{6.19}$$

to calculate τ_ν. Equation 3.19 may then be solved for the emission measure. Because the HII region is resolved, we can measure its angular size, and if we know the distance, we can calculate its characteristic physical dimension, s. Combining the various steps, and assuming the source width and depth are the same, we obtain

$$n_e = \sqrt{\frac{EM}{s}}.$$

The above calculations are frequently made for individual sight lines through the HII region, and as such give line-of-sight average physical parameters instead of integrated values for the entire nebula.

The value adopted for s requires some assumption about the geometry of the source — often that the depth of the source (along the line-of-sight) is similar to its transverse dimension on the plane of the sky. In an early radio study of galactic HII regions, Peter Mezger and A. Henderson[3] developed a similar, but somewhat more rigorous formulation that gives integrated values for the entire nebula, valid for several different geometries. They assumed a uniform density, isothermal nebula, and calculated the necessary integrals for spherical, cylindrical, and Gaussian geometries. Here, we present their results for a spherical region; their article may be consulted for the other two geometries. For the mean electron density of the nebula, they obtain

$$n_e = 878\,\mathrm{cm}^{-3}\left(\frac{T_e}{10^4\ \mathrm{K}}\right)^{0.175}\left(\frac{\nu}{\mathrm{GHz}}\right)^{0.05}\left(\frac{F_\nu}{\mathrm{Jy}}\right)^{0.5}\left(\frac{d}{\mathrm{kpc}}\right)^{-0.5}\left(\frac{\Theta}{\mathrm{arcmin}}\right)^{-1.5}, \tag{6.20}$$

where d is the distance to the HII region and Θ is its angular diameter.

[3]1967, Astrophysical Journal, vol. 147, p. 471.

Once the average electron density is known, the ionized gas mass within the sphere is calculated by assuming that for each electron there is a proton of mass 1.67×10^{-24} g and multiplying by the volume of the HII region. In terms of the observable units, for a uniform spherical region, Mezger and Henderson give

$$M = 0.28\,\mathrm{M}_\odot \left(\frac{T_e}{10^4\ \mathrm{K}}\right)^{0.175} \left(\frac{\nu}{\mathrm{GHz}}\right)^{0.05} \left(\frac{F_\nu}{\mathrm{Jy}}\right)^{1/2} \left(\frac{d}{\mathrm{kpc}}\right)^{5/2} \left(\frac{\Theta}{\mathrm{arcmin}}\right)^{3/2}. \tag{6.21}$$

Another useful parameter to derive from the observations is the minimum number of ionizing photons required to maintain the ionization of the nebula. Starting with Equation 6.11, we can substitute in for the Strömgren radius (as the angular size times the distance) and use Equation 6.20 for n_e (and setting $n_p = n_e$). The recombination coefficient is temperature-dependent, which we express as

$$\alpha_B = 2.59 \times 10^{-13}\ \mathrm{cm}^3\ \mathrm{s}^{-1} \left(\frac{T_e}{10^4\ \mathrm{K}}\right)^{-0.85}.$$

Combining these steps, we obtain

$$Q \geq 7.6 \times 10^{46}\ \mathrm{s}^{-1} \left(\frac{\nu}{\mathrm{GHz}}\right)^{0.1} \left(\frac{T_e}{10^4\ \mathrm{K}}\right)^{-0.5} \left(\frac{F_\nu}{\mathrm{Jy}}\right) \left(\frac{d}{\mathrm{kpc}}\right)^{2}. \tag{6.22}$$

The value of Q given by Equation 6.22 is the minimum value for the Lyman continuum photon rate because it neglects the fact that some photons are absorbed by dust grains within the HII region and do not contribute to the ionization. Moreover, real HII regions will not be perfectly spherical or uniform. Some sight lines from the star may not have sufficient hydrogen to absorb all of the ionizing photons passing in their direction; the stellar photon flux in these directions would not fully contribute to the ionization. Hence, this value of Q leads to a lower limit to the temperature of the ionizing star, as listed in Table 6.1.

6.2.2 Radio Recombination Line Emission from HII Regions

The recombination process not only results in optical emission lines, which cool the plasma, but also in the emission of radio frequency lines, called *radio recombination lines* or RRLs (see Section 4.2.2). Although RRLs are much weaker than optical recombination lines, and so are not a major coolant, they do not suffer the dust extinction that optical lines do. For younger HII regions, still deeply embedded within molecular clouds, RRLs are often the only means to obtain kinematic information about the ionized gas.

As explained in Section 6.1.1, the recombination coefficients reflect the probability that a recombining electron will go into a particular atomic level. Because α_n decreases with increasing n, recombinations to very high n levels are not common, but they do occur: emission of hydrogen RRLs has been detected in HII regions with $n > 300$, while astronomical carbon RRLs have been detected with $n > 600$.

What happens after a recombination to such a high level? The electron will naturally decay toward the ground state of $n = 1$. However, it will almost never go *directly* to the ground state. As we discussed in Section 4.2.2, the Einstein A coefficients are largest for transitions with $\Delta n = 1$, followed by $\Delta n = 2$, and so on. Hence, an electron that recombines to the $n = 110$ level will have the highest probability of decaying to the $n = 109$ level, producing an H109α line. Transitions of larger Δn can occur, giving rise to the lines H108β, H107γ, and so forth, but these transitions have a lower probability of occurrence. Once in the new level, the excited electron will decay again and again, dropping level by level toward the ground state. Because the energy differences between these very high n levels are so small, the resulting radiation is at radio wavelengths. Radio recombination lines are produced when the transitions satisfy $n \gtrsim 30$ and $\Delta n \ll n$. Because of this condition on Δn, a convenient approximation to the frequency (given precisely in Section 4.2.2) is

$$\nu_{\rm n} \approx \frac{2\,R\,c\,\Delta n}{n^3}, \tag{6.23}$$

where R is the Rydberg constant. The frequency spacing between adjacent α lines, with $\Delta n = 1$, is $\Delta\nu \approx 3\nu_n/n$. Thus, at lower frequencies (higher quantum numbers) RRLs are more closely spaced.

In the following discussion we will assume that RRL emission is in thermodynamic equilibrium. Thus, the excitation temperature is equal to the electron temperature (see Section 4.1.5). This is a fairly good assumption because atoms in high-n states suffer many collisions, partly because they have large cross-sections (the Bohr orbit goes as n^2) and partly because they have fairly small Einstein A coefficients, so the excited atoms have relatively long lifetimes, sufficient to suffer collisions with other particles. We discuss departures from thermodynamic equilibrium in Section 6.2.3.

The ratios of brightness temperatures of the line and continuum emission (or, alternatively, their flux densities) can be used to determine the electron temperature of an HII region. The basis of the calculation is that for optically thin emission, the line brightness temperature, T_L, and the continuum brightness temperature, T_C, both depend on their respective optical depths and the electron temperature, T_e, via

$$T_L = \tau_\nu^L\, T_e \quad \text{and} \quad T_C = \tau_\nu^C\, T_e.$$

Because of our assumption of thermodynamic equilibrium, we use T_e rather than T_{ex} to estimate the brightness temperature of the line emission. Previously, we only considered one emission mechanism at a time, so it was sufficient to use τ_ν without a superscript. Because the line and continuum emission arise from different physical processes, we must use the corresponding τ_ν for each of them. As we will see, the optical depths for the line and the continuum have different dependencies on T_e. Therefore, the ratio

$$T_L/T_C = \tau_\nu^L/\tau_\nu^C$$

can be solved for T_e.

The optical depth of the continuum emission, which is produced by the bremsstrahlung process, is given by Equation 3.19 as

$$\tau_\nu^C \approx 0.0824 \left(\frac{T_e}{\rm K}\right)^{-1.35} \left(\frac{\nu}{\rm GHz}\right)^{-2.1} \left(\frac{EM}{{\rm cm}^{-6}\,{\rm pc}}\right). \tag{6.24}$$

To find the optical depth of the line emission, we return to the discussion in Section 4.1. The absorption coefficient for the line is given by Equation 4.5 as

$$\kappa_\nu^L = \frac{h\nu}{4\pi}[n_l\, B_{lu} - n_u\, B_{ul}]\,\phi_\nu.$$

To obtain the optical depth, we follow a similar procedure to that used in the derivation of Equation 4.18, but instead of expressing τ_ν^L in terms of B_{lu} and N_l we use Equations 4.6 and 4.7 to express it in terms of A_{ul} and N_u, giving us

$$\tau_\nu^L = \frac{c^2}{8\pi\nu^2}\, N_u\, A_{ul}\, \left(e^{h\nu/kT_e} - 1\right) \phi_\nu,$$

where we have used $\Delta E_{ul} = h\nu$. For these radio recombination lines $h\nu \ll kT_e$, so we can make the approximation

$$\left(e^{h\nu/kT_e} - 1\right) \approx \frac{h\nu}{kT_e}.$$

We can also substitute for the line profile function, ϕ_ν, at the line center (see Section 4.1.2) with $2(\ln 2)^{1/2}\pi^{-1/2}\Delta\nu^{-1}$, where $\Delta\nu$ is the full-width at half maximum. Thus we have for the optical depth at the line center of a radio recombination line

$$\tau_o^L \approx \frac{c^2\sqrt{\ln 2}}{4\pi^{3/2}\nu^2} N_u A_{ul} \frac{h\nu}{kT_e}\frac{1}{\Delta\nu}. \tag{6.25}$$

A general expression for the Einstein A coefficient for electric dipole transitions is given by

$$A_{ul} = \frac{64\pi^4}{3hc^3}\nu_{ul}^3 \mid d_{ul}\mid^2,$$

where d_{ul} is the electric dipole matrix element. The electric dipole matrix elements are related to the dipole moment associated with the transition between the two states. For large n recombination lines in hydrogen and for $\Delta n = 1$ (α transitions), we can approximate these electric dipole matrix elements by the mean electric dipole moment of the atom. Thus

$$\mid d_{n+1\to n}\mid \approx \frac{e a_n}{2},$$

where a_n is the Bohr radius of the nth orbit. The Bohr radius is given by

$$a_n = \frac{h^2}{4\pi^2 m_e e^2} n^2,$$

where m_e and e are the electron mass and charge, so that

$$\mid d_{n+1\to n}\mid \approx \frac{h^2}{8\pi^2 m_e e} n^2.$$

Using the approximation of Equation 6.23 for ν_{ul} with $\Delta n = 1$ (for α transitions) and substituting in for the various physical constants, the Einstein A coefficient for α-transitions is

$$A_{n+1\to n} \approx \frac{5.35\times 10^9\ \mathrm{s}^{-1}}{n^5}. \tag{6.26}$$

For the column density in the upper level, N_u, we use $n_u L$. To obtain n_u, since we are assuming thermodynamic equilibrium, we can use the Boltzmann equation (Equation 4.14 but with the excitation temperature replaced by the electron temperature). We can express the population in the nth level of hydrogen to that in the ground state ($n = 1$) by

$$\frac{n_n}{n_1} = n^2\, e^{-\Delta E/kT_e},$$

where ΔE is the energy difference between the ground state ($n = 1$) and excited state (n) and the statistical weight for the nth level in hydrogen is $g_n = 2n^2$. Be careful of the notation in this equation, as n indicates both number density and also the principle quantum number.

For typical HII region temperatures, nearly all of the neutral hydrogen atoms are in the ground state and very few are in any of the excited states. Therefore, n_1 is approximately equal to the density of neutral hydrogen atoms, n_H. However, most of the hydrogen in an HII region is ionized, so we need a relation between the number of neutral hydrogen atoms and the number of ionized hydrogen atoms. For this we use the generalized Saha equation, which, for thermodynamic equilibrium, gives the ratio of the number density of atoms or ions in two adjacent ionization states ($i+1$ and i) as a function of temperature. The Saha equation is

$$\frac{n_e\, n_{i+1}}{n_i} = \left(\frac{2\pi m_e k T_e}{h^2}\right)^{3/2} \frac{2Z_{i+1}}{Z_i}\, e^{-\chi/kT_e}, \tag{6.27}$$

where Z is the partition function for each ionization state (see discussion in Section 4.2.3 as applied to molecular rotational states) and χ is the ionization energy for ionization state i. If we apply this equation to hydrogen, then the $i+1$ ionization state refers to ionized hydrogen and the i ionization state applies to neutral hydrogen. Since ionized hydrogen is just a proton, its partition function is simply $Z_{i+1} = 1$. Assuming that nearly all of the neutral hydrogen is in the ground state, the partition function Z_i is just the statistical weight of the $n = 1$ state or $Z_i = 2$. Therefore, for hydrogen we have

$$\frac{n_e\, n_{H^+}}{n_H} \approx \left(\frac{2\pi m_e k T_e}{h^2}\right)^{3/2} e^{-\chi/kT_e},$$

where χ is 13.6 eV and $\chi/k = 158,000$ K.

Combining the results from the Boltzmann and Saha equations, we solve for the number density of neutral hydrogen atoms in the nth level in terms of the temperature and the product $n_e n_{H+}$. Note that for these high-n radio recombination lines, the energy ΔE in the Boltzmann equation is nearly equal to χ in the Saha equation, and so the exponential terms cancel. We find

$$n_n \approx n^2 \left(\frac{h^2}{2\pi m_e k T_e}\right)^{3/2} n_e n_{H^+}. \tag{6.28}$$

We can now obtain the expression for the line center optical depth of a hydrogen recombination line by starting with Equation 6.25 and substituting for ν using Equation 6.23, A_{ul} from Equation 6.26, and for N_u using Equation 6.28 ($n_n = N_u/L$). Inserting the numerical values for the various constants we get

$$\tau_o^L \approx 5.44 \times 10^{-13} \left(\frac{T_e}{\mathrm{K}}\right)^{-5/2} \left(\frac{\Delta\nu}{\mathrm{Hz}}\right)^{-1} n_e n_{H^+} L.$$

The quantity $n_e n_{H^+} L$ for a pure hydrogen nebula is equal to the emission measure, EM. Although the ionization of helium contributes to both the number density of electrons and ions, since the ratio of hydrogen to helium by number is about 10 and because only a small fraction of the helium is ionized, we can neglect the contribution of helium to the emission measure.

Instead of expressing the line width in frequency units, we can express it in velocity units using

$$\Delta\nu = \nu \frac{\Delta v}{c}.$$

If we also express the frequency in GHz, the emission measure in units of cm^{-6} pc, and the line width in km s^{-1} we find

$$\tau_o^L \approx 503 \left(\frac{T_e}{\mathrm{K}}\right)^{-5/2} \left(\frac{\nu}{\mathrm{GHz}}\right)^{-1} \left(\frac{\Delta v}{\mathrm{km\ s^{-1}}}\right)^{-1} \left(\frac{EM}{\mathrm{cm^{-6}\ pc}}\right), \tag{6.29}$$

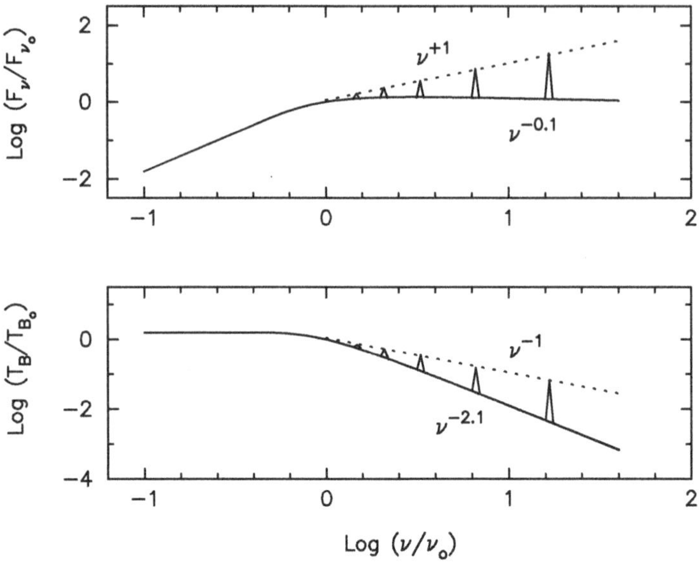

Figure 6.3: Radio recombination lines only appear in the optically thin part of the spectrum, above the turnover frequency. The separation between consecutive α lines (of different n levels) increases with the frequency, so at lower frequencies the lines are more closely spaced with respect to one another. The intensity of the lines relative to the background continuum rises as $\nu^{1.1}$.

where ν is the frequency of the RRL being observed, ν_{ul}. As before, assuming the line and continuum are both optically thin, we have

$$\frac{T_L}{T_C} = \frac{\tau_o^L}{\tau_\nu^C}.$$

Using Equations 6.24 and 6.29 for the optical depths, we can write this line-to-continuum ratio as

$$\frac{T_L}{T_C} = 6110 \left(\frac{T_e}{\mathrm{K}}\right)^{-1.15} \left(\frac{\nu}{\mathrm{GHz}}\right)^{1.1} \left(\frac{\Delta v}{\mathrm{km\ s^{-1}}}\right)^{-1}. \tag{6.30}$$

We see that the line-to-continuum ratio depends on temperature, as we alluded to earlier, and thus provides an observational means to determine T_e. Equation 6.30 may be solved for the electron temperature, T_e, yielding

$$T_e = 1960\ \mathrm{K} \left(\frac{T_L}{T_C}\right)^{-0.87} \left(\frac{\nu}{\mathrm{GHz}}\right)^{0.96} \left(\frac{\Delta v}{\mathrm{km\ s^{-1}}}\right)^{-0.87}. \tag{6.31}$$

Because the line and continuum brightness temperatures convert to flux densities with the same frequency dependence, we may use either T_L/T_C or F_ν^L/F_ν^C to calculate T_e (see Figure 6.4 and Example 6.5). We caution that this derivation assumes thermodynamic equilibrium. In some cases, RRL intensities show the effects of stimulated emission, resulting in an underestimate of T_e and other quantities that are derived from it.

Also noteworthy in Equation 6.30 is the $\nu^{1.1}$ dependence. This means that higher frequency RRLs (from lower n transitions) are stronger relative to the continuum than lower frequency (higher n transitions) lines. This is illustrated in Figure 6.3, where higher frequency RRLs have brightness temperatures increasing as $\nu^{1.1}$ with respect to the continuum.

Again, the conversion of brightness temperature to flux density has the same frequency dependence for both line and continuum, so in the optically thin regime, where the flux density of the continuum is nearly constant, RRLs have steadily rising flux densities at higher frequencies. Thus, other considerations aside, high frequency RRLs are more easily detected than low frequency RRLs.

It is instructive to relate the trends seen in Figure 6.3 to our discussions of optical depth throughout this book. An understanding of opaque radiation provides physical insight regarding the decrease of the ratio of the line-to-continuum intensities with decreasing frequency. As discussed in Chapter 3, at large optical depths the continuum intensity approaches the Planck function for the gas temperature and, as discussed in Chapter 4, the spectral line intensity approaches the Planck function for the excitation temperature. For HII regions, in thermodynamic equilibrium, these temperatures are equal. Thus, at large optical depths the line and continuum intensities are the same, and the spectral lines are not evident against the continuum. In short, Figure 6.3 demonstrates that the brightness temperature of opaque thermal radiation equals the Planck function for both continua and spectral lines. This is consistent with the idea that high optical depths lead to thermal equilibrium between the gas and the radiation.

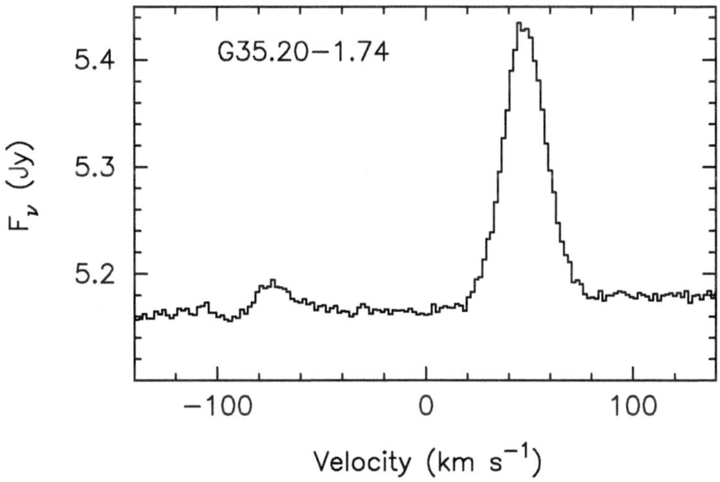

Figure 6.4: The H110α spectrum of the G32.5−1.74 HII region. The hydrogen line is centered at an LSR velocity of 47.7 km s^{-1} and its full-width at half-maximum is 23.8 km s^{-1}. Weak He110α emission is seen at about −74 km s^{-1}. The 4.87-GHz continuum flux density of the HII region is 5.17 Jy, while the combined flux density of the hydrogen line plus the continuum at line center is 5.43 Jy. The data shown here are from observations made with the Arecibo 305-m meter telescope, reported in Araya et al., 2002, Astrophysical Journal Supplement, vol. 138, p. 63. Data for the spectrum were kindly provided by Esteban Araya.

Example 6.5:

The G35.20−1.74 HII region is at a distance of 3.3 kpc and has an angular size of about 2 arc minutes (see Figure 6.17 in Volume I). Based on its H110α spectrum, shown in Figure 6.4,

(a) calculate the electron temperature, assuming a pure hydrogen nebula, and

(b) find it's electron density, assuming the region is spherical, isothermal and uniform.

Answer:
(a) The rest frequency of the H110α line is found from Equation 4.31 to be 4,874.15 MHz. We substitute this value into Equation 6.31, along with the parameters mentioned in the caption of Figure 6.4, and use the fact that $T_L/T_C = F_\nu^L/F_\nu^C$ to obtain

$$T_e = 1960\ \mathrm{K}\left(\frac{0.26\ \mathrm{Jy}}{5.17\ \mathrm{Jy}}\right)^{-0.87}\left(\frac{4.874\ \mathrm{GHz}}{\mathrm{GHz}}\right)^{0.96}\left(\frac{23.8\ \mathrm{km\ s^{-1}}}{\mathrm{km\ s^{-1}}}\right)^{-0.87} = 7670\,\mathrm{K}.$$

(b) Using Equation 6.20 with the 7,670 K electron temperature from part (a), we have

$$n_e = 878\ \mathrm{cm^{-3}}\left(\frac{7670}{10^4\ \mathrm{K}}\right)^{0.175}\left(\frac{4.874\ \mathrm{GHz}}{\mathrm{GHz}}\right)^{0.05}\left(\frac{5.17\ \mathrm{Jy}}{\mathrm{Jy}}\right)^{1/2}\times$$
$$\left(\frac{3.3\ \mathrm{kpc}}{\mathrm{kpc}}\right)^{-1/2}\left(\frac{2\ \mathrm{arcmin}}{\mathrm{arcmin}}\right)^{-3/2} = 400\ \mathrm{cm^{-3}}.$$

6.2.3 Gas Density and Temperature from RRLs: Non-Equilibrium Effects

In our derivation of the electron temperature in Equation 6.31, we assumed thermodynamic equilibrium conditions. Unfortunately, this assumption is not always valid in relation to RRL emission. Non-equilibrium conditions will affect the calculated value for T_e. RRL can also be used to infer the density of the ionized gas, although the procedure is more complicated than obtaining the electron temperature. Morever, the density calculation is even more susceptible to non-thermodynamic equilibrium effects. A complete description of the accurate determination of T_e and n_e in the general case is outside the scope of our discussion. Instead, we will explain in general physical terms how non-thermodynamic equilibrium models may be used to infer the electron temperature and density.

To describe the non-equilibrium effects, *departure coefficients* are introduced, that quantify how far the conditions are from thermodynamic equilibrium. In particular, $b_i \equiv n_i/n_i^*$, where n_i^* represents the density of atoms in the ith state if thermodynamic equilibrium holds, and n_i is the true density of atoms in the ith state. To consider one non-equilibrium effect, imagine a 'bucket analogy', where buckets are strung one above another, each bucket representing an atomic level. Each bucket has a hole to let water drain into the lower buckets, but buckets lower down have bigger holes, so they drain out more quickly, leading to fuller buckets higher up and emptier buckets lower down. This produces a population inversion and gives rise to a mild *masing* effect in the RRL (see Sections 4.1.1.3 and 4.2.3).

Outside of thermodynamic equilibrium, there is a combination of effects that determines the line intensities. One is the masing effect, which tends to increase the line strength. Another is *pressure broadening* of the lines, which tends to decrease the peak line strength. The pressure broadening arises from interactions between the excited atoms, and is sensitive to the electron density and to the principle quantum number of the atom — atomic radius is proportional to n^2, so highly-excited atoms have much larger collision cross-sections. As a result of this coupling between collisions and radiative transitions, the non-thermodynamic equilibrium version of Equation 6.30 has a dependency on gas density and principle quantum number as well as T_e. For RRLs with high principal quantum numbers, pressure broadening often dominates over thermal broadening, especially for high-density HII regions. The ratio of the pressure broadening to the thermal broadening (the latter is discussed in Section 4.1.2) is given by

$$\frac{\Delta v_{\rm pressure}}{\Delta v_{\rm thermal}} \approx 0.14 \left(\frac{n}{100}\right)^{7.4} \left(\frac{T_e}{10^4\ {\rm K}}\right)^{-1/2} \left(\frac{n_e}{10^4\ {\rm cm}^{-3}}\right). \tag{6.32}$$

The observed line width is a combination of several mechanisms. Thermal and turbulent motions can both be described by Gaussian line profiles, so the convolution of these two effects is given by

$$(\Delta v_{\rm Gaussian})^2 = \Delta v_{\rm thermal}^2 + \Delta v_{\rm turbulent}^2.$$

Pressure broadening, on the other hand, presents a *Lorentzian profile*, which has a narrower central region and more extended line wings. The convolution of a Gaussian profile with a Lorentzian profile is called a *Voigt profile*. Because the different line broadening mechanisms have differing dependencies on the density, temperature, and principle quantum number, the change in line width and intensity of several different RRLs, spanning a range of principle quantum number, can be used to constrain the temperature and density of the HII region. The procedure, then, is to plot T_L/T_C as a function of principle quantum number, and adjust T_e and n_e to obtain the best fit to the observed ratios.

Example 6.6:

(a) For the electron temperature of 7670 K calculated in Example 6.5, what thermal line width is expected for the H110α line?

(b) Based on this thermal line width and the electron density calculated in Example 6.5, what is the pressure broadening line width?

Answer:
(a) Using Equation 4.9 for the Doppler broadening parameter, we find that

$$\Delta\nu_D = \left(\frac{4.874\times10^9\ {\rm Hz}}{3.0\times10^{10}\ {\rm cm\ s^{-1}}}\right)\left[\frac{(2)(1.38\times10^{-16}\ {\rm erg\ K^{-1}})(7670\ {\rm K})}{(1.67\times10^{-24}\ {\rm g})}\right]^{1/2}$$
$$= 1.83\times10^5\ {\rm Hz}.$$

The full width at half maximum in velocity units is given by

$$\Delta v_{\rm FWHM} = \left(\frac{3.0\times10^{10}\ {\rm cm\ s^{-1}}}{4.874\times10^9\ {\rm Hz}}\right) 2\sqrt{\ln 2}\,(1.83\times10^5\ {\rm Hz}) = 18.7\ {\rm km\ s^{-1}}.$$

(b) Using Equation 6.32, we have

$$\Delta v_{\rm pressure} \approx 18.7\ {\rm km\ s^{-1}} \times 0.14 \left(\frac{110}{100}\right)^{7.4}\left(\frac{7670}{10^4\ {\rm K}}\right)^{-1/2}\left(\frac{400}{10^4\ {\rm cm^{-3}}}\right)$$
$$= 0.242\ {\rm km\ s^{-1}}.$$

Example 6.6(a) shows that thermal broadening (18.7 km s^{-1}) cannot explain the full observed line width (23.8 km s^{-1}). Moreover, part (b) shows that pressure broadening is not responsible for the additional line width. Rather, the additional line width is due to turbulent line broadening, arising from internal gas motions within the HII region. We note that at electron densities of order a few hundred per cubic centimeter, the H110α line is not substantially pressure broadened. At higher densities, typical of ultracompact HII regions, the 110α line could show pressure broadening on the same order as the thermal broadening.

6.3 THE CLASSIFICATION AND EVOLUTION OF HII REGIONS

Strictly speaking, an "HII region" simply means a region of ionized hydrogen. However, there are multiple ways to ionize a gas, and in practice the term "HII region" is reserved for plasmas that are photo-ionized by the UV radiation from massive stars. Astronomical plasmas with other sources of ionization are generally referred to by other names. In Section 6.4, we will discuss plasmas known as planetary nebulae, which are photo-ionized by white dwarf stars.

6.3.1 Classification of HII Regions

Even with the agreement that "HII region" indicates a photo-ionized plasma around a young massive star (or stellar cluster), there is still an enormous range in the properties of HII regions. These nebulae were first identified from their optical spectral line emission, which resulted in a selection effect. In particular, the HII regions that were first discovered and studied, often called *classical HII regions*, are large, low density regions. They are old enough to have expanded and dispersed the molecular cloud from which the massive star(s) formed, and can now be seen in optical emission lines. As explained in Section 4.2.2, the lower n hydrogen recombination lines from a plasma are much brighter than high n (radio) recombination lines. Because classical HII regions are not significantly obscured by dust, they are extensively studied by means of these bright, optical hydrogen lines and the optical forbidden lines of OIII, NII and OII.

In some locations, clusters of massive stars form almost simultaneously, leading to extremely high ionizing photon rates. These massive clusters are able to form substantially larger nebulae, known as *giant HII regions*. Even more extreme massive clusters have been found in starburst galaxies (see Section 8.3.2) which form even larger nebulae than have been found in the Milky Way; these are sometimes referred to as super-giant HII regions. Typical ranges for some physical parameters of these regions are shown in Table 6.2. As with classical HII regions, the giants and super-giants are usually studied by means of optical and infrared recombination lines. Such studies provide useful information on the environments where these stellar clusters form and on the range of stellar types that form coevally.

Table 6.2: Physical Parameters of HII Regions

Class of Region	Size (pc)	Density (cm^{-3})	Emission Measure (pc cm^{-6})	Ionized Mass ($M_\odot$)
Super-Giant	$\gtrsim 100$	~ 10	$\gtrsim 10^4$	$10^6 - 10^8$
Giant	~ 100	~ 30	10^5	$10^4 - 10^6$
Classical	~ 10	~ 100	10^5	$10^2 - 10^4$
Compact	$\lesssim 0.5$	$\gtrsim 5{,}000$	10^6	~ 1
Ultracompact	$\lesssim 0.1$	$\gtrsim 10^4$	$\gtrsim 10^7$	$10^{-3} - 10^{-1}$

Perhaps more interesting to radio astronomers are the *compact* and *ultracompact* HII regions. These regions are smaller and denser than classical HII regions; they are still deeply embedded within molecular clouds where massive stars have recently formed, and may still be accreting matter. The dust extinction of the surrounding clouds is quite high in the optical and near-IR; hence these regions can only be studied at radio and far-IR wavelengths. They were first discovered as bright point sources in centimeter-wave continuum surveys of the Galactic plane, made with single-dish telescopes in the 1960s. Subsequent interferometric surveys had much higher angular resolution and led to the discovery of an entire class of smaller and denser HII regions. Regions less than about a parsec in size, and electron densities of several thousand cm^{-3} came to be called compact HII regions, while even smaller, denser regions are called ultracompact (UC) HII regions. An example of such a region, W3(OH), is shown in Figure 6.5. The ionized gas region is slightly less than 2 arc seconds in diameter, corresponding to a linear distance of 0.020 pc at the source distance of 2.1 kpc.

These UC HII regions are one of the earliest observable manifestations of recently formed massive stars. The simplest model is that of a star or stars forming inside a dense core of a molecular cloud. When ionizing photons are produced by the star, they ionize the surrounding gas, forming a very small, dense HII region. As we will see in the following section, this high-pressure UC HII region will quickly expand, reaching sizes and densities that correspond to compact, and eventually, classical HII regions.

In reality, the situation is rather more complicated. Various studies show that a single high-mass star forms about once every 100 years in the Milky Way. If so, then given the expansion rate of the UC HII regions we would only expect to see a few dozen such objects in the entire Galaxy. Nevertheless, studies by Ed Churchwell and Douglas Wood[4] suggested that the Milky Way hosts hundreds of these UC regions. Over the following years, many studies — both observational and theoretical — led to the conclusion that these UC HII regions stay in their very compact state for much longer than the simple expansion model suggests. There are multiple reasons for their longevity, but there are two principal causes. One is that various phenomena act to increase the ambient pressure on the UC HII regions, thus slowing their expansion. The other is that various mechanisms serve to provide a new source of neutral hydrogen that can be ionized by the star.

It is worth noting that radio observations of compact and UC HII regions do not detect the star itself. As we will see in Chapter 7, the direct radio detection of stars is usually rather difficult. Nevertheless, because they form so quickly after the birth of the star, UC HII regions provide a means to study the earliest stages of stellar evolution.

6.3.2 Evolution of HII Regions

The initial formation of the Strömgren sphere is quite rapid. If one imagines "turning on" the UV photon flux from the star instantaneously, then the surrounding gas will become ionized and heated to a temperature of order 10^4 K on very short timescales. Once the initial Strömgren sphere forms around a massive star, this bubble of plasma will be highly over-pressured with respect to the ambient molecular cloud material. Hence, if there are no other factors involved, the HII region will begin to expand. As the gas expands, the density will drop, making recombinations less frequent. Thus, fewer photons will be required to maintain the ionization of the gas, and the extra photons can go towards ionizing additional gas. The time evolution of an HII region, then, is characterized by growth, both in size and in mass

[4] 1989, Astrophysical Journal Supplement Series, vol. 69, p. 831 and 1989, Astrophysical Journal, vol. 340, p. 265.

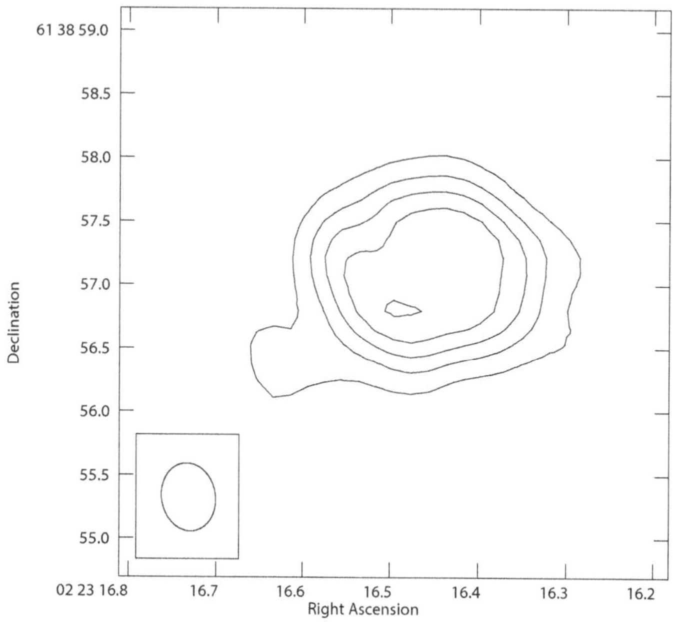

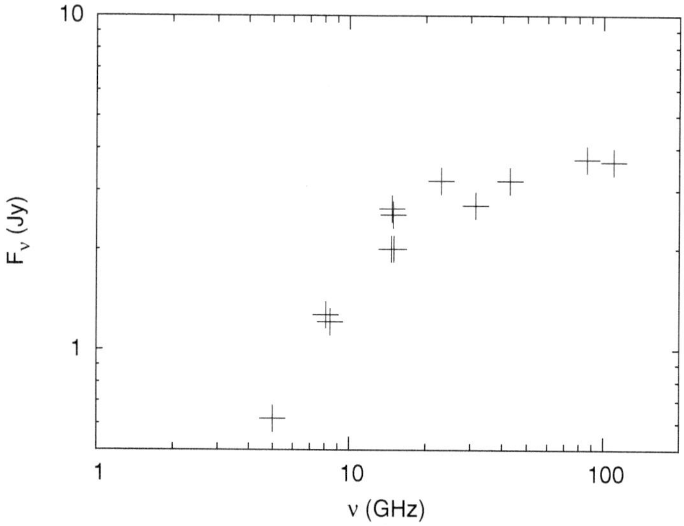

Figure 6.5: Top: A 15 GHz continuum image of the W3(OH) HII region. The synthesized beam size of the observations (about 0.5 arcseconds) is shown in the lower left corner. Bottom: The radio flux density distribution for the W3(OH) HII region. At frequencies below about 15 GHz the region is optically thick, and the spectrum turns optically thin above about 25 GHz.

of ionized gas. The simplest treatment of this expansion is to assume that it will continue until pressure equilibrium with the surrounding cloud is reached. For simplicity, we assume a neutral atomic hydrogen cloud. In this case, the ideal gas law, $P = nkT$, provides a means to evaluate the final state of the expansion. In particular, if the molecular cloud has density n_c and temperature T_c, it will have a pressure $P_c = n_c k T_c$ and the ionized gas will reach pressure equilibrium when $P_i = 2n_e k T_e = n_c k T_c = P_c$. Here n_e and T_e represent the electron density and temperature, respectively, and the factor of 2 arises because the particle density doubles upon ionization. Solving this equality for the equilibrium electron density, we have

$$n_e = n_c \left(\frac{T_c}{2T_e} \right).$$

As shown in Equation 6.12, the Strömgren radius depends on density as $R_S \propto n^{-2/3}$ and so the ratio of the initial and final radii of the expansion is given by

$$r_f = r_i \left(\frac{2T_e}{T_c} \right)^{2/3}.$$

Typical ionized gas temperatures are 10^4 K while cloud temperatures are of order 20 K. Thus, one expects a final radius about 100 times larger than the initial Strömgren radius. If the pressure of the ambient medium is not very high, the time to reach pressure equilibrium will be quite large. In fact, the final, pressure-equilibrium radius may never be reached in the lifetime of the star.

A somewhat more sophisticated treatment of HII region expansion[5] considers that the high-pressure bubble of ionized gas will produce a shock front that propagates outward into the cloud, independently of the ionized gas. The shock front is basically a sound wave, traveling outward at the sound speed of the gas. The edge of the HII region — called the *ionization front* — travels outward at a speed determined by the recombination rate and the availability of Lyman continuum photons. The ionization front initially travels outward at fairly high (super-sonic) speed, without disturbing the gas, in the sense of setting it into motion. But as more gas becomes ionized, more Lyman continuum photons from the star are needed to maintain the ionization against the recombinations. Thus, fewer photons are available to extend the ionization front outward into the neutral gas, and its speed slows, eventually reaching the order of the sound speed in the ionized gas. Meanwhile, the highly over-pressured plasma has begun to expand, sending a shock front into the neutral gas of the cloud. To properly follow the expansion of the HII region, one needs to model the expansions of both the shock front and the ionization front.

The size of the HII region as a function of time is found from the velocity of the advancing ionization front. The result of this analysis is that radius of the HII region as a function of time is given by

$$R(t) = R_S \left(1 + \frac{7}{4} \frac{c_i t}{R_S} \right)^{4/7}, \tag{6.33}$$

where c_i is the sound speed in the ionized gas, typically of order 10 km s^{-1}, and R_S is the initial Strömgren radius. This functional form implies a rapid initial growth, followed by slower, but continued, expansion at later times.

This analytical model provides a reasonable first approximation to the expansion of HII regions. Numerical calculations are required to properly treat the hydrodynamics of the expanding shock wave along with the radiative transfer leading to the ionization front. Such calculations can also include many complicating factors that arise in real HII regions, such as inhomogeneous density distributions, motion of the star through the cloud, the presence of high-velocity stellar winds, and numerous others.

---

[5]For example, Lyman Spitzer Jr., 1978, *Physical Processes in the Interstellar Medium,* Chapter 12, John Wiley & Sons: New York or John Dyson and David Williams *The Physics of the Interstellar Medium,* Chapter 7, Institute of Physics Publishing: Bristol.

6.4 PLANETARY NEBULAE

Massive stars, of about 8 $M_\odot$ or more, are not the only stars to emit large numbers of UV photons. In the final stages of the evolution of intermediate mass stars, when they become white dwarfs, surface temperatures can reach 10^5 K — substantially hotter than young O-stars — and hence the stars emit copious amounts of Lyman continuum photons. During the asymptotic giant phase of their stellar evolution, these stars eject a significant fraction of their mass into the surrounding interstellar medium. Because the stars are bloated at this stage, the pull of gravity on their outer layers is lower; moreover, their cool atmospheres permit the formation of dust grains, which are then pushed outward by radiation pressure. The combined effect can lead to mass loss rates as high as 10^{-4} $M_\odot\,yr^{-1}$. This material is ejected at fairly low velocities, so it does not travel too far before the star passes through the post-AGB phase on its way to becoming a white dwarf. The hot stellar core is exposed during this period, giving rise to the high energy photons that ionize the surrounding ejected material, producing an emission nebula called a *planetary nebula.* Such nebulae were first observed in the 18th century, and their somewhat misleading name dates from that time, because in early telescope observations they resembled planets. The distribution of gas around the star, both in terms of density and chemical composition, can result in complicated and intriguing shapes and emission patterns when the gas is ionized.

Strictly speaking, planetary nebulae (PNe) *are* HII regions, in the sense that both are plasmas, or regions of ionized gas. Because of their very different origins, however, it is customary to reserve the term HII region for plasmas produced by young, massive stars, and use the term PNe for plasmas produced by white dwarf stars illuminating their shroud of ejected material.

Although the origin of planetary nebulae is quite different from that of HII regions, the nebular properties are not so different. Typical PNe electron densities are $10^2 - 10^4$ cm^{-3}, while sizes are a few tenths of a parsec. Although there are morphological differences between the two, these are only evident at very high angular resolution. Many radio continuum observations are insufficient to distinguish between the two types of nebulae.

Perhaps the chief difference in nebular properties is that PNe are hotter than HII regions. The latter are typically in the 6,000 to 12,000 K range, while the former have T_e from about 8,000 to 20,000 K. This is a direct consequence of white dwarfs being hotter than young O stars, hence heating the plasma to a higher temperature, as discussed in Section 6.1. For the same reason, PNe are found to contain a substantial amount of HeII (and more highly ionized states of carbon, oxygen, and nitrogen), that is largely absent from HII regions. The higher stellar temperatures of white dwarfs result in more high-energy photons, many with energies greater than 24.4 eV — the energy required to ionize helium. Massive stars produce very few of these very energetic photons, and hence there is very little HeII in HII regions. Thus, if an RRL from HeII is detected, the object is probably a planetary nebula.

The planetary nebula gas is in a state of fairly rapid expansion with velocities of order 30 km s^{-1}. Due to this expansion — along with the fact that they are not being fed with new material — the density of these nebulae falls fairly rapidly and they become undetectable after only a few tens of thousands of years; astronomically speaking, they are just a flash in the pan. During their short lives though, their optical appearance produces striking astronomical images.

QUESTIONS AND PROBLEMS

1. Estimate the initial Strömgren radius for an HII region produced by an O5 star embedded in a pure hydrogen cloud of density 10^4 cm^{-3}. Assume an electron temperature of 10^4 K. If one-third of the ionizing photons from the star are absorbed by dust within the cloud, what would the Strömgren radius be?

2. In Example 6.1 we assumed that the plasma temperature was 10^4 K. Qualitatively, describe how the Strömgren radius would change if the plasma temperature were 8,000 K? Justify your answers based on physical processes within the nebula. Hint: See Section 6.1.1.

3. Why is α_B used in the ionization-recombination equilibrium calculation instead of α_A? Does this mean that no H atoms recombine to the ground state?

4. Calculate the frequency and Einstein A coefficient of the H42α line.

5. In Section 6.2.2 we gave an approximation for the RRL frequency as

$$\nu_{\rm n} \approx \frac{2\,R\,c\,\Delta n}{n^3}.$$

   This approximation is useful to see the dependence of line frequencies on n and Δn, but it is too imprecise to use when planning observations. Using the H90α line as an example, show that the approximation lacks the necessary precision. Hint: See Section 4.2.2.

6. (a) For an HII region with a temperature of 10^4 K, what thermal linewidth would you expect for hydrogen recombination lines?
   (b) What value would you expect for helium lines?
   (c) We observe the HII region in the H167α recombination line that has a pressure-broadened line width of 25 km s^{-1} and no contribution from turbulent line broadening. What is the density of the HII region?

7. For a given value of the principle quantum number n, the H$n\alpha$ and He$n\alpha$ lines will have different frequencies. Which line will have the higher frequency, the H$n\alpha$ or He$n\alpha$? Demonstrate that regardless of the value of n, the frequency difference between the H$n\alpha$ and the He$n\alpha$ lines — when expressed in velocity units — is 122 km s^{-1}.

8. Show that a sphere that emits with a uniform intensity I_ν has a surface flux given by πI_ν. Hint: See the fundamental relation between flux and integral given in Section 1.2.4.

9. Figure 6.5 shows a 15-GHz image of the ultracompact HII region W3(OH) along with a radio spectrum of the source. Assuming a distance of 2.0 kpc, an electron temperature of 10^4 K, and a uniform, spherical density distribution, estimate
   (a) the electron density,
   (b) the total mass of ionized gas, and
   (c) the spectral type of the ionizing star.

10. The Orion HII region is located at a distance of 420 pc. This HII region is approximately spherical with an angular diameter of 12 arcminutes. The spectrum from 70 MHz to 70 GHz is shown in Figure 6.6.
    (a) What is the linear size of the HII region?

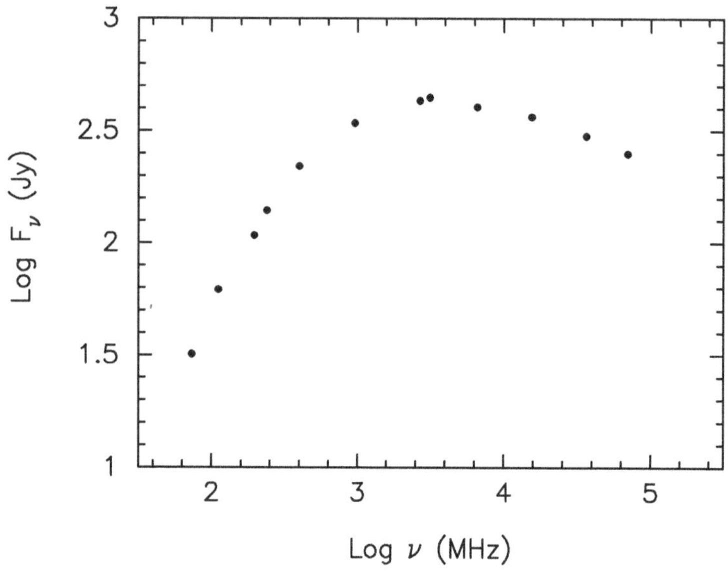

Figure 6.6: Radio spectrum of the Orion HII region; see Problem 10.

(b) What is the solid angle of the HII region?
(c) What is the temperature of the HII region?
(d) Assuming that the HII region is a uniform density sphere, what is the number density of free electrons?
(e) What is the optical depth of the HII region at 100 MHz? At 1.0 GHz? At 10 GHz?

11. In Example 6.1(*a*), we found that the initial Strömgren radius for an HII region produced by an O6 star in 10^4 cm^{-3} gas is 0.183 pc. Assuming a sound speed of 10 km s^{-1}, how long would it take for the HII region to expand to a size ten times larger?

12. In Section 6.3.2 we suggested that an HII region will reach pressure equilibrium after it has expanded by about a factor of 100 from its initial Strömgren radius. How long would this take for the HII region in the previous problem? Is this likely to happen (consider the main sequence lifetime of the exciting star)?

CHAPTER 7

Radio Emission from Stellar Objects

RADIO emission has been detected from stars in various stages of their stellar evolution. The first stellar object detected was our own Sun, and radio observations have provided very important diagnostics of the Sun's outer atmosphere (the chromosphere and corona) and of its magnetic activity. However, only recently have we had the sensitivity to detect other Sun-like stars. Earlier detections of stellar radio emission were either from more evolved stars with very extended atmospheres, from young stars blowing powerful winds, or from stars with much more intense magnetic activity than the Sun (flare stars). We can also detect radio emission from the end products of stellar evolution. Stars more massive than the Sun will end their lives in violent core-collapse supernova explosions, and the ejecta from these explosions can be readily detected by their synchrotron radiation at radio wavelengths. These supernovae explosions leave behind stellar remnants, either neutron stars or black holes. The remnant neutron stars are rapidly rotating and produce one of the most remarkable radio emitting stellar objects — pulsars. In the following, we discuss the radio emission from the Sun, stars and pulsars.

7.1 SOLAR RADIO EMISSION

The Sun has been known to be a source of radio emission since 1942. Its radio emission was first detected as interference with radar being used by England during the Second World War. At first it was thought that the radar was being jammed by the Germans; but James Hey realized that the interfering radio signals were from the Sun. The first detection of solar radio emission occurred at 20 and 70 MHz — the frequencies used for radar at that time — and was non-thermal emission resulting from the Sun's magnetic activity. The Sun's radio emission is rather complicated; it is usually divided into three components: (1) the quiet Sun (when there is no sunspot activity), (2) the slowly varying component (radio emission associated with the sunspot cycle) and (3) radio bursts (due to strong magnetic activity that produces radio emission).

7.1.1 The Quiet Sun

We discuss first the thermal radio emission produced by the Sun at the time of minimum sunspot activity, called the quiet Sun. This emission is primarily free-free emission arising from layers of ionized gas spanning a wide range of electron densities and temperatures.

These layers constitute the Sun's atmosphere, so before we can discuss the radio emission, we need to briefly describe the density and temperature structure of the Sun's atmosphere.

When we observe the Sun in optical light we see emission arising from the Sun's photosphere. The photosphere is essentially the layer when looking into the Sun at which the optical depth reaches unity at optical wavelengths. At optical wavelengths the primary source of opacity is the H^- ion, while at radio wavelengths the primary source is free-free opacity. Although the Sun is entirely gaseous, we often describe the photosphere as the "surface" of the Sun and refer to the temperature at the photosphere as the Sun's surface temperature, which is about 5800 K. The gas at this point produces a nearly perfect blackbody spectrum.

Above the photosphere the gas density falls dramatically, and its temperature decreases slightly; this cooler layer of more tenuous gas produces absorption lines that are seen in the Sun's optical spectrum. The temperature reaches a minimum of about 4400 K at a height of approximately 500 km above the photosphere. However, higher up in the Sun's atmosphere, in layers called the chromosphere and the corona, the temperature begins to rise. The chromosphere extends to about 3000 km above the photosphere where the temperature is about 10,000 K. Above the chromosphere there is a transition zone where the temperature rises sharply to over 1 million K and the electron density drops precipitously; this extended hot region is called the corona. The hot corona was first discovered through radio wavelength observations and was unexpected. The corona extends to several solar radii above the photosphere. The source of heating of the coronal gas is still somewhat of a mystery, but is likely related to the magnetic activity of the Sun, either by magnetohydrodynamic waves or by many small magnetic reconnection events.

Although the gas in the chromosphere and corona is very tenuous, at long radio wavelengths it can have a significant optical depth. This optical depth is almost entirely due to the free electrons, producing free-free absorption (see Section 3.2.4). The free-free optical depth increases with increasing wavelength, so when looking down onto the Sun it becomes opaque (optical depth greater than unity) at greater heights above the photosphere for longer wavelengths. In Figure 7.1 we plot the height above the photosphere where the optical depth is unity as a function of frequency.

Because of the complex temperature and density structure of the Sun, its radio brightness temperature is a strong function of wavelength. The measured radio emission is dominated by the emission at the height in the atmosphere where the optical depth due to free-free absorption is about unity. Figure 7.2 shows the trend of the brightness temperature of the Sun with frequency. At frequencies higher than 10,000 MHz, the detected radio emission arises in the chromosphere, and the brightness temperature varies from about 6000 K at 100,000 MHz to about 10,000 K at a frequency of 20,000 MHz. However, at frequencies lower than 10,000 MHz, the location in the atmosphere where the free-free optical depth is about unity is in the transition zone where the temperature is much higher. Therefore at frequencies lower than about 10,000 MHz, the brightness temperature is greater than 10,000 K. At a frequency of about 100 MHz, the Sun is opaque nearly at the height of the corona and the brightness temperature at these very long wavelengths is about 800,000 K.

Also plotted in Figure 7.1, the dashed line indicates the plasma frequency at each height above the photosphere. As we showed in Section 2.2.1, radiation cannot propagate if its frequency is lower than the plasma frequency. Consulting Figure 7.1, we see that below frequencies of about 80 MHz, as we look down onto the Sun we reach a height where the observing frequency equals the plasma frequency before we reach the height where the optical depth due to free-free absorption is unity. A consequence is that the radiation follows a curved path as it propagates outward from the Sun. As we show in Appendix D, the

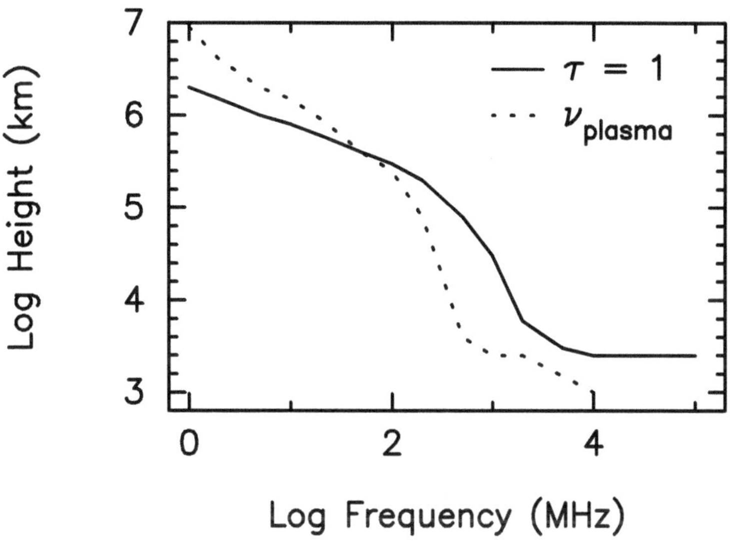

Figure 7.1: The solid line shows the height above the photosphere where the optical depth along the path going in towards the Sun is unity as a function of frequency. The dashed line shows the plasma frequency as a function of height above the photosphere. This figure is based on data in the on-line lecture notes of Dale Gary (https:/web.njit.edu/ gary/728/Lecture10.html).

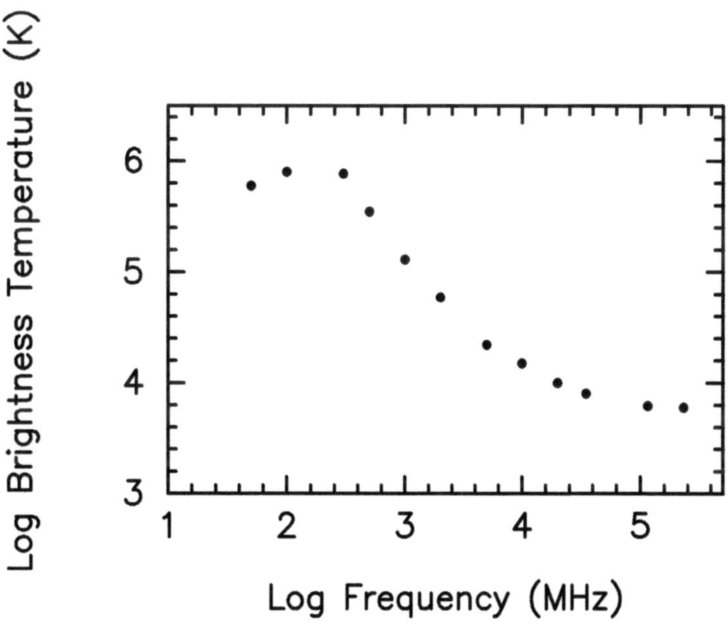

Figure 7.2: The variations of the measured brightness temperature of the Sun at sunspot minimum as a function of wavelength. The figure is based on data from the book by Mukul Kundu (*Solar Radio Astronomy*, 1965, John Wiley & Sons) and the review by E. Fürst (in *Radio Physics of the Sun*, 1980, Dordrecht and Reidel Publishing Co., pg. 25).

presence of free electrons makes the refractive index of the medium a function of frequency; the curved path followed by the radiation (i.e., its refraction) is a manifestation of this effect. This bending can be significant at frequencies below 100 MHz. The trajectories of light rays at these low radio frequencies through the Sun are illustrated in the book by Mukul Kundu[1]. One effect of this bending is that at low frequencies the path through the Sun's atmosphere never has sufficient column density to produce an optical depth of unity or greater. Thus, at these low frequencies the brightness temperature is less than the gas temperature in the corona, which explains why the measured brightness temperatures plotted in Figure 7.2 decrease at frequencies below 100 MHz.

Not only does the brightness temperature of the Sun vary with wavelength, but so does its apparent radius. At a frequency of 500 MHz, the Sun becomes opaque at a height of about 80,000 km above the photosphere; therefore the Sun will have a radius 80,000 km larger than its photospheric radius of about 700,000 km or about 10% larger. At even lower frequencies, the Sun can have a radius as much as twice its photospheric radius. Also, due to the temperature gradient in the Sun's atmosphere, at frequencies lower than about 3000 MHz (or about 10 cm wavelength), the Sun will be limb brightened, which means that the intensity or brightness temperature at the edge of the Sun is greater than that at the center and is a direct consequence of the temperature profile in the Sun's chromosphere and corona.

Since the plasma frequency is a function of the electron density and the free-free optical depth is a function of both the electron density and the electron temperature, radio wave propagation through the solar atmosphere is highly dependent on the Sun's electron density and temperature structure. Both of these parameters vary by three orders of magnitude or more, at different heights in the solar atmosphere, which has profound effects on the radiative transfer at radio wavelengths.

7.1.2 Slowly Varying Component of the Sun

The quiet Sun radio emission is altered when there is increased sunspot activity, which is a result of magnetic activity of the Sun. A sunspot is a region near the photosphere of the Sun where the magnetic field is greatly enhanced — sunspots have magnetic field strengths of 2,000 to 4,000 gauss, much greater than the magnetic field of the quiet Sun of about 1 gauss. The enhanced magnetic field suppresses convection currents in the Sun's interior, reducing the energy transfer to the surface, thereby causing sunspots to be slightly cooler than the surrounding photosphere, so they appear dark at optical wavelengths. However, in sunspot regions, the electron density at heights of 10,000 to 100,000 km above the Sun's surface can be an order of magnitude higher than in regions outside the sunspots or at times of the quiet Sun. At wavelengths longer than a few centimeters, the higher electron density produces a larger free-free optical depth, and the height at which the optical depth is unity can be much higher above the photosphere than normal and thus at locations where the temperature is much higher. Above a sunspot, the brightness temperature of the radio emission at wavelengths of 10 to 20 cm can be as large as 10^6 K. During sunspot maximum, regions above sunspots can contribute more to the total radio flux density of the Sun than the much larger area of the Sun outside the sunspots.

The flux density of the Sun at a wavelength of 10.7 cm is regularly monitored to follow the sunspot activity. Figure 7.3 shows the monthly averaged 10.7-cm flux density of the Sun over the past 60 years; the most obvious feature of this plot is the 11-year sunspot cycle. The flux density of the Sun at centimeter wavelengths is greatest at sunspot maximum and smallest at sunspot minimum. Note that the minimum flux density of the Sun is typically about 6×10^{-18} erg s^{-1} cm^{-2} Hz^{-1} or 60 solar flux units (a solar flux unit is 10^4 Jy).

[1] *Solar Radio Astronomy*, 1965, John Wiley & Sons.

This flux density corresponds to a disk-averaged brightness temperature of about 37,000 K, in agreement with what would be predicted for the quiet Sun shown in Figure 7.2. The flux density at sunspot maximum is usually about 200 solar flux units, corresponding to a solar-disk-average brightness temperature of about 123,000 K, although the precise value varies between sunspot cycles.

Example 7.1:

Using the solar flux densities mentioned above, we see that Sun's flux density, and hence its disk-averaged brightness temperature, is about three times larger during solar maximum than during solar minimum. The brightness temperature above sunspots, measured at 10.7 cm, is about 10^6 K. Estimate the fraction of the solar surface covered by sunspots during solar maximum.

Answer:
Let f represent the fraction of the Sun's surface covered by sunspots and which presents a brightness temperature of 10^6 K. Thus $1 - f$ is the fraction of the surface without sunspots, and presents a brightness temperature of 37,000 K. We can write the following equality:

$$(f)10^6 \text{ K} + (1 - f)37,000 \text{ K} \approx 123,000 \text{ K}.$$

Solving, we find that $f = 0.089$; thus about 9% of the Sun's surface is covered by sunspots during solar maximum.

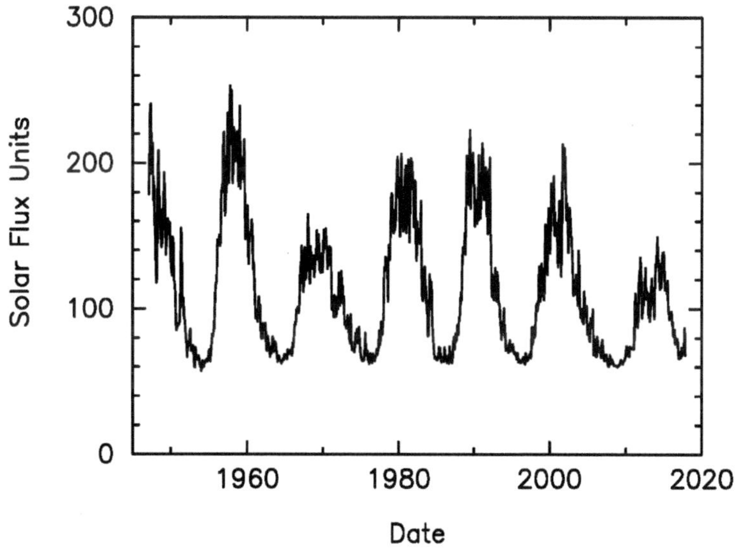

Figure 7.3: The measured flux density of the Sun at a wavelength of 10.7 cm as a function of time. Each point represents the monthly average flux density in solar flux units. One solar flux unit is equivalent to 10^4 Jy. The data are from the solar weather monitoring website of the National Research Council of Canada. See www.spaceweather.gc.ca

7.1.3 Radio Bursts

It was the occurrence of a radio burst at meter wavelengths associated with a large sunspot group that caused the jamming of the English radar in 1942 that we alluded to in Section 7.1. The radio burst emission follows the appearance of solar flares at optical wavelengths. These radio bursts occur over a range of wavelengths from centimeters to decameters and vary in duration from a few minutes to nearly an hour. In solar flares, which are a result of the intense magnetic fields in sunspots, magnetic reconnection can cause a sudden re-arrangement of the magnetic field lines into a lower energy configuration. Magnetic energy is released in the process and transferred to the surrounding plasma, heating the gas.

The release of magnetic energy can also accelerate electrons to very high speeds and the interaction with the intense magnetic fields produces non-thermal radiation. When interacting with a magnetic field, non-relativistic electrons can produce cyclotron (or gyromagnetic) radiation, mildly relativistic electrons produce what is called gyrosynchrotron radiation, while highly relativistic electrons can produce synchrotron radiation (see Section 3.3). In addition to these non-thermal emission mechanisms, the radio emission can also be produced by a coherent process called plasma radiation, where the electrons act in unison. In this case the electrons accelerated in a flare interact with the surrounding plasma driving charge oscillations, or plasma waves, called Langmuir waves. The plasma oscillations, under the right circumstances, can excite electromagnetic waves that can escape from the plasma. The electromagnetic waves are produced at the plasma frequency or at harmonics of this frequency. This mechanism can produce extremely high brightness temperature emission (even greater than that produced by the synchrotron mechanism) and is the dominant emission mechanism at long radio wavelengths. The details of the radio emission from the Sun during these radio bursts are complex, and we refer the reader to *Solar Radio Astronomy* by Mukul Kandu[2], or to *Solar Radiophysics* by McLean and Labrum[3], and to the review article by George Dulk[4] for more details.

7.2 RADIO EMISSION FROM STARS

Until recently, Sun-like stars were not detectable at radio wavelengths. With the improved sensitivity of telescopes such as the Jansky Very Large Array (JVLA) and the Atacama Large Millimeter/submillimeter Array (ALMA), though, the study of stars at radio wavelengths is now feasible. We start with a discussion of the expected radio flux density due to the thermal emission from stars assuming that they have no magnetic activity, similar to the emission of the quiet Sun.

7.2.1 Thermal Radio Emission

We can approximate the thermal radio emission from stars as blackbody emission and at radio wavelengths and stellar temperatures we can safely use the Rayleigh-Jeans approximation for the Planck function (see Section 3.2.2). The flux density of a star (see Equation 1.9) can be approximated as the product of its intensity and the solid angle it subtends, where the solid angle is given by Equation 1.7. A star's flux density is

$$F_\nu \;=\; I_\nu\,\Omega \;=\; \frac{2kT}{\lambda^2}\,\frac{\pi R^2}{d^2},$$

[2]1965, John Wiley & Sons.

[3]1985, Cambridge University Press.

[4]1985, Annual Reviews of Astronomy and Astrophysics, vol. 23, p. 169.

where T is the gas temperature at the radius, R, where the line of sight into the star becomes optically thick at the observing wavelength, λ, and d is the distance of the star. We can rewrite this expression using the properties of the emission of the quiet Sun at 1 cm wavelength as a reference. At 1 cm wavelength, the detected radio emission from the Sun originates slightly above the photosphere at a radius of about 7×10^{10} cm, where the temperature is about 9,000 K. Therefore we can rewrite the 1-cm flux density of a star, using the Sun for scaling, as

$$F_\nu = 400\,\mu\text{Jy}\left(\frac{T}{9,000\,\text{K}}\right)\left(\frac{1\,\text{cm}}{\lambda}\right)^2\left(\frac{R}{7\times10^{10}\,\text{cm}}\right)^2\left(\frac{1\,\text{pc}}{d}\right)^2. \tag{7.1}$$

7.2.1.1 Main Sequence Stars

With the JVLA, at a wavelength of 1 cm, it is possible to detect sources with flux densities as low as 10 μJy. Despite this amazing sensitivity, using Equation 7.1 we see that stars like the quiet Sun can only be detected out to a distance of a little over 6 pc. Several solar-type stars within 5 pc of the Sun have been detected with the JVLA. For example, the stars Tau Cet, Eta Cas A, and 40 Eri A were detected by Jackie Villadsen and colleagues[5] at a frequency of 34.5 GHz or a wavelength of 0.87 cm, with flux densities ranging from 16 to 25 μJy. The flux densities they measured for these stars correspond to disk averaged brightness temperatures of about 10,000 K, similar to the brightness temperature of the Sun at this wavelength.

Main sequence stars of later spectral types are expected to have smaller radii and lower surface temperatures than the Sun, and thus the detection of the thermal emission from such stars would be very difficult unless they were unusually close. On the other hand, main sequence stars of earlier spectral types (O, B, A, and F stars) with larger radii and higher surface temperatures are more easily detected.

Example 7.2:

A spectral type B0 main sequence star has a radius about 8 times larger than the Sun and a surface temperature of about 20,000 K. Assume this B0 star has a brightness temperature at 1 cm of 20,000 K. How far away could we detect such a star at a wavelength of 1 cm if the detection threshold was a flux density of 10 μJy?

Answer:
We can use Equation 7.1 and solve for the distance to obtain

$$d^2 = \left(\frac{T}{9,000\text{ K}}\right)\left(\frac{1\text{ cm}}{\lambda}\right)^2\left(\frac{R}{7\times10^{10}\text{ cm}}\right)^2\left(\frac{400\mu\text{Jy}}{F_\nu}\right)\text{pc}^2.$$

For the B0 star we find

$$d = \left(\frac{20,000\text{ K}}{9,000\text{ K}}\right)^{\frac{1}{2}}\left(\frac{1\text{ cm}}{1\text{ cm}}\right)\left(8\right)\left(\frac{400\ \mu\text{Jy}}{10\ \mu\text{Jy}}\right)^{\frac{1}{2}}\text{pc} = 75.4\text{ pc}.$$

Therefore the distance to which we could detect such a B0 star is about 75 pc.

---

[5] 2014, Astrophysical Journal, vol. 788, p. 112.

The flux density of stars will be much larger at higher radio frequencies, because $F_\nu \propto \nu^2$ on the Rayleigh-Jeans side of the black-body spectrum. The ALMA observing bands range from 8.6 to 0.32 mm. Although sensitivity is lower at shorter wavelengths, at a wavelength of 0.87 cm ALMA can easily detect flux densities as low as 50 μJy — sufficient to detect solar-type stars at distances up to about 100 pc. Alpha Centauri A, a star similar to the Sun and at a distance of only 1.3 pc, was easily detected by Rene Liseau and colleagues[6] with ALMA at wavelengths of 0.44, 0.87 and 3.1 mm. The brightness temperature inferred from these measurements was slightly higher than the photospheric temperature of the star as would be expected if Alpha Centauri A has a similar atmosphere to the Sun. The companion to Alpha Centauri A, Alpha Centauri B, is a K main sequence star and this star was also detected in these ALMA observations.

Can we measure the sizes of stars like our Sun at radio wavelengths? The closest Sun-like star, Alpha Centauri A, has an angular diameter (see Section 1.2.4) of only 3.5×10^{-8} radians or 7.2 mas (milli-arcsecond). This would require a resolution just beyond the capabilities of current radio interferometers, such as the JVLA and ALMA. Higher angular resolution can be achieved with very long baseline interferometry. However, to obtain detections, such techniques require sources with much higher brightness temperatures than the thermal emission from the photospheres of stars. Thus, the angular sizes of stars like our Sun cannot currently be measured at radio wavelengths. More massive main sequence stars have larger angular diameters; however these stars are generally much more distant, so they have even smaller angular sizes, and are also impossible to measure.

7.2.1.2 Giant and Supergiant Stars

As stars end their main sequence phase (core hydrogen burning phase) they evolve into giant and supergiant stars. Although, as giant and supergiant stars they usually have slightly lower photospheric temperatures, they are very much larger than their main sequence size. For a star like our Sun, we believe it will expand into a red giant star with a radius about 100 times larger than present and have a photospheric temperature of about 3500 K. Using Equation 7.1, if the detection threshold is a flux density of 10 μJy at a wavelength of 1 cm, then the Sun as a red giant star could be detected out to a distance of 400 pc. The nearest red giant star is only about 30 pc away, and there are hundreds of red giant stars that could be readily detectable at radio wavelengths.

Stars much more massive than the Sun evolve into supergiant stars. Some of the largest stars are red supergiant stars and have radii one thousand times larger than the Sun. These stars have similar photospheric temperatures as red giant stars and could be detected out to a distance of 4 kpc. One of the nearest examples of a red supergiant star is Betelgeuse at a distance of about 222 pc and with a photospheric temperature of about 3600 K and radius of 1044 $R_\odot$. Using Equation 7.1, we would predict a flux density for Betelgeuse at a wavelength of 7 mm of about 7 mJy. Although such stars are rare, the thermal radio emission from red supergiant stars should be easy to detect. Betelgeuse was observed by Jeremy Lim and collaborators[7] at a wavelength of 7 mm to have a flux density of 28 mJy. Note that this flux density is four times larger than our prediction.

Since these supergiant stars have such large diameters, if they are relatively nearby they may have resolvable angular diameters. For example, at optical wavelengths Betelgeuse has an angular diameter of 44 milli-arcseconds (mas). As stars go, this is a very large angular

[6]2015, Astronomy and Astrophysics, vol. 573, p. L4.
[7]1998, Nature, vol. 392, p. 575.

diameter, but this star is only barely resolved with interferometers such as the JVLA or ALMA. The paper by Jeremy Lim and collaborators also measured the angular diameter of Betelgeuse at a wavelength of 0.7 cm and measured a size about twice that of its optical diameter. Thus, at radio wavelengths, we are observing thermal emission from an extended atmosphere. The inferred brightness temperature at 0.7 cm was similar to the photospheric temperature, but the larger size explains why the measured flux density was much larger than predicted. Lim and colleagues observed at 1.3, 2, 3.6 and 6 cm wavelengths, and at these longer wavelengths the size of Betelgeuse was even larger, but the brightness temperature decreased. Unlike the Sun, the temperature higher up in Betelgeuse's atmosphere is lower than the temperature at its photosphere.

7.2.2 Winds from Asymptotic Giant Branch Stars

Solar-type stars, very late in their lives, become asymptotic giant branch stars or AGB stars. AGB stars are often long period variable stars and have significant mass loss in the form of dense stellar winds. These stars are shedding their outer envelope, evolving toward planetary nebulae. The cool, dense winds from these stars make an ideal environment for the formation of both dust and molecules. In fact, we believe AGB star winds are a significant source for the dust found in the interstellar medium.

The spectral-line emission from molecules and the continuum emission from dust in the expanding envelopes surrounding these stars can be readily detected at millimeter wavelengths. One of the primary molecules formed in these winds is carbon monoxide. In Figure 7.4 we show a spectrum of the CO $J = 1-0$ emission line toward the star NSV 11225 (also called TMSS +20370). Observations of molecular line emission provide important information concerning the kinematics and mass loss in these stars' winds. The solid line in Figure 7.4 is a fit assuming the star has a spherically symmetric expanding envelope in which the temperature is decreasing outward and assuming that the CO emission is optically thick. From this fit, the expansion velocity for the envelope of NSV 11225 was determined to be 15 km s^{-1} and the mass loss rate to be about 10^{-7} M$_\odot$ yr^{-1}. Such stars evolve into planetary nebulae in tens of millions of years.

As in molecular clouds, the most abundant molecule in AGB envelopes after H_2 is CO. Although most of the molecular species detected in AGB envelopes are also detected in molecular clouds (see Section 5.2), some molecules found in AGB envelopes, such as KCl, NaCl, AlCl, AlF, NaCN, and MgCN, are nearly unique to these envelopes. Clearly the chemistry in these stellar envelopes is of interest and is slightly different than the chemistry occurring in molecular clouds. Finally, we note that some AGB stars have OH, SiO or H_2O maser emission (see Section 4.2.3) from their envelopes. The brightness temperature of the maser emission is sufficient for VLBI techniques, and such high-resolution images of these maser emission regions provide a powerful tool for inferring the size and kinematics of the extended envelopes in these stars.

Toward the end of a star's AGB phase, they become so obscured by dust at visible wavelengths that they can only be studied at infrared and radio wavelengths. This final stage represents a short-lived transition from the AGB phase to a planetary nebula phase, and is referred to as a protoplanetary nebulae (PPN). As in their AGB phase, protoplanetary nebulae can be studied at radio wavelengths via both the dust continuum emission and molecular line emission from their very extended envelopes.

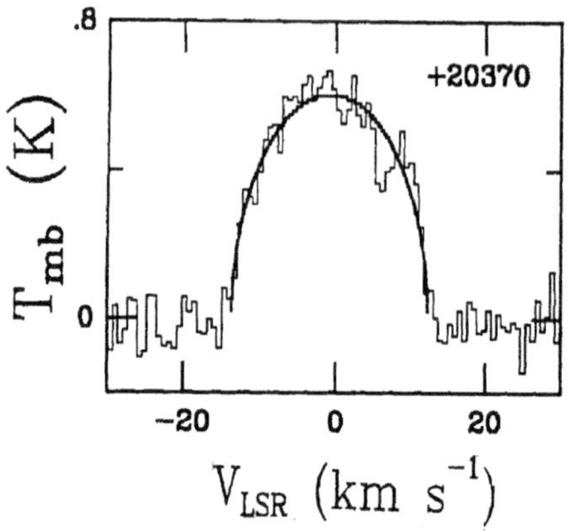

Figure 7.4: The spectrum of the $J = 1 - 0$ transition of CO toward the star NSV 11225 (also called TMSS +20370). The x-axis is the local standard of rest velocity and the y-axis is main-beam antenna temperature (the antenna temperature corrected for telescope losses). This spectrum was obtained with the Five College Observatory 14-m telescope by Michael Margulis and collaborators (1990, Astrophysical Journal, vol. 361, p. 673). The solid line shows a model fit to the spectrum.

7.2.3 Flare Stars

Flare stars were first detected in the early part of the 20th century by their sudden brightening at visible wavelengths. These stars can brighten by a factor of a few to greater than ten, and the brightening can last for a few minutes to nearly an hour. Nearly all flare stars are spectral-type M main sequence stars, also called dwarf M stars (or dM stars). Most of these flare stars also have hydrogen emission lines in their optical spectrum; these emission-line dwarf M stars are denoted as dMe stars. Therefore the terms 'flare stars' and 'dMe stars' are often used interchangeably. Many of these flare stars are also X-ray sources, even when not flaring, indicating that these stars have very hot coronae. The coronae in flare stars are likely hotter and denser than the Sun's corona.

Flare stars have intense magnetic fields (up to thousands of gauss) covering much of their surface with magnetic activity that produces the hot corona and the flaring events. The intense magnetic activity that produces radio bursts from the Sun is likely similar to the mechanism responsible for producing the visible flaring of these stars. Radio flares from these stars were detected in the 1960s and typically have durations of only minutes. The radio flares on these stars are scaled-up versions of the radio bursts seen in the Sun, as their radio flares can produce a thousand times more radio power than solar radio bursts. The flare emission can have both broad-band and narrow-band emission, and the inferred brightness temperatures of the radio flares often exceed 10^9 K. As with solar radio bursts, the radio emission is produced by a combination of non-thermal processes. Despite their high brightness temperatures, these flare stars are only detectable out to distances of about 20 pc because of their small sizes.

Radio emission can also be detected from some flare stars even when they are not flaring. This quiescent radio emission varies very slowly with time, but unlike the slowly varying component of the Sun, it is also almost certainly non-thermal in origin. To produce this non-thermal radio emission a source of energetic electrons is needed and in many cases with energies greater than 100 keV. Since these electrons lose energy rapidly, it is not clear how a population of energetic electrons can be maintained in these flare stars. One idea is that these flare stars are continually undergoing microflares, which accelerate and maintain a population of high energy electrons in these stars' coronae.

7.3 YOUNG STARS

Young stars, recently formed or still forming, frequently present radio emission. Although such emission can arise in the magnetosphere of the star itself, much more common is radio emission from the circumstellar material or the interaction of the star with its immediate surroundings. Here, we describe two common manifestations of such emission: disks and jets.

7.3.1 Proto-stellar Disks

In general terms, stars form by the gravitational collapse of molecular cloud material within dense cloud cores (see Section 5.2.3). This material, falling into the potential energy well of the proto-stellar object, carries with it both gravitational energy and angular momentum. The collapsing motion is a combination of radial infall and orbital rotation about the young stellar object (YSO). Because the collapse motion is inhibited radially by the angular momentum of the material, but not inhibited vertically, these structures become flattened, circular disks. As a consequence, an accretion disk is formed around the YSO. As the infalling material approaches the proto-star, it first falls onto the disk, and then is gradually accreted onto the YSO, building up the stellar mass as accretion proceeds. The remnants of the accretion disk can later form a planetary system around the young star.

These accretion disks are ubiquitous in the formation of low-mass stars of a few solar masses or less; they are also thought to occur during the formation of more massive stars, although the data are more scarce and less compelling in the latter case. Typical diameters for low-mass proto-stellar disks are a few hundred astronomical units (AU) and their thickness is generally a tenth (or less) of their diameter.

The disk is primarily composed of gas, both molecular and atomic, and dust grains. The nature of these two components — their composition, size, and temperature — is key to understanding the radio emission from the disks. Dust grains within the disk are heated to temperatures of several hundred kelvins, and hence are strong radiators of continuum emission (see Section 5.3). Molecules, which are abundant in this high-density environment, present emission in both low-level and highly-excited rotational states. Both continuum and spectral line emissions are important for determining the disk properties.

Continuum emission from the disk depends critically on the temperature of the dust grains. Unfortunately, the precise temperature structure of the disk is rather difficult to predict; it depends on many factors, including the density and composition of the disk, the size distribution of the dust grains, the radiation field of the star and its radiative transfer, and even the fluid dynamics within the disk material. The main sources of heating are the stellar radiation and friction in the accretion disk. Both are more intense closer to the star; hence the inner parts of the disk tend to be substantially hotter than the outer parts. Thus the radio brightness of the disk will be a maximum near the center, and fall off towards the outer parts. Although a range of temperatures is present, a typical average temperature is on the order of a few hundred kelvins. Consulting Figure 3.3, we see that the peak of the emission lies in the infrared, but there is still significant emission at millimeter wavelengths. An example of such emission from the disk surrounding HL Tau, at $\lambda = 1.3$ mm, is shown in Figure 7.5. HL Tau is a T-Tauri star still actively accreting material from its surrounding disk with an accretion rate, found by Nuria Calvet and collaborators[8], of about 4×10^{-6} $M_\odot$ per year.

Figure 7.5: ALMA observations of the $\lambda = 1.3$ mm continuum emission from the disk surrounding HL Tau. The solar-type star HL Tau is about 140 pc distant from Earth; its disk is inclined at 46° with respect to the plane of the sky, and thus internal structure within the disk is readily visible at the 0.028 arcsecond resolution of this image. The angular diameter of the disk is about 1.5 arcseconds corresponding to a physical diameter of about 200 AU. Image courtesy of ALMA(ESO/NAOJ/NRAO); C. Brogan, B. Saxton (NRAO/AUI/NSF).

Information about the disk structure can be inferred from multi-frequency continuum observations and modelling of the spectral energy distribution of this emission. Spectral-line observations of the gaseous component of the disk make a powerful complement to the continuum observations. There are a wealth of molecular species present in the gas phase of the disk material. The relatively warm environment and presence of a strong infrared radiation field provide excellent conditions for astro-chemical reactions. As described in Chapter 4, spectral-line observations can provide information about the density, temperature, and composition of the molecular gas, but most crucially, they provide information on the disk rotation.

Measurements of the disk kinematics provide valuable information on the mass of the disk and the central young star. For example, in HL Tau, observations of the emission from the $J = 2 - 1$ rotational transitions of ^{13}CO and $C^{18}O$ were made by Hsi-Wei Yen and collaborators[9] using ALMA. They found disk kinematics consistent with Keplerian

[8]1994, Astrophysical Journal, vol. 434, p. 330.

[9]2017, Astronomy and Astrophysics, vol. 608, p. 134.

rotation, implying that the mass of the system is dominated by the central star and the mass of the disk is unimportant in the disk rotation. They measured a disk rotation speed (after correction for the disk inclination) given roughly by

$$V_{\rm rot} \sim 2.4 \text{ km s}^{-1} \left(\frac{R}{140 \text{ AU}}\right)^{-1/2},$$

where $V_{\rm rot}$ is the rotational velocity at radius R.

Example 7.3:

Using the kinematic information for the HL Tau disk and assuming Keplerian rotation, estimate the mass of HL Tau.

Answer:
The rotation speed at 140 AU is 2.4 km s^{-1} or 2.4×10^5 cm s^{-1}. For Keplerian motion, we assume that the centrifugal force of gas in the disk is supplied by the gravitational attraction to the central mass, thus

$$\frac{V_{\rm rot}^2}{R} = \frac{GM}{R^2}.$$

Therefore

$$M = \frac{V_{\rm rot}^2 R}{G} = \frac{(2.4 \times 10^5 \text{ cm s}^{-1})^2 \,(140 \text{ AU})\,(1.5 \times 10^{13} \text{ cm AU}^{-1})}{6.67 \times 10^{-8} \text{ cm}^3 \text{ g}^{-1} \text{ s}^{-2}} = 1.8 \times 10^{33} \text{ g}.$$

Thus, the mass of HL Tau is 0.9 $M_\odot$.

In the case of HL Tau, since the rotation is Keplerian, the disk mass must be much smaller than the mass of the central star. Although HL Tau is still accreting material and its mass may grow somewhat larger with time, it will ultimately form a star much like our Sun.

7.3.2 Thermal Radio Jets

A well-known issue in the process of star formation is the loss of angular momentum. A cloud that collapses to form stars has an angular momentum per unit mass, called the specific angular momentum, of order $10^{21} - 10^{23}$ cm^2 s^{-1}. The newly formed stars, however, have specific angular momenta of order 10^{18} cm^2 s^{-1}. Clearly, some mechanism must exist to carry away a large fraction of the angular momentum. Collimated stellar winds are a likely means by which the forming star sheds angular momentum. Although the engine to drive these winds is still not well-understood, it likely involves the interaction of the rotating disk and the stellar magnetic field. The result is supersonic flows of partially ionized gas that are ejected from YSOs. These streams of outflowing plasma are most commonly observed at centimeter wavelengths by their thermal bremsstrahlung emission (see Section 3.2.4). Non-thermal jets have also been discovered in some massive YSOs. Their spectra are seen to decline with increasing frequency and linear polarization has been detected, two characteristics of synchrotron radiation (see Section 3.3). The means to accelerate electrons to the needed relativistic speeds is still not well-understood, but may be related to shocks where the thermal jets impact the ambient medium. Here, we focus on the much better understood phenomenon of thermal radio jets.

The velocity of the outflowing gas in these jets is typically several hundred kilometers per second and their linear extent can be as large as a parsec. These jets are often traced by their optical and near-infrared emission that result from internal shocks that excite a variety of atomic and molecular emission lines. At radio wavelengths, the emission is primarily thermal bremsstrahlung emission from the ionized component of these jets. The spatial extent of the thermal radio jets is much smaller than the optical jets, typically only a few tens to a few thousands of AU. The jets can be highly collimated or have quite wide opening angles. There is some evidence that the degree of collimation is correlated with the jet age, but this result has yet to be firmly established. In Figure 7.6 we show the thermal radio jets associated with the young binary star L1551 IRS5, in which the radio jets are relatively well-collimated.

These thermal jets present the interesting property that both their flux density and angular size are frequency dependent. Theoretical considerations show that for an isothermal jet with constant velocity and constant ionization fraction, the flux density is given by

$$F_\nu \quad \propto \quad \nu^{1.3-0.7/\epsilon},$$

and the angular size is given by

$$\Theta_\nu \quad \propto \quad \nu^{-0.7/\epsilon},$$

where ϵ is the power index giving the jet half-width (w) as a function of distance from the driving source, $w \propto r^\epsilon$. Spherically symmetric flows, along with conical flows of uniform opening angle, will have $\epsilon \sim 1$, so that $F_\nu \propto \nu^{+0.6}$ and $\Theta_\nu \propto \nu^{-0.7}$. High angular resolution, multi-frequency observations confirm these relations, as illustrated in Figure 7.7. The dependence of jet size on frequency can be understood in terms of the optical depth of the free-free emission. Essentially, one looks into the jet until the optical depth is of order unity. At higher frequencies, the $\tau = 1$ condition is met closer to the jet axis (where the density is higher), and so the observed size decreases with increasing frequency. The topic of protostellar jets has been recently reviewed by Guillem Anglada and collaborators[10].

7.3.3 Molecular Outflows

Young stellar objects are surrounded by their parental molecular gas; the properties of these molecular clouds were discussed in Section 5.2.1. The collimated winds from young stars collide with this surrounding cloud material and can accelerate some of this gas to higher velocities. Since the primary tracer of the molecular gas is emission from CO, the spectra of CO were the first to show evidence for this perturbation. Due to the Doppler effect (see Section 1.4), the gas accelerated by these winds shows up as redshifted and blueshifted emission. The presence of oppositely directed collimated flows of molecular gas, called bipolar molecular outflows, were first detected by Ronald Snell, Robert Loren and Richard Plambeck[11] associated with the young star L1551 IRS5. Spectra of the CO emission from the L1551 IRS5 bipolar molecular outflow are shown in Figure 7.8. These spectra show emission from both blueshifted and redshifted gas. The distribution of the blueshifted and redshifted CO emission in L1551 IRS5 is shown in the upper-right panel in Figure 7.6. In the L1551 outflow, one collimated stream of molecular gas is moving toward us and to the south-west, while the opposite stream is moving away from us and to the north-east. As can be seen from the spectra in Figure 7.8, the molecular gas is accelerated to a velocity of only about 5 to 10 km s^{-1} relative to the undisturbed gas, although some correction may

[10]2018, Astronomy and Astrophysics Reviews, vol. 26, p. 3.
[11]1980, Astrophysical Journal Letters, vol. 239, p. L17.

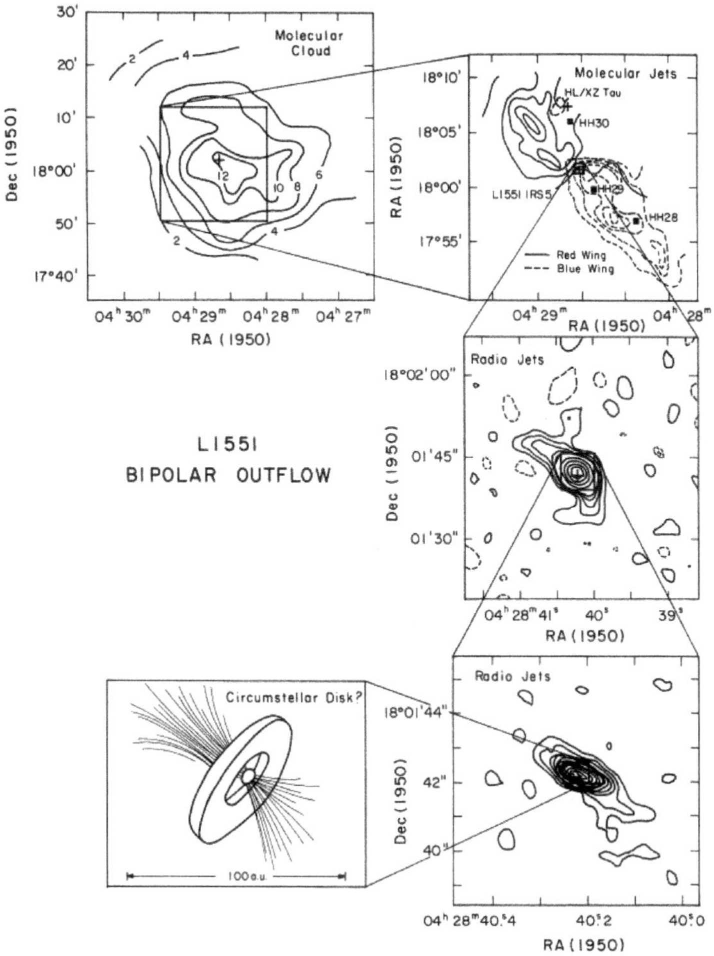

Figure 7.6: The outflow phenomenon occurs on multiple size scales and in the form of both molecular and ionized gas. Here we illustrate the phenomenon for the YSO L1551 IRS5. The lower-left panel shows a sketch of the disk surrounding the YSO. The lower-right panel shows the radio emission obtained with the VLA from the thermal radio jets; these jets have an extent of a few hundred AU. The middle panel shows a more sensitive, but lower resolution, VLA image of the radio jets tracing their emission out to 1500 AU. The upper-right panel shows the associated molecular outflow. In this image the integrated intensity of the CO emission for gas moving toward us (blueshifted) is shown in dashed contours and the integrated intensity of the CO emission for gas moving away (redshifted) is shown in solid contours. Finally, the upper-left panel shows the CO emission from the surrounding molecular cloud. This figure is from the paper by Ronald Snell and colleagues (1985, Astrophysical Journal, vol. 290, p. 587).

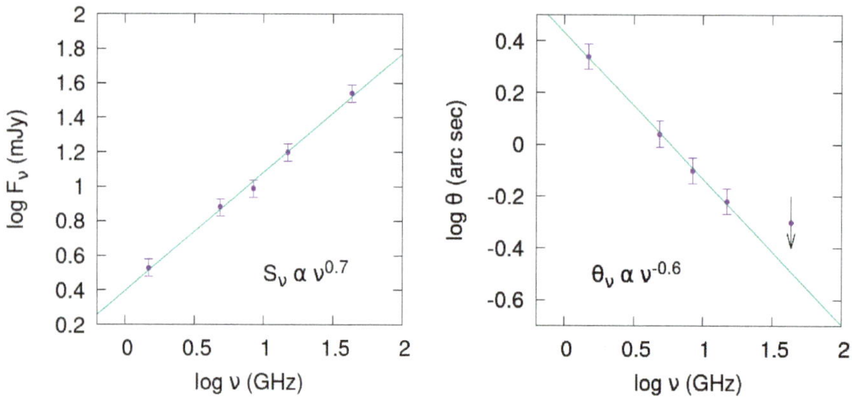

Figure 7.7: Multi-frequency Very Large Array observations of the HW2 thermal jet show that both the flux density of the jet and its angular size follow power-law relations that are consistent with $\epsilon = 1.2 \pm 0.2$. Image courtesy of Joel Hartsough, based on data from the paper by L. F. Rodríguez and collaborators (1994, Astrophysical Journal, vol. 427, p. L103).

be needed for the inclination of the molecular outflow. The molecular gas is moving much slower than the wind material. For the L1551 molecular outflow the spatial extent is about one parsec and the total mass of accelerated gas is of order 6 $M_\odot$.

These molecular outflows provided some of the first evidence that YSOs were producing high speed collimated winds. Today we know that these molecular outflows are ubiquitous and are associated with newly forming stars of all masses. In regions where dust obscuration prevents the detection of the optical or infrared emission from the shocked winds, these molecular outflows are the only means for determining that such activity is occurring. The molecular outflows are also detected in a variety of molecules in addition to CO, and the abundance of these other molecular species provides good evidence that the chemistry in this outflow gas has been altered by the collision with the wind material.

Molecular outflows provide valuable information concerning the energetics of winds from these young stars. The total momentum in the outflowing molecular gas in L1551 IRS5 was measured by Stojimirovic and collaborators[12] to be at least 20 $M_\odot$ km s^{-1} and the total energy to be at least 1.5×10^{45} erg. Since the total gravitational binding energy of the surrounding molecular cloud is of order 2×10^{45} erg, such an outflow has the potential to disperse much of the surrounding parental cloud gas. Assuming that momentum is conserved in the interaction between the stellar wind and the surrounding cloud gas, estimates of an outflow age along with the wind velocity imply that the mass loss rate in the stellar wind is of order 10^{-6} $M_\odot$ yr^{-1}. This rate is somewhat smaller than the rate of disk accretion; however it is still a significant fraction of the accreting matter is ejected in winds, and these winds are responsible for carrying away excess angular momentum from the disk.

7.4 RADIO PULSARS

In stars more massive than about 8 $M_\odot$, in their post-main sequence evolution they can fuse heavier and heavier elements, building up an iron core. When this iron core mass exceeds the Chandrasekhar mass limit of about 1.4 $M_\odot$ (this mass limit is slightly dependent on the composition of the core), the mass of the core can no longer be supported by electron

[12] 2006, Astrophysical Journal, vol. 649, p. 280.

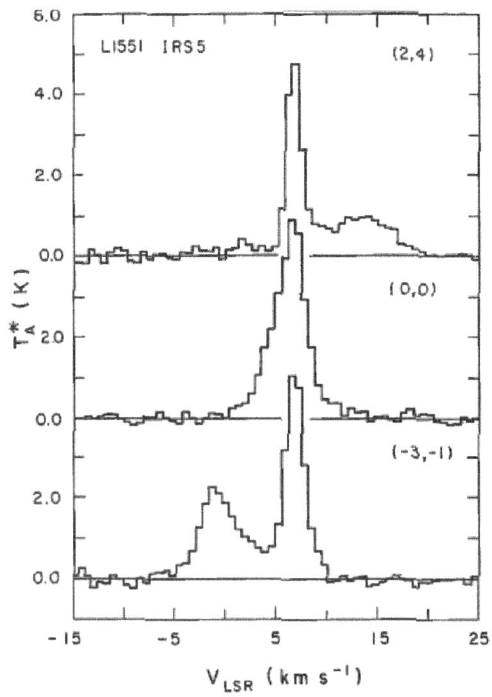

Figure 7.8: CO spectra taken in three locations in the bipolar molecular outflow from L1551 IRS5. The top spectrum is towards a position to the north-east of the central source and shows redshifted emission. The middle spectrum is toward the central source and shows primarily the unperturbed gas, although the line wings are evidence for both blueshifted and redshifted emission. The bottom spectrum is towards a position to the south-west of the central source and shows blueshifted emission.

degeneracy pressure and the core collapses catastrophically, leading to a supernova explosion. For some stars, the remnant core can ultimately be supported by neutron degeneracy pressure, resulting in the formation of a stable neutron star; if the mass exceeds the limit for support by neutron degeneracy pressure (a mass between 2 and 3 $M_{\odot}$), then the collapse results in a black hole. In this section we discuss neutron stars and their radio frequency emission.

The angular momentum of the core is conserved as it collapses, causing a very large increase in angular velocity. Through some combination of magnetic flux conservation and the generation of new magnetic flux, the collapsed core also attains very high levels of magnetic field intensity. The end product of the collapse, then, is a rapidly-rotating, highly-magnetized, neutron star. Typical parameters for such stars include a mass of about 1.5 $M_{\odot}$, a radius of about 10 km, a magnetic field of order 10^{12} gauss, and spin rates ranging from about 0.1 Hz to 700 Hz. The existence of neutron stars was predicted by Walter Baade and Fritz Zwicky in 1934; at the time it was thought they would be undetectable. With the advent of X-ray astronomy in the early 1960s, numerous X-ray point sources were detected and it was speculated that some of these sources could be accreting neutron stars in close binary systems. In 1967 Jocelyn Bell and colleagues detected a source with rapidly varying, but very regular, radio emission. Similar sources were also detected forming a class of objects that were named pulsars and the first source detected was pulsar PSR B1919+21. Pulsars were quickly recognized to be rapidly rotating neutron stars. The radio emission as a function of time emitted by pulsar PSR 03291+54 is illustrated in Figure 7.9. This pulsar has a period of slightly less than 1 second.

The model of a pulsar is a rapidly rotating neutron star, whose magnetic field axis is mis-aligned with its rotation axis. Radiation is emitted along the magnetic axis of the star, in the form of a "lighthouse" beam, extending outward from the magnetic poles. The radiation mechanism responsible for this emission is discussed in Section 7.4.2. A pulsar does not 'pulse' in the sense of turning 'on' and 'off'. Rather, as its radio beam sweeps out a path on the sky, if the beam periodically points toward Earth, we detect a pulse of emission as the beam sweeps by; see Figure 7.10. There are expected to be many more pulsars in existence than we detect, but their orientation is such that their emission beams never sweep over the Earth, and so we do not see them.

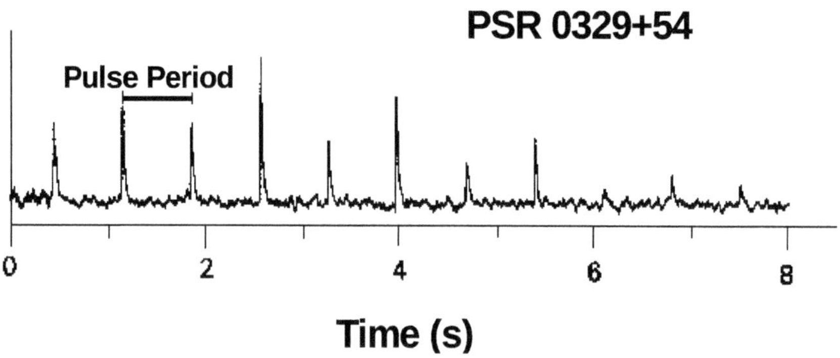

Figure 7.9: Plot of amplitude versus time for the radio emission from the pulsar PSR 0329+54. Individual pulses vary in amplitude and shape, but the average of many pulses produces a very stable pulse profile. Plot provided by Joseph Taylor.

Although the first neutron star to be clearly recognized was a radio pulsar, it is important to realize that not all neutron stars are radio pulsars. There is a veritable zoo of neutron stars, many of which do not emit radio pulses.

7.4.1 Pulsar Mechanics

A pulsar spin rate decreases with time due to energy loss by magnetic dipole radiation. This radiation is emitted at a frequency below the plasma frequency of the interstellar medium and so is undetectable. As the star rotates, the component of the dipole aligned with the spin axis remains constant (remember that the magnetic field axis is misaligned relative to the spin axis), but the component perpendicular to the spin axis rotates and thus emits magnetic dipole radiation. The power emitted can be inferred from Larmor's formula (Equation 3.1) where now instead of using the acceleration of the electric dipole (qa), we use the acceleration of the magnetic dipole, d^2m/dt^2,

$$L = \frac{2}{3c^3}\left(\frac{d^2\vec{m}}{dt^2}\right)^2 .$$

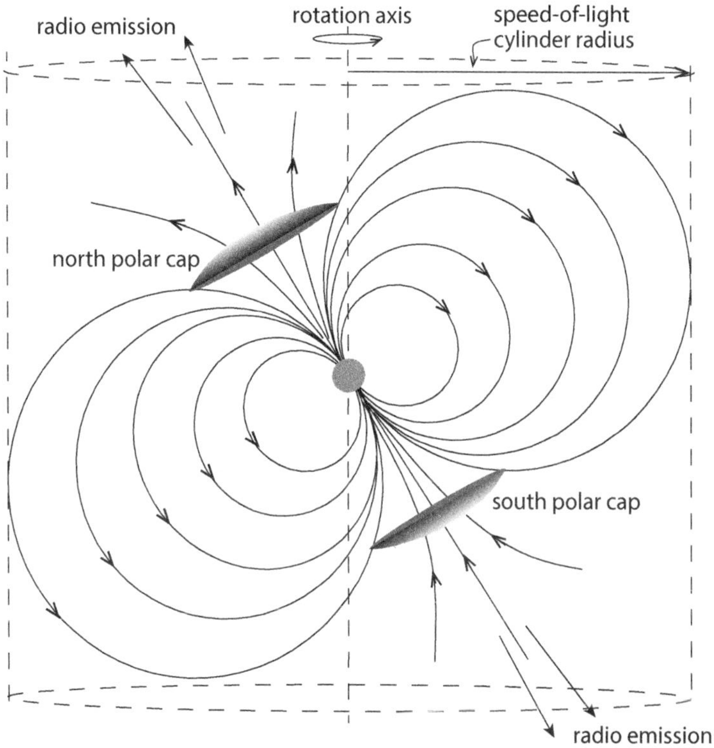

Figure 7.10: The speed-of-light cylinder of a pulsar, of radius r_c, is defined by the radius at which the tangential velocity, given by ωr, equals the speed of light. Charged particles moving along the closed magnetic field lines within this radius will return to the stellar surface. Particles moving on all other magnetic field lines do not return to the stellar surface.

Since only the perpendicular component of $\vec{m}$ is varying, and it rotates with angular velocity ω, the acceleration of the magnetic dipole is

$$\frac{d^2\vec{m}}{dt^2} = \omega^2 m \sin\alpha,$$

where α is the angle between the rotation axis and the magnetic dipole axis. Therefore, the power radiated is

$$L = \frac{2}{3c^3} m^2 \omega^4 \sin^2\alpha. \tag{7.2}$$

We use L for luminosity instead of P for power to avoid confusion with the pulsar rotation period. It is worth noting that radiation loss by relativistic particles, while not included in Equation 7.2, can also be important.

The energy radiated away by magnetic dipole radiation comes at the expense of the kinetic energy of rotation of the pulsar. Thus, with time, the pulsar angular velocity will gradually decrease and its rotational period will increase. The rotational kinetic energy is given by $K = \frac{1}{2} I\omega^2$ where I is the moment of inertia of the pulsar and ω is its angular velocity. The energy lost by the pulsar is given by the time rate of change of K, or $\dot{K} = dK/dt = I\omega\dot{\omega}$. Here we have used the 'dot-notation' to indicate differentiation with respect to time. In practice, we measure the rotational period rather than the frequency, so it is customary to use $P = 2\pi/\omega$ to express K in terms of the period:

$$K = \frac{2\pi^2 I}{P^2}. \tag{7.3}$$

The slowing of the pulsar is given by the spin-down rate, expressed as $\dot{P} = dP/dt$. In terms of period rather than frequency, $\dot{K}$, known as the spin-down luminosity, is given by

$$\dot{K} = -4\pi^2 I \frac{\dot{P}}{P^3}. \tag{7.4}$$

Note that $\dot{K}$ is negative, because there is a *loss* of kinetic energy. It is also useful to note that the time-derivative of a period, $\dot{P}$, is a dimensionless quantity.

The moment of inertia for any given pulsar is usually not well-determined, but it is unlikely to vary greatly from one pulsar to another. The general form for a moment of inertia is $I = kMR^2$; in the case of a uniform density sphere, $k = 2/5$. The internal density profile of a star depends on the equation of state that describes its interior, so the value of k will deviate somewhat from the nominal value of 0.4 for uniform density. However, k is unlikely to vary by more than a factor of 2 from the uniform density value. It is common practice to estimate I by assuming $k = 2/5$, assuming the Chandrasekhar limit for the mass ($M \sim 1.4$ $M_\odot$) and adopting a radius of $R = 10$ km. Thus, the canonical value for I is 10^{45} g cm^2, within a factor of a few.

Example 7.4:

Typical values for P and $\dot{P}$ of a pulsar are 1 second and 10^{-15}, respectively. What is the spin-down luminosity of such a pulsar?

Answer:
Using Equation 7.4 and assuming the canonical value of the moment of inertia, $I = 10^{45}$ g cm^2, we obtain

$$\dot{K} = -4\pi^2 \left(10^{45}\text{ g cm}^2\right) \left[\frac{10^{-15}}{(1\text{ s})^3}\right] = -3.95 \times 10^{31}\text{ erg s}^{-1}.$$

This is almost exactly 0.01 solar luminosities.

If the dominant energy loss is magnetic dipole radiation, then the rate of energy loss given in Equation 7.4 is approximately equal to the magnetic dipole luminosity given in Equation 7.2. By combining these equations we can solve for $m \sin\alpha$. And since a magnetic dipole moment is related to its magnetic field, roughly, by

$$B \sim m/r^3,$$

at the surface of the neutron star we expect

$$m \sim B_{surf}\, R^3.$$

We can, then, estimate the strength of the neutron star's magnetic field. Equating L from Equation 7.2 and the absolute value of $\dot{K}$ from Equation 7.4 yields

$$B_{surf}\ \sin\alpha \sim \left(\frac{3c^3 I}{8\pi^2 R^6}\right)^{1/2} (P\dot{P})^{1/2}. \tag{7.5}$$

Using P and $\dot{P}$ from Example 7.4 and the canonical value for the moment of inertia, we find a surface magnetic field of order 10^{12} gauss.

As pulsars age and radiate away their rotational kinetic energy, the increase in their rotation period can be used to estimate their age. A pulsar age can be defined by noting that $K/\dot{K}$ has dimensions of time. Using Equations 7.3 and 7.4 we can write this ratio as

$$\frac{K}{\dot{K}} = -\frac{P}{2\dot{P}}. \tag{7.6}$$

Note that $\dot{K}$ is negative because the pulsar is losing energy, while $\dot{P}$ is positive because the rotation period is increasing. This fractional change in pulsar rotational kinetic energy is then used to define the *characteristic age*, or *spin-down age*, τ_{sd}, of the pulsar. The derivation of τ_{sd} follows from the assumption that magnetic braking is slowing the pulsar rotation, in accordance with the expression

$$\dot{\omega} = -k\omega^n, \tag{7.7}$$

where n is called the braking index and k is some constant (not to be confused with the k constant of the moment of inertia). Considering an angular velocity ω_0 at time $t = 0$ and angular velocity ω at time $t = \tau$, we have

$$\int_{\omega_0}^{\omega} \frac{d\omega'}{(\omega')^n} = \int_0^{\tau} -k\, dt, \tag{7.8}$$

which integrates to

$$\frac{1}{-n+1}\left[\omega^{1-n} - \omega_0^{1-n}\right] = -k\,\tau.$$

Provided that $n > 1$ (for magnetic dipole braking, $n = 3$), and assuming that the original angular velocity was much greater than the current value (i.e., $\omega_0 \gg \omega$), then the second term in square brackets can be neglected, giving us

$$\frac{1}{1-n}\,\omega^{1-n} = -k\,\tau.$$

Making the substitution

$$\omega^{1-n} = \frac{\omega}{\omega^n}$$

and using Equation 7.7 we have

$$\omega^{1-n} = -k\frac{\omega}{\dot{\omega}}$$

and solving for τ gives us

$$\tau = \frac{1}{1-n}\frac{\omega}{\dot{\omega}}.$$

It is convenient to express this result in terms of the observable rotation period rather than the angular velocity. To make this conversion we use $\omega = 2\pi/P$ and $\dot{\omega} = -2\pi\dot{P}/P^2$. For magnetic dipole braking, with $n = 3$, this gives us a spin-down age of

$$\tau_{sd} = \frac{P}{2\dot{P}}. \tag{7.9}$$

We stress that even if the assumptions that the spin-down is due entirely to magnetic dipole braking and that the pulsar has slowed substantially from its original spin rate are true, this expression for τ_{sd} is only indicative of the pulsar age; many other factors can affect P and $\dot{P}$, during the life of the pulsar. For the handful of pulsars in which the braking index has been measured the values of n are all less than 3 (see paper by Oliveira and collaborators[13]), suggesting that other energy loss mechanisms in addition to magnetic dipole radiation may be important.

Example 7.5:

The pulsar of Example 7.4 currently has a period of 1 second and a $\dot{P}$ of 10^{-15}. Assuming that the pulsar experiences constant magnetic dipole braking, how old is the pulsar? How long will it take for its spin period to increase to 10 seconds?

Answer:
The assumptions implicit in Equation 7.9 are met, so we can use that expression to estimate the pulsar age.

$$\tau_{sd} = \frac{P}{2\dot{P}} = \frac{1 \text{ s}}{2\left(10^{-15}\right)} = 5 \times 10^{14} \text{ s},$$

which is about 16 million years.

Because the problem is posed in terms of period rather than angular velocity, it is convenient to make the conversion from ω to P, before evaluating the integral in Equation 7.8. Thus, adopting $n = 3$, we write Equation 7.7 as

$$\dot{P} = k\left(\frac{4\pi^2}{P}\right).$$

To find the value of k, we substitute in the known values for P and $\dot{P}$:

$$k = \frac{P\dot{P}}{4\pi^2} = \frac{(1 \text{ s})\left(10^{-15}\right)}{4\pi^2} = 2.5 \times 10^{-17} \text{ s}.$$

To get the age, then, we can write the following equation

$$\int_1^{10} P\, dP = 4\pi^2 k \int_{5\times10^{14}\text{ s}}^{\tau_{10}} dt$$

or

$$\frac{99}{2}\text{s}^2 = 4\pi^2\left(2.5 \times 10^{-17}\text{s}\right)\left[\tau_{10} - 5 \times 10^{14} \text{ s}\right],$$

giving

$$\tau_{10} = 4.9 \times 10^{16} \text{ s} = 1.6 \times 10^9 \text{ yr}.$$

The combination of P with $\dot{P}$ provides a useful means to describe the age and evolution of pulsars, in analogy to the Hertzsprung-Russell diagram. Such a "P–P-dot diagram" is shown in Figure 7.11. Immediately evident are two distinct populations of pulsars. The majority of pulsars have periods in the range of 0.1 to 10 seconds, and are slowing at rates of about 10^{-15} seconds per second. But a second population of pulsars is evident, with periods of about 1 to 10 milliseconds, and spin-down rates about five orders of magnitude slower than the majority of pulsars.

The first category, to which most pulsars belong, follows the "normal" pattern of pulsar creation in a supernova event, in which the neutron star forms with an initial spin period

[13] 2018, Journal of Cosmology and Astroparticle Physics, vol. 11, p. 25

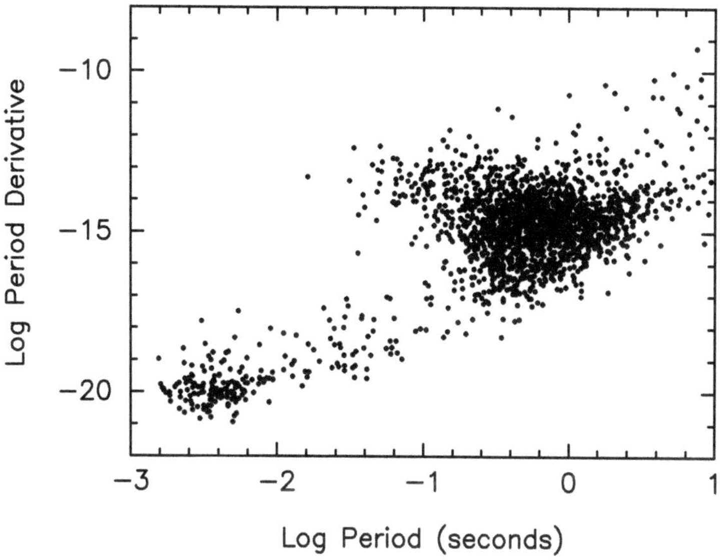

Figure 7.11: Plot of P versus $\dot{P}$ for pulsars in the ATNF pulsar catalog (see URL www.atnf.csiro.au/research/pulsar/psrcat). Units of the P axis are seconds; the $\dot{P}$ axis is dimensionless (i.e., seconds per second). Two distinct groupings of pulsars are evident: millisecond pulsars, with periods of order a few milliseconds, and normal pulsars, with periods from about 0.1 to 10 seconds.

somewhat below 100 ms. The well-known (and young) Crab and Vela pulsars are in this category, with periods of 33 and 89 ms, respectively. As these normal pulsars age, slowing at a rate of order 10^{-15} seconds per second, their rotation periods gradually lengthen, and they move toward the right in Figure 7.11.

The second category, called "millisecond pulsars", contrary to what you might expect, are much older pulsars. They have already radiated away much of their rotational kinetic energy, and have also lost much of their magnetic field strength. Because of their smaller magnetic moment, we see from Equation 7.2 that their magnetic dipole radiation is greatly reduced, and thus they lose energy more slowly. This is the reason for their much smaller $\dot{P}$ values, compared to normal pulsars. Almost all millisecond pulsars are found to have close binary stellar companions; in contrast, most normal pulsars do not have close companions. This result provides the key to understanding the very high rotation rates of the millisecond pulsars. If the binary companion is sufficiently close to the neutron star, it can transfer mass and angular momentum to the latter. A common scenario is for the companion to be a red giant, swollen in size to exceed its Roche lobe. The resulting mass transfer will impart angular momentum to the pulsar, causing it to"spin-up" to very high rotation rates. Such pulsars are sometimes referred to as "recycled pulsars".

7.4.2 Pulsar Emission Mechanisms

Magnetic dipole radiation is thought to be responsible for much of the loss of rotational kinetic energy of the pulsar. Nevertheless, this is not the radiation we detect from pulsars with our radio telescopes. The magnetic dipole radiation is emitted at the rotational frequency of the pulsar, which is in the kHz range or lower. In fact, although pulsars were discovered via their radio emission, the electromagnetic waves they emit can be of much higher frequency — including visible light and even X- and gamma-rays. Many details of the emission processes are still under investigation, but some general aspects of the mechanisms involved are becoming well-understood.

The radiation emitted by pulsars spans over 10 orders of magnitude in frequency. Different frequencies of radiation arise from different physical regions surrounding the pulsar. Before describing the location where each type of radiation originates, it is useful to consider the geometry of the neutron star, along with its atmosphere, a plasma envelope surrounding the neutron star. Because the charged particles of the plasma are confined to move along the magnetic field lines, the magnetosphere is whirled around in co-rotation with the star. The speed of a particle co-rotating at a distance r from the star's rotation axis is given by ωr. The co-rotation of the plasma must break down at some distance from the star, or else the tangential speed would exceed the speed of light. The distance from the rotation axis at which this happens defines the speed-of-light cylinder, whose radius is given by $r_c = c/\omega$ (see Figure 7.10). The co-rotating magnetosphere is defined by the region of closed magnetic field lines; that is, where the field lines return to the neutron star surface without going outside the speed-of-light cylinder. Thus, charged particles streaming along the field lines in this region of the magnetosphere will return to the stellar surface while still co-rotating with the magnetosphere, at velocities less than c. In contrast, there is a solid angle surrounding each magnetic pole, called a *polar cap*, where the field lines do not close but instead continue out into space. Charged particles streaming along the field lines from the polar cap regions are eventually forced out of co-rotation.

Example 7.6:

What radius is the speed-of-light cylinder for a pulsar of period 1 second?

Answer:
The radius of the speed-of-light cylinder is given by

$$r = \frac{c}{\omega} = \frac{c}{1/2\pi P} = 2\pi P c.$$

Using $P = 1$ s and $c = 3 \times 10^{10}$ cm s^{-1}, we obtain $r = 1.9 \times 10^{11}$ cm, which is 2.7 $R_\odot$.

The charge distribution within the magnetosphere produces strong electric fields. This is a consequence of mechanical equilibrium: if free charges feel a net, non-zero force, they will move until they achieve a zero-force configuration. That is, in order for the Lorentz force on a charged particle to vanish, we require that an electric field be induced such that

$$\vec{F} = 0 = q\vec{E} + \frac{q}{c}\vec{v} \times \vec{B} = q\vec{E} + \frac{q}{c}(\vec{\omega} \times \vec{r}) \times \vec{B}.$$

This electric field, which can reach strengths of order 10^{10} statvolt cm^{-1}, accelerates charged particles within the magnetosphere. These charged particles then emit radiation by a variety of mechanisms, covering the full electromagnetic spectrum. If the radiation emitted is sufficiently energetic, pair-production can occur within the magnetosphere, and the newly produced electron-positron pairs can also emit radiation.

The motion of the charged particles is a combination of streaming along the magnetic field lines, caused by the acceleration due to the electric field, together with gyration about the magnetic field lines. Because of the relativistic particle energies, the gyrating motion gives rise to synchrotron emission (see Section 3.3). Because of the relativistic effects related to synchrotron radiation, these photons can achieve very high energies, and the optical through gamma-ray emission is thought to arise from this process. However, the mechanism for the radio emission must be different.

Example 7.7:

The pulsar J2145−0750 is at a distance of 0.5 kpc, with a period of 16 milliseconds, and has a measured pulse flux density of 100 mJy at 436 MHz ($\lambda = 68.8$ cm). Estimate the brightness temperature of the emission by assuming that an upper limit to the size of the emitting region is given by the light travel distance between consecutive pulses. This assumption is plausible because whatever is causing the pulse cannot propagate at a speed faster than c, so the phenomenon must be restricted to a dimension less than c times the pulse period.

Answer:
The upper limit to the size of the emitting region is 0.016 s $\times 3 \times 10^{10}$ cm s^{-1} or 4.8×10^8 cm. Beginning with Equation 3.13, we express the brightness temperature as

$$T_B = \left(\frac{\lambda^2}{2\,k}\right) I_\nu .$$

The intensity, I_ν, is the flux density per unit solid angle, F_ν/Ω. For a circular source, the solid angle may be written as $(\pi/4)\,\theta^2$, where θ is in radians (see Equation 1.8). Thus we have

$$T_B = \left(\frac{\lambda^2}{2\,k}\right) I_\nu = \left(\frac{\lambda^2}{2\,k}\right)\frac{F_\nu}{\Omega} = \left(\frac{\lambda^2}{2\,k}\right)\frac{4}{\pi}\frac{F_\nu}{\theta^2} .$$

Replacing θ with D/d, the angular size of the assumed emitting region (given here as the ratio of the linear source size D to the source distance d) and simplifying terms, we have

$$T_B = \frac{2\lambda^2}{\pi k}\left(\frac{d}{D}\right)^2 F_\nu .$$

Substituting in numerical values, this gives

$$T_B = \frac{2\,(68.8\,\text{cm})^2}{\pi(1.38\times10^{-16}\,\text{erg}\,\text{K}^{-1})}\left(\frac{1.54\times10^{21}\ \text{cm}}{4.8\times10^{8}\ \text{cm}}\right)^2\left(10^{-21}\frac{\text{erg s}^{-1}}{\text{cm}^2\,\text{Hz}}\right) = 2.2\times10^{23}\text{K}.$$

This is a lower limit to the brightness temperature because the size of the emitting region is likely to be even smaller than we have assumed. The extremely high brightness temperature tells us that the radiation is emitted by a coherent process.

The preceding example provides an important clue to the nature of the pulsar radio emission: the extremely high brightness temperature — much higher than a physically possible temperature — indicates that some coherent process is involved in the emission process, similar to the case of maser emission, discussed in Section 4.2.3.

The mechanism responsible for the radio emission is still an open question. However, there is general agreement that the mechanism must be coherent involving relativistic electrons and positrons and that the radiation is produced in the regions above the polar caps (see Figure 7.10) where the field lines are open. It is also generally accepted that the electrons and positrons are secondary particles produced by pair production by gamma-ray photons produced by synchrotron emission. One mechanism often mentioned to produce the radio emission is *curvature radiation.* Curvature radiation is produced when the highly relativistic electrons and positrons stream along the curved magnetic field lines. The electrons and positrons must be produced in bunches by pair production, and these bunched particles radiate coherently. The coherent emission of a large number of charges results in a greatly amplified radiation intensity, which can explain the extremely high brightness temperatures that are detected.

7.4.3 Pulsar Searches

At present, about 2600 pulsars have been discovered within the Milky Way, which represents about 1% of the number that are estimated to exist. Not surprisingly, there are many active searches for these as yet undiscovered pulsars. The progenitors of pulsars are massive O- and B-type stars, so the obvious place to search for pulsars is the Galactic plane, where massive stars form and spend their relatively short lives. The distribution of pulsars extends somewhat further out of the plane than massive stars, presumably because the former can be given a "kick" during the supernova process which forms them, imparting an extra velocity that, over time, will carry the pulsar away from its Galactic plane birthplace. This detail notwithstanding, a sensible place to search for normal pulsars is where massive stars recently formed. Searches for millisecond pulsars must consider a larger region of the sky, because these older objects may have moved farther from the Galactic plane.

Even knowing where to look, pulsar searches present some special challenges that other astronomical objects do not. Chief among these are their unknown periods and their unknown dispersion measures (see Section 2.2.2). We will discuss the problem with periods first.

Unlike the discovery pulsar, PSR B1919+21, or the pulsar B0329+54, shown in Figure 7.9, most pulsars are too weak for individual pulses to be detected. Typically, hundreds or thousands of pulses must be observed and averaged by aligning the data with the correct periodicity, before the pulse profile can be detected. The averaging interval for the data is given by the pulse period, but this is unknown *a priori.* A straight-forward, but computationally expensive, approach is to average the data stream many different times, using a different averaging interval each time. Each average is then searched for a pulsar signal.

A much more efficient alternative is to Fourier transform the data, and search for peaks in the resulting power spectrum. The fundamental pulsar frequency, as well as its harmonics, will have more power and hence the Fourier transform will show peaks at those values. The pulsar data can then be folded at these frequencies, (i.e., averaged using an interval set by the inverse of the frequency component) to see if a pulse profile emerges.

A more significant problem is the frequency dispersion of the pulsar signal, discussed in Section 2.2.2. Because the lower frequencies within a pulse travel more slowly than the higher frequencies, the pulse arrives smeared out in time. An example is shown in Figure 7.12 in which the pulse arrival time as a function of frequency is shown for one pulse from the pulsar J1819–1458. For large dispersion measures and short pulsar rotation periods, distinct pulses will overlap with one another: the lower frequencies of one pulse will arrive later than the higher frequencies of the following pulse. To correct for this, the signal must be de-dispersed; that is, different frequencies must be shifted in time, with the higher frequency channels retarded so as to arrive synchronously with the lower frequency channels. A simple-minded (but effective) way to do this is to observe the pulsar with a spectrometer, and into each frequency channel introduce an appropriate time lag, Δt_{tr}, as given by Equation 2.21.

When searching for undiscovered pulsars, a substantial problem is that the dispersion measure (DM) is not known *a priori.* Hence, either a brute force trial-and-error method

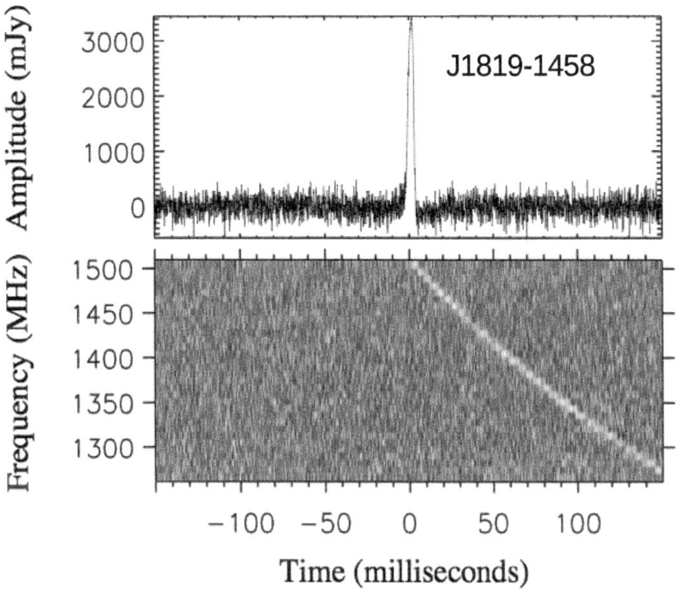

Figure 7.12: The different frequencies of a pulse signal travel at different speeds through the ionized gas of the interstellar medium, with lower frequencies traveling slower than higher frequencies. As a result, the pulse arrives "smeared-out" in time. The bottom figure shows time on the horizontal axis, frequency on the vertical axis, and intensity as the gray scale. The top figure shows the pulse corrected for this dispersion effect. Figure provided by Maura McLaughlin.

must be used, testing many different values of the DM, or, as with the period, we may use the Fourier transform to search for the desired information. If the data are considered as two-dimensional, with arrival time as one dimension and signal frequency as the other, then a two-dimensional Fourier transform will convert the time axis to a frequency and the frequency axis to a time — in particular, to a delay time. The slope of the transformed data in the frequency-delay plane gives the delay as a function of frequency, which is determined by the dispersion measure. Fitting the slope of the line gives an approximate value for the DM, which can be further refined by various methods.

The de-dispersion is limited by the spectral resolution of the frequency sampling, i.e., the channel width. All frequencies within a single spectral channel are corrected by the same delay. Such de-dispersion is termed incoherent de-dispersion. A more sophisticated technique, known as coherent de-disperion, avoids this problem, and provides a precise de-dispersion at all frequencies. The technique is based on digital signal processing concepts that are beyond the scope of the present discussion.

In the early years of pulsar studies, the computationally-intensive operations of de-dispersion and period searching were carried out with hardware, using purpose-built "pulsar backends" that contained high-speed electronic circuits to make the calculations. With the exponential growth in computing power since that time, nowadays most pulsar searches are done in software, sometimes in combination with special hardware. The incoming signal is digitized, and high-speed computers execute digital signal processing algorithms on the data to search for the pulsar signals.

Pulsar searches are characterized by their need to explore a large parameter space of rotation period and dispersion measure. Once these parameters are determined, the observations can be optimized to study other pulsar phenomena, such as their pulse profiles, pulse time-of-arrival, inter-pulse emission, giant pulses, glitches, polarization, and so forth.

7.4.4 Binary Pulsars

In recent years a number of binary pulsar systems have been discovered. Such systems are unique laboratories to study fundamental physical processes — particularly theories of gravity. The first such system discovered, PSR B1913+16, was observed with the Arecibo telescope in 1974 by Russell Hulse and Joseph Taylor[14], and earned them the 1993 Nobel Prize in Physics. Subsequent observations showed that the orbital decay of the system corresponded to the expected energy loss due to gravitational radiation, as predicted by the general theory of relativity.

Such binary pulsar systems are the equivalent of two point masses in mutual orbit about their center of mass. The orbital parameters, then, provide for numerous tests of gravitational theories. Relatively few such systems are known, and alternative means for testing gravity theories are being developed. Most notable are pulsar timing arrays, in which a large number of isolated millisecond pulsars are observed, and the time-of-arrival of their pulses is precisely measured and correlated with those of other pulsars in the network. Such pulsar timing arrays can be used to detect the passage of gravitational waves.

In 2017 LIGO/VIRGO detected gravitational waves produced by the merger of two neutron stars (see paper by the LIGO Scientific Collaboration and the VIRGO Collaboration[15]). The orbital decay seen in binary pulsars like B1913+16 is a very early stage of what ultimately becomes the inspiral and merger event that produced detectable gravitational waves. It may take 300 million years for a system like B1913+16 to merge.

7.4.5 Radio Pulsars as Probes of the Interstellar Medium

The pulse characteristics from radio-pulsars provide astronomers with tools for measuring two important parameters of the interstellar medium. First, the short time duration of a pulse along with its continuous spectrum facilitates a measure of the average density of free electrons along the line of sight to the pulsar. The interstellar medium contains free electrons, and as explained in Section 2.2.2, such free electrons can affect the pulse travel time toward the Earth, with the delay in the arrival time being frequency-dependent (see Equation 2.21). In Section 7.4.3, we described how, in the process of detecting pulsars, we must correct the pulse arrival time for the dispersion of the interstellar medium, thus determining the dispersion measure for each pulsar. What poses a problem for the detection of the pulsars can be converted into a tool for studying the interstellar medium. The dispersion measure, DM (see Equation 2.23), depends on the free electron column density along the line of sight. If the distance to the pulsar is known by other means, we can calculate the average free electron density along the given sight line. This procedure has been applied to a large of number of pulsars, producing a map of the free electron density of the interstellar medium; see, for example, the paper by James Cordes and Joseph Lazio[16]. The free electrons being

[14] 1974, Astrophysical Journal Letters vol. 195 p. 51,
[15] 2017, Physical Review Letters, vol. 119 p. 161101-1.
[16] 2002, arXiv:astro-ph/0207156.

modeled are primarily in the Warm Ionized Medium; see the introduction to Chapter 5. Observations have shown that the interstellar free electron density is non-uniform. Locally, i.e. in the neighborhood of the Sun, the free electron density is about 0.025 cm^{-3}.

Example 7.7:

(a) Determine the dispersion measure for the pulsar J1819−1458 shown in Figure 7.12.

(b) Assuming this pulsar is at a distance of 3.5 kpc, what is the average electron density along the line of sight?

Answer:
(a) From Figure 7.12 we see that the arrival of the pulse at 1300 MHz is detected 130 milliseconds later than its arrival at 1500 MHz. Substituting these frequencies and time delays into Equation 2.24 we have

$$DM = 2.41 \times 10^{-4}\ \text{cm}^{-3}\ \text{pc} \left(\frac{0.130\ \text{s}}{\text{s}}\right)\left(\frac{1}{(1300\ \text{MHz}/\text{MHz})^2} - \frac{1}{(1500\ \text{MHz}/\text{MHz})^2}\right)^{-1}$$
$$= 213\ \text{cm}^{-3}\ \text{pc}.$$

(b) The mean electron density can be calculated by dividing the DM by the distance.

$$n_e = \frac{213\ \text{cm}^{-3}\ \text{pc}}{3500\ \text{pc}} = 0.061\ \text{cm}^{-3}$$

Second, pulsar radiation has a net linear polarization, which allows a measurement of the interstellar magnetic field. As the linearly polarized radio wave propagates through the interstellar medium, it will suffer Faraday rotation (see Section 2.2.3) due to the Galactic magnetic field. The total amount of rotation depends on the wave's frequency and the integrated value of the product of the electron density and the component of the magnetic field along the line of sight. This effect is described by the rotation measure, RM, given by Equation 2.27. Typical values of the RM for Galactic pulsars range from 10^{-5} to 0.1 rad cm^{-2}.

The similarity between DM and RM is evident: they differ only by the factor of the magnetic field along the line of sight. If we are content with obtaining the average value of $B_{\|}$ along the sight line, then, as described in Section 2.2.3, we can take the magnetic field outside the integral, and express it in terms of the ratio of RM to DM. The interstellar magnetic field is typically about 5 μgauss.

Example 7.8:

The pulsar J1819−1458, described in Example 7.7, has a rotation measure of 0.033 rad cm^{-2} (see paper by Aris Karastergiou and collaborators[17]). What is the average line-of-sight parallel component of the magnetic field toward this pulsar?

[17]2009 Monthly Notices of the Royal Astronomical Society, vol. 396, p. L95.

Answer:
Using Equation 2.30, we have

$$\langle B_{||} \rangle = 0.0124 \text{ gauss} \left(\frac{0.033 \text{ rad cm}^{-2}}{\text{rad cm}^{-2}} \right) \left(\frac{213 \text{ cm}^{-3} \text{ pc}}{\text{cm}^{-3} \text{ pc}} \right)^{-1} = 1.9 \times 10^{-6} \text{ gauss.}$$

Since we measure only the magnetic field component along the line of sight, the total magnetic field is larger. This inferred value, then, is consistent with the typical value of 5 μgauss stated above.

7.4.6 Supernova Remnants

Figure 7.13: VLA image of the Crab supernova remnant at 6 cm wavelength. The image field of view is 0.1167 × 0.1167 degrees. Image courtesy of NRAO/AUI and M. Bietenhol.

As stated in the introduction of Section 7.4, pulsars and all neutron stars are formed when massive stars explode in core-collapse supernovae (most are designated type II supernovae). Not surprisingly, then, pulsars are often found in *supernova remnants* (SNR), which are large nebulae produced by the supernova blast waves expanding into the interstellar medium. Figure 7.13 shows an image of the supernova remnant known as the Crab nebula (also called Taurus A), inside of which is the Crab pulsar, also designated PSR B0531+21. The Crab pulsar was the first pulsar found to be associated with a supernova remnant. The supernova remnants, themselves, can be very bright radio sources. In fact, the supernova remnant Cas A, shown in Figure 7.14, was one of the first radio sources detected (see Volume I, Section 1.1). At X-ray wavelengths a neutron star was detected at the center of Cas A, but this neutron star does not show radio pulses, so either it is not a pulsar or its beams do not sweep over the Earth.

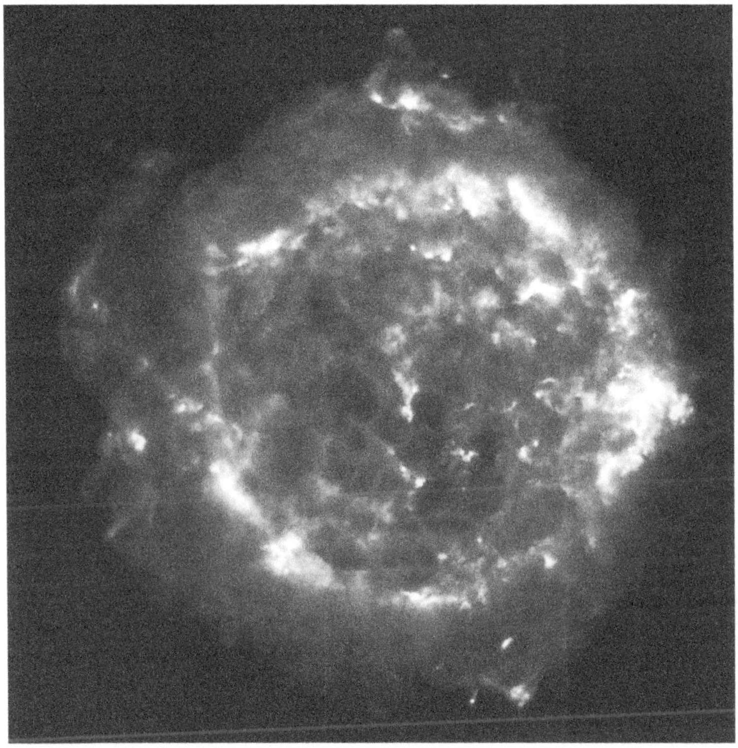

Figure 7.14: VLA image of the Cas A supernova remnant at 6-cm wavelength. The displayed image is 0.1×0.1 degrees. Image courtesy of NRAO/AUI. Credit: P. E. Angerhofer, R. Braun, S. F. Gull, R. A. Perley, and R. J. Tuffs.

In addition to massive star core-collapse supernovae, there are supernovae that result from white dwarf stars in binary systems in which the white dwarf accretes a sufficient amount of mass to put it over the Chandrasekar mass limit leading to a catastrophic collapse. This collapse heats up the white dwarf to temperatures sufficient to produce a thermonuclear denotation resulting in another type of supernova, usually designated type Ia. These supernovae explosions are not believed to leave any remnant star. Tycho's supernova, studied by Tycho Brahe in 1572, is an example of a supernova from the detonation of a white dwarf and its remnant is still radio bright today.

All supernova remnants can be a significant source of radio emission. SNRs can also be bright in all realms of the electromagnetic spectrum. The large power emitted by a SNR at radio wavelengths is due to synchrotron radiation (see Section 3.3). Figure 3.7, which includes the radio-frequency spectra of Cas A and the Crab (Tau A), shows that the SNR emission is a declining power-law, similar to the synchrotron emission seen in radio galaxies, like Virgo A, also shown in Figure 3.7. The relativistic electrons needed to produce the synchrotron radiation in these supernova remnants can be produced either by the central pulsar, as is the case for the Crab nebula, or can be accelerated in the shock front produced by the explosion as is likely the case for Cas A.

As a supernova remnant ages, its magnetic field will decrease and its energetic particles will cool, leading to a decreasing radio flux density. The flux density of Cas A, in fact, has been confirmed to be decreasing measurably over the past decades. Old SNRs, then, will not be detectable and the very luminous ones must be young. The Crab nebula is believed to be due to a supernova that occurred in 1054 and so is less than 1000 years old, while Cas A is believed to be about 350 years old. Despite the large radio fluxes of SNRs, we will not discuss these objects in any further detail here.

QUESTIONS AND PROBLEMS

1. The average flux density of the Sun at a frequency of 2800 MHz (wavelength 10.7 cm) in April 2018 was 70 SFU. What was the disk-averaged brightness temperature of the Sun that month?

2. The solar maximum of solar cycle 24 was in 2014. This solar cycle was not as active as previous cycles. Using the data in Figure 7.3, estimate the average fraction of the solar surface covered by sunspots during the 2014 maximum.

3. The solar radio burst on January 18, 2011 lasted for 4 minutes and at a frequency of 245 MHz had a peak flux density of 750 SFU. What was the Sun's disk-averaged brightness temperature at the peak of this burst?

4. The main-sequence star Tau Ceti is a solar-type star at a distance of 3.65 pc. It has a radius about 0.8 $R_\odot$ and a disk-averaged brightness temperature of about 10,000 K at a frequency of 34.5 GHz. Estimate its flux density at this frequency.

5. The brightness temperature of the quiet Sun increases dramatically from frequencies around 100,000 MHz to frequencies around 300 MHz, but does the Sun's flux density also increase? Compute the flux density of the Sun at 100,000 and 300 MHz using the data shown in Figure 7.2. Assume the Sun has a diameter given by its photospheric radius at 100,000 MHz and at 300 MHz assume the Sun is about 20% larger.

6. Solar-type stars have been detected at 34.5 GHz. If solar-type stars were like the quiet Sun, do you think they would be more detectable at lower frequencies? Make use of your result from Question 5.

7. The nearest flare star is at a distance of 2.7 pc. If it experienced a flare like that which occurred on the Sun in January 18, 2011 (see Question 3), would it be detectable at 245 MHz? Assume a detection limit of 100 μJy.

8. Based on $C^{18}O$ ALMA observations of VLA 1623A, Nadia Murillo and collaborators[18] found that the inner disk of this very young protostar exhibits Keplerian rotation. At a distance of 50 AU, they find a rotation velocity of 2 km s^{-1}. What is the mass of the central protostar?

9. In Section 7.3.3 we suggested that the collimated wind from L1551 IRS5 had a mass loss rate of 10^{-6} $M_\odot$ yr^{-1}. If the dynamical age of the outflow is 10^5 years and the wind velocity is 200 km s^{-1}, confirm this mass loss rate assuming conservation of momentum between the wind and the accelerated molecular gas.

10. We showed that the radio free-free emission from an isothermal, homogeneous sphere of plasma has $F_\nu \propto \nu^{+2}$ and $F_\nu \propto \nu^{-0.1}$ in the optically thick and thin regimes, respectively. Consider now an isothermal ionized stellar wind, whose density decreases as r^{-2} with distance r from the star. Show that in this case, regardless of the value of

[18] 2013, Astronomy and Astrophysics, vol. 560, p. A103.

τ_ν, $F_\nu \propto \nu^{+0.6}$. Hint: Assume that the gas emits as a black-body whose size is defined by the position where $\tau_\nu = 1$. The following integral will be useful:

$$\int_{-\infty}^{+\infty} \frac{dx}{(x^2+a^2)^2} = \frac{\pi}{2}\frac{1}{a^3}.$$

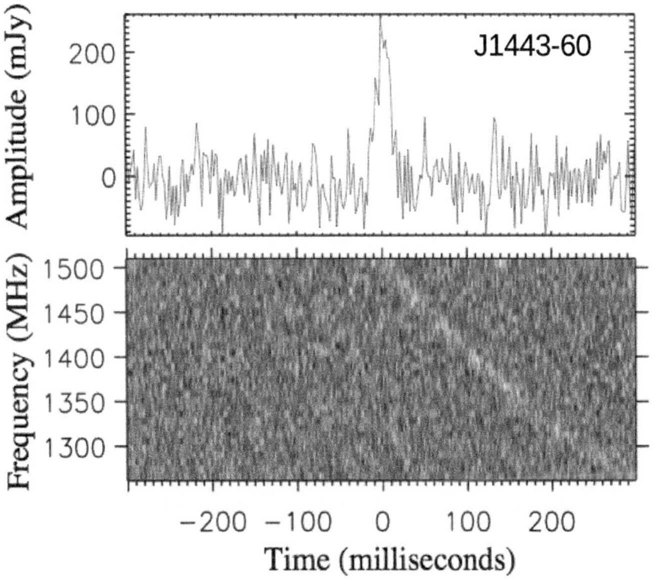

Figure 7.15: The top figure shows a pulse profile corrected for dispersion from the pulsar J1443−60. The bottom figure shows time on the horizontal axis, frequency on the vertical axis, and intensity as the gray scale. Figure provided by Maura McLaughlin.

11. Figure 7.15 shows a plot of frequency versus time for one pulse from the pulsar J1443−60. What is the dispersion measure of this pulsar? If the average electron density along the path to this pulsar is 0.025 cm^{-3}, what would be the distance to this pulsar?

12. Millisecond pulsars typically have periods of about 0.003 seconds and period derivatives of 10^{-20} seconds per second. What is the spin-down luminosity and spin-down age of such a pulsar? Assume these pulsars have the canonical value of 10^{45} g cm^2 for their moment of inertia.

13. The pulsar at the center of the Crab Nebula has a period of 0.033 seconds and is at a distance of about 2200 pc. An average pulse at a frequency of 333 MHz has a peak flux density of about 25 Jy (see paper by David Moffett and Timothy Hankins[19]). The radio emission likely arises from a region with a radius smaller than 10^9 cm. What is the minimum brightness temperature of the radio emission?

[19] 1996, Astrophysical Journal, vol. 468, p. 779.

CHAPTER 8

Galaxies at Radio Wavelengths

THE local universe abounds with galaxies that have a variety of shapes and sizes. Galaxies are often categorized by their morphology using a classification scheme developed by Edwin Hubble in the early 20th century. In the Hubble classification, which is based on their optical appearance, galaxies are classified as spiral, barred spiral, S0 (also called lenticular), elliptical and irregular. A schematic of the Hubble galaxy classification is shown in Figure 8.1. Both the spiral and S0 galaxies have flattened, disk-shaped morphologies with central bulges, but S0 galaxies do not have the spiral arms that are characteristic of spiral galaxies. The elliptical galaxies have an ellipsoidal shape with a smooth distribution of light, and vary from being nearly spherical to being highly oblong. Finally, irregular galaxies, unlike the spiral and elliptical galaxies, do not have any well-defined shape. Astronomers often refer to elliptical and S0 galaxies as early-type galaxies and spiral and irregular galaxies as late-type galaxies, based on an old and incorrect belief that galaxies evolved with time from elliptical morphologies to spiral or irregular morphologies.

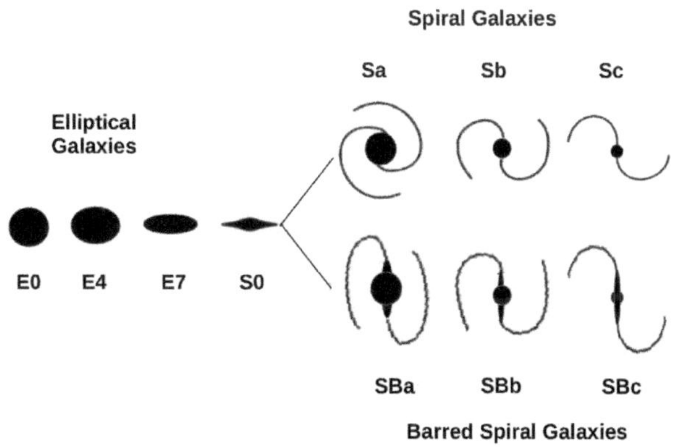

Figure 8.1: A schematic representation of the morphology of galaxies and the Hubble classification scheme. The elliptical galaxies are given the designation E0 through E7, where E0 galaxies are spherical and E7 galaxies are the most flattened. The S0 galaxies are disk galaxies with no obvious spiral arms. The spiral-armed galaxies are divided into spiral galaxies (Sa, Sb, and Sc) and barred spiral galaxies (SBa, SBb. and SBc). Sa and SBa galaxies have more tightly wound arms and large central bulges, while Sc and SBc have more loosely wound arms and smaller central bulges. Galaxies that do not fit any of these morphologies are classifed as irregular galaxies.

Galaxies of all morphologies have a broad range of sizes and masses. Elliptical galaxies vary from dwarf ellipticals with sizes less than 1 kpc in diameter and stellar masses (total mass in stars) less than 10^7 $M_\odot$ to giant ellipticals with diameters greater than 100 kpc and stellar masses greater than 10^{12} $M_\odot$. Spiral galaxies also have a broad range of sizes and masses; the largest have diameters of 100 kpc with stellar masses greater than 10^{11} $M_\odot$, while the smallest have diameters of only 5 kpc and stellar masses of only 10^9 $M_\odot$. The Milky Way is a barred spiral galaxy, with a stellar mass of about 6×10^{10} $M_\odot$, making it a larger mass spiral galaxy.

The most numerous types of galaxies are small, low-mass galaxies, referred to as dwarf galaxies. These dwarf galaxies have either irregular or elliptical morphologies. Although dwarf galaxies are much more numerous than massive galaxies, the galaxies that contribute most to the total stellar mass and starlight in the local universe are the spiral and giant elliptical galaxies.

Galaxies are not distributed uniformly in space, but instead form a hierarchical structure. Galaxies are often found in groups and these groups can vary from having only a few galaxies as members to well over a thousand galaxies. The larger groups are called galaxy clusters, and are often classified as either poor, if they contain only tens of members, to rich, if they have hundreds to thousands of galaxies. These galaxy clusters are also clustered together, forming what is referred to as superclusters. The Milky Way is a member of a poor galaxy cluster called the Local Group. The Milky Way galaxy, together with another large spiral galaxy, Andromeda (also called M31), are the most massive galaxies in the Local Group, while most of the remaining members are dwarf galaxies. Galaxies not in groups or clusters are known as field galaxies.

Studies of galaxies in clusters and field galaxies reveal that the morphologies of galaxies depend on their environment. Elliptical galaxies are more prevalent in rich clusters, while spiral galaxies are more likely found in poor clusters or in the field. In fact, the giant elliptical galaxies are almost exclusively found at the center of rich galaxy clusters. Since the density of galaxies (number of galaxies per unit volume) in rich clusters is much higher than in poor clusters or in the field, this observed morphology difference is termed the density-morphology relation, and was recognized by Alan Dressler in 1980. The interactions between galaxies is much more frequent in rich clusters than in the field, and these interactions can lead to the disruption of disk structures in spiral galaxies and change a galaxy's morphology. Thus, in rich galaxy clusters, the frequent interactions result in the conversion of spiral galaxies into elliptical galaxies producing the observed density-morphology relation.

Because galaxies evolve with time, we have been qualifying our discussion with the term "local" universe, because we see local galaxies as they are today. If we observe distant galaxies, we see the Universe as it was in the past. In Section 8.4 we will discuss radio observations of distant galaxies, which are less-evolved and have quite different properties than galaxies today.

The stellar and gaseous contents of local galaxies vary enormously and these properties are often related to a galaxy's morphology. In general, elliptical and S0 galaxies are composed of mostly old stars and contain less gas and dust than spiral galaxies of the same mass. Since these galaxies have less raw material to form new stars, the star formation activity in some of these galaxies has nearly come to a halt. On the other hand, spiral and irregular galaxies are generally composed of both young and old stars and have relatively more gas and dust, allowing these galaxies to continue forming stars. The central bulges in spiral galaxies contain mostly old stars that are similar to the stars present in elliptical galaxies.

The colors of different galaxy types at optical wavelengths are also different. Elliptical galaxies are red, because they have mostly old red stars, while spiral and irregular galaxies are blue, because they are still forming stars and the most massive young stars are extremely luminous and blue. Because they are so luminous their blue light can dominate the total light emitted by these galaxies. Because of this dichotomy, we often talk about red and dead galaxies (not forming stars) and blue galaxies (star-forming galaxies).

In star-forming galaxies, the rate at which stars are forming varies significantly between galaxies. Some galaxies have exceptionally high rates of star formation, and these are termed *starburst galaxies*. The rate of star formation in these galaxies is so high that it cannot be maintained for very long as all of the gas would be consumed; therefore the period of rapid star formation must be relatively brief. These starburst galaxies are often interacting galaxies and these interactions are believed to trigger the onset of rapid star formation. The light from the newly formed stars is generally absorbed by the surrounding dust and then reradiated at infrared wavelengths. Such galaxies can be very bright in the infrared, and if their infrared luminosity is greater than 10^{12} $L_{\odot}$ they are referred to as ultraluminous infrared galaxies (ULIRGs).

In addition to differences in the stellar and gas content, the kinematics of galaxies also differ and are related to their morphology. In spiral galaxies, like the Milky Way, most of the stars (except those in the bulge and halo) along with the gas and dust are confined to a thin plane, and they orbit the center of the galaxy in nearly circular orbits all in the same direction. In elliptical galaxies, the stars orbit in highly eccentric orbits with random orientations.

Most large galaxies, including the Milky Way, have super-massive black holes at their centers. If these black holes are accreting copious amounts of material, they can produce an enormous luminosity detectable at all wavelengths from gamma-rays to radio. Galaxies containing actively accreting super-massive black holes are called Active Galactic Nuclei (AGN). Quasars, radio galaxies, and Seyfert galaxies are all examples of AGN, identified in different ways, and in Chapter 9 we discuss the radio emission from these objects in detail. Galaxies without AGN activity are often referred to as "normal" galaxies; in this chapter we will discuss the radio emission from these normal galaxies.

We now turn to what we can learn about normal galaxies (non-AGN) by radio wavelength studies. We know most about the galaxies in the local universe as they are sufficiently close to permit detailed studies, and much of what we will discuss in this chapter pertains to these nearby galaxies. We should not be surprised that the radio emission from many of the local galaxies has much in common with the radio emission of the Milky Way discussed in Chapter 5. Studying other galaxies has both drawbacks and advantages compared with studies of the Milky Way. The major drawback is that the galaxies are much further away, making the radio emitting regions fainter and of smaller angular size, and thus more difficult to study than similar regions in the Milky Way. However, our view of the Milky Way is from the inside, making it much more difficult to get a good grasp on its global properties, whereas our view of other galaxies usually provides us with a panoramic view of their structure. As in the Milky Way, radio observations provide an important means for studying the gas content of galaxies, and we start our discussion with observations of the 21-cm line of atomic hydrogen.

8.1 21-CM HI OBSERVATIONS

The physics of the 21-cm spectral line of atomic hydrogen is described in Section 4.2.1 and observations of this line in the Milky Way are described in Section 5.1. Observations of this spectral line in other galaxies have also been instructive, yielding constraints on several important properties of galaxies. As we discuss 21-cm observations of other galaxies, it is worth remembering that the diffraction-limited angular resolution of single-dish telescopes at a wavelength of 21 cm is quite poor (see discussion in Volume I, Chapter 3). For example, the angular resolution of even a large telescope, such as the 100-m diameter Green Bank Telescope, operating at a wavelength of 21 cm is approximately 21 cm/10^4 cm = 0.0025 radians or about 9′. As a comparison, observations with even small optical telescopes typically achieve angular resolutions of order 1″ or about 500 times better angular resolution. Almost all galaxies have angular diameters less than 9′, so single-dish telescopes usually provide little information about how the HI emission is distributed within the galaxy; rather they only give the total flux in the spectral line. However, radio interferometers, such as the Jansky Very Large Array (JVLA), can achieve high angular resolution. The JVLA can achieve an angular resolution at 21 cm of order 1″. However, due to the low brightness temperature of the HI emission, it is only practical to obtain HI images with the JVLA with resolutions of between 5″ and 60″. Nevertheless, interferometers can be used to map the distribution of 21-cm spectral line emission within nearby galaxies.

Determination of the total atomic gas content of a galaxy does not require resolving the galaxy and so single-dish observations are adequate. Furthermore, understanding of the atomic gas content of galaxies in general requires measurements of the HI gas mass in hundreds of galaxies. Studies involving observations of large numbers of galaxies are largely accomplished using single-dish telescopes. We start our discussion of 21-cm HI observations showing how to use the measured line flux to calculate the total mass of atomic hydrogen gas.

8.1.1 HI Mass of Galaxies

An example of an HI 21-cm spectrum, obtained at the Arecibo Observatory of the galaxy UGC 93, is shown in Figure 8.2. UGC 93 is a spiral galaxy at a distance of about 70 Mpc and has an optical diameter of 2.1 × 1.5 arcminutes. The Arecibo Observatory telescope has a diameter of 305 m, and therefore an angular resolution at 21 cm of about 3.4 arcminutes. Even with Arecibo, the HI emission of this galaxy is still unresolved; however such unresolved measurements can still be used to obtain the total mass of atomic hydrogen in this galaxy.

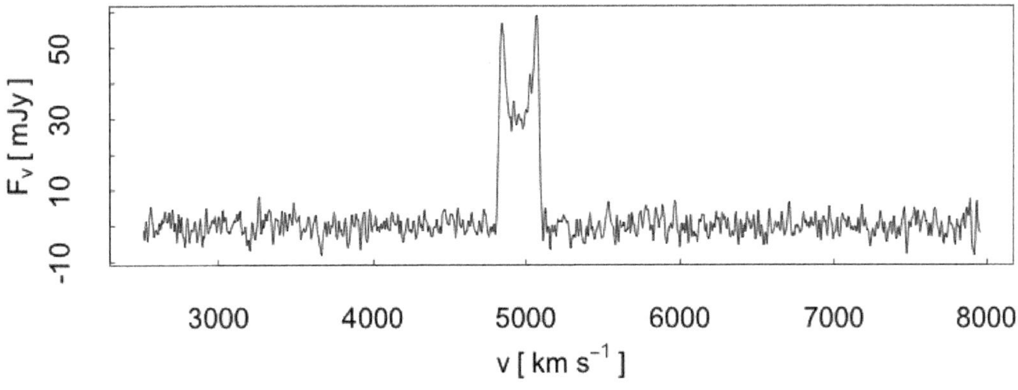

Figure 8.2: The 21 cm HI spectrum of the galaxy UGC 93 obtained at the Arecibo Observatory as part of the first data release from the ALFALFA survey (see Haynes et al. 2011, Astronomical Journal, vol. 142, p. 170). The flux density (in milli-jansky, mJy) is plotted as a function of velocity (in km s^{-1}). The plot is from the NASA/IPAC Extragalactic Database (NED) which is operated by the Jet Propulsion Laboratory, California Institute of Technology, under contract with the National Aeronautics and Space Administration.

We discussed in Section 4.2.1 that the 21-cm spectral line of HI is usually optically thin, so there is a simple relation between the integrated 21-cm line intensity and the atomic hydrogen column density. The atomic hydrogen column density, $N_{\rm H}$, will likely vary across a galaxy, but the total number of hydrogen atoms in the galaxy can be computed from the integral of the column density distribution over the area of the galaxy. Thus, the mass of atomic hydrogen in a galaxy, $M(\rm HI)$, is

$$M({\rm HI}) = \int N_{\rm H}\, m_{\rm H}\, dA,$$

where $m_{\rm H}$ is the mass of a hydrogen atom. Taking the derivative of the relationship between area, A, distance, d, and solid angle, Ω, from Equation 1.6, we have $dA = d^2 d\Omega$. Therefore we can rewrite the integral over the area of the galaxy into an integral over the solid angle of the galaxy, and thus

$$M({\rm HI}) = \int N_{\rm H}\, m_{\rm H}\, d^2 d\Omega = m_{\rm H}\, d^2 \int N_{\rm H}\, d\Omega.$$

The relation between the atomic hydrogen column density and the integrated line intensity is given in Equation 4.29. Converting the integral of intensity over frequency to an integral over velocity and expressing the velocity in km s^{-1}, we can write the hydrogen column density as

$$N_{\rm H} = 2.92 \times 10^{36}\ {\rm cm}^{-2} \left(\frac{\int I_\nu\, dv}{{\rm erg\ s^{-1}\ cm^{-2}\ Hz^{-1}\ sr^{-1}\ km\ s^{-1}}}\right).$$

Using this expression, we can rewrite the total mass of atomic hydrogen in grams as

$$M({\rm HI}) = 4.88 \times 10^{12}\ {\rm g} \left(\frac{d}{\rm cm}\right)^2 \left(\frac{\int\int I_\nu\, d\Omega\, dv}{{\rm erg\ s^{-1}\ cm^{-2}\ Hz^{-1}\ km\ s^{-1}}}\right),$$

where we have included the value of the mass of a hydrogen atom in the numerical coefficient. The quantity $\int I_\nu\, d\Omega$ is the flux density, F_ν, of the 21-cm spectral line, so the mass of HI can be rewritten as

$$M({\rm HI}) = 4.88 \times 10^{12}\ {\rm g} \left(\frac{d}{\rm cm}\right)^2 \left(\frac{\int F_\nu\, dv}{{\rm erg\ s^{-1}\ cm^{-2}\ Hz^{-1}\ km\ s^{-1}}}\right),$$

where $\int F_\nu\, dv$ is the flux density integrated over the 21-cm line. Expressing mass in solar masses, flux density in janskys and distances in Mpc, we can rewrite the atomic hydrogen gas mass as

$$M({\rm HI}) = 2.36 \times 10^5\ {\rm M_\odot} \left(\frac{d}{\rm Mpc}\right)^2 \left(\frac{\int F_\nu\, dv}{{\rm Jy\ km\ s^{-1}}}\right). \tag{8.1}$$

In summary, assuming the 21-cm emission is optically thin, we can use the integrated flux density of the 21-cm line to determine the total mass of atomic hydrogen in a galaxy. Taking into account all the other atoms, we need to multiply the HI mass by a factor of 1.36 to determine the total atomic gas mass.

Example 8.1:

The HI spectrum of the galaxy UGC 93 is shown in Figure 8.2 and the HI line has an integrated flux density of 10.7 Jy km s^{-1}. UGC 93 is at a distance of 70 Mpc. What is the total mass of atomic hydrogen?

Answer:
We can use Equation 8.1 that relates the HI mass to the integrated flux density of the 21-cm line and the distance to the galaxy. We find

$$M(\mathrm{HI}) = 2.36 \times 10^5 \ \mathrm{M}_\odot \left(\frac{70 \ \mathrm{Mpc}}{\mathrm{Mpc}}\right)^2 \left(\frac{10.7 \ \mathrm{Jy\ km\ s^{-1}}}{\mathrm{Jy\ km\ s^{-1}}}\right) = 1.2 \times 10^{10} \ \mathrm{M}_\odot.$$

The mass of HI gas in galaxies from single-dish observations ranges from values greater than 10^{10} $\mathrm{M}_\odot$ to values less than 10^6 $\mathrm{M}_\odot$. The gas mass depends largely on the physical size of the galaxy, but there is also a dependence on the nature of the galaxy. To separate the two effects, it is useful to normalize the gas mass by a measure that reflects the relative size or mass of the galaxy. To characterize the gas content of galaxies, it is common to normalize the gas mass by the stellar mass (M_*, the total mass in stars) of the galaxy. We denote the ratio of the total HI gas mass to the total stellar mass as f_{HI} given by M_{HI}/M_*. In Figure 8.3 the values of f_{HI} for galaxies in two separate samples are plotted versus their stellar mass. The two samples are (1) a sample of early-type, or red, galaxies, and (2) a sample of late-type, or blue, galaxies. The value of f_{HI} varies from as high as 3, for gas-rich galaxies with three times more mass in gas than in stars, to values less than 0.001 in gas-poor galaxies with more than 1000 times more mass in stars than in gas.

There are two important trends to note in the hydrogen gas content of the galaxies plotted in Figure 8.3. First, for both the early-type and late-type galaxies, the ratio of HI mass to stellar mass decreases with increasing stellar mass. Thus, larger values of f_{HI} are found for the galaxies with smaller stellar mass. The other trend is that f_{HI} varies with galaxy morphology. Lower values of f_{HI} are found for early-type galaxies (red galaxies), while larger values are found for late-type galaxies (blue galaxies). This trend with morphology confirms what we stated earlier — that late-type, or blue, galaxies are more gas rich than early-type, or red galaxies. The Milky Way is a massive spiral galaxy, with an HI gas mass of about 8×10^9 $\mathrm{M}_\odot$ (see Chapter 5) and a stellar mass of 6×10^{10} $\mathrm{M}_\odot$; therefore $f_{\mathrm{HI}} = 0.13$. We can see from Figure 8.3 that the atomic gas content of the Milky Way is typical of a massive late-type galaxy.

Most galaxy catalogues are based on optical images, and thus are biased toward galaxies that are optically bright. The galaxy samples shown in Figure 8.3 are all optically selected bright galaxies. Recently there have been a number of "blind" HI surveys in which large regions of the sky are systematically searched for HI emission. Such surveys are called "blind" because they are not based on any previous information about the presence or absence of galaxies. Note that these blind surveys are still biased, but instead of having an optical bias, they are biased toward HI gas-rich galaxies, and therefore such surveys are ideal for finding the most atomic hydrogen gas-rich galaxies in the local universe. One of the largest of these blind surveys is called the Arecibo Legacy Fast ALFA Survey (ALFALFA) and uses the Arecibo 305-m radio telescope with the Arecibo L-band Feed Array (ALFA) to map approximately 7000 square degrees of sky (see the paper by Riccardo Giovanelli and collaborators[1]). The HI spectrum of UGC 93 shown in Figure 8.2 is from the ALFALFA survey. Based on about 9400 galaxies from the ALFALFA survey, Shan Huang and colleagues[2] found galaxies with HI masses as large as 10^{11} $\mathrm{M}_\odot$ and galaxies with values of f_{HI} as large

[1] 2005, Astronomical Journal, vol. 130, p. 2598.
[2] 2012, Astrophysical Journal, vol. 756, p. 113.

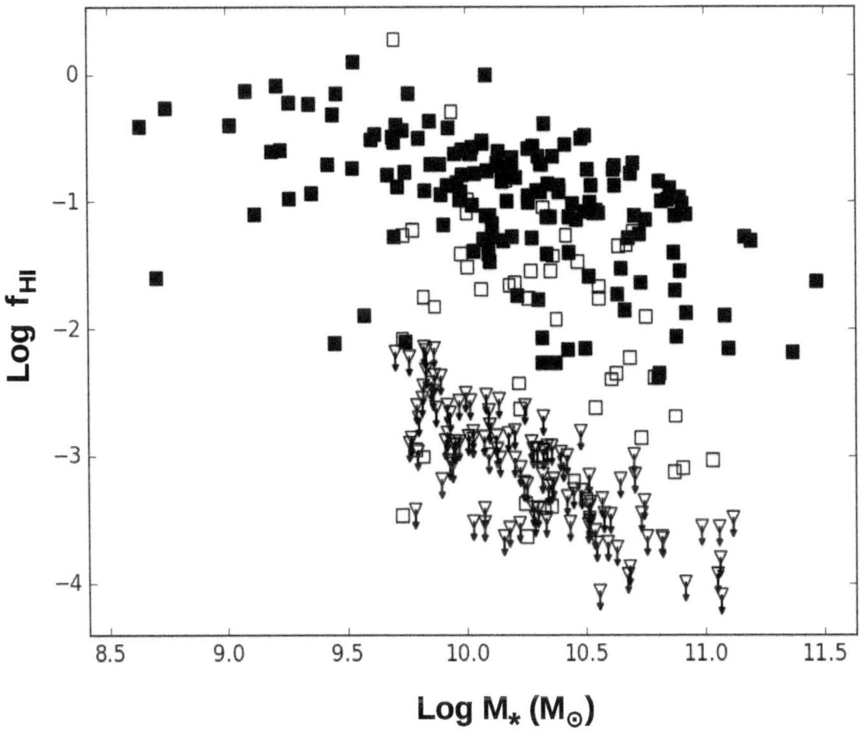

Figure 8.3: Plot of the ratio of HI gas mass to total stellar mass ($f_{\rm HI}$) versus the stellar mass for two samples of galaxies. One is a sample of late-type galaxies from the FCRAO Extragalactic CO Survey (see paper by Judith Young and collaborators, 1995, Astrophysical Journal Supplement, vol. 98. p. 219), indicated in the plot by filled symbols. The second is a sample of early-type galaxies from the ATLAS 3D survey (see paper by Michele Cappellari and colleagues, 2011, Monthly Notices of the Royal Astronomical Society, vol. 413 p. 813), indicated by open symbols. Many of the early-type galaxies were not detected in 21-cm HI emission and therefore only upper limits to the HI gas mass could be determined; these limits are indicated by triangles with downward pointing arrows. The data for these two samples were compiled by Michael McCrackan, who kindly provided this plot.

as 100 — both values are much larger than what is found for the optically selected galaxies shown in Figure 8.3.

These blind HI surveys have uncovered many optically faint galaxies that have a large mass of atomic gas. These galaxies are called low-surface-brightness (LSB) galaxies, and have ratios of HI masses to stellar masses of order unity or larger, so these galaxies are gas-dominated. Some LSB galaxies are large spiral galaxies, but unlike the spiral galaxies in most optically selected galaxy catalogues, these galaxies have low intensity at optical wavelengths. Because these galaxies are difficult to detect at optical wavelengths they are often under-represented in optical galaxy catalogues. An example of an LSB galaxy is UGC 93 (whose spectrum is shown in Figure 8.2), which is approximately 50 kpc in diameter. Such LSB galaxies may make up a large fraction of spiral galaxies. It is unclear why these gas-dominated galaxies are not forming stars more rapidly.

8.1.2 Imaging HI in Galaxies

With radio interferometers it is possible to obtain high resolution images of the HI emission in nearby galaxies. One of the large projects undertaken with the Jansky Very Large Array (JVLA) is called The HI Nearby Galaxy Survey (THINGS), which imaged the emission from the 21-cm line in 34 galaxies (see the paper by Fabian Walter and colleagues[3]). One of the spiral galaxies in this survey is NGC 3184 which is relatively nearby, at a distance of 11.1 Mpc, and is nearly face-on, only inclined by about 16 degrees with respect to the plane of the sky. An image of the integrated intensity of the HI 21-cm line from this galaxy is shown in Figure 8.4, accompanied by an optical image of the galaxy. Since the integrated intensity of the 21-cm line is proportional to the HI column density, this image reveals the distribution of atomic hydrogen across the galaxy. As in many spiral galaxies (including the Milky Way, see Section 5.1.3), the spatial extent of the HI gas in NGC 3184 extends far beyond the stars as traced by the optical light. Also like many spiral galaxies, including the Milky Way, NGC 3184 has a paucity of HI gas near its center. The HI emission in NGC 3184 shows spiral structure in the atomic gas, similar to the spiral arm pattern seen in the optical light. What is striking in the HI image is the presences of HI holes, regions devoid of HI emission. Ioannis Bagetakos and collaborators[4] identified 40 HI holes in this image ranging in diameter from 500 to 1400 pc. These holes are called HI super-shells. These are believed to be produced by the stellar winds and supernovae explosions associated with massive stars in OB associations and super star clusters that have swept away the gas in these regions. Hence, these massive stars have had a major impact on the distribution of gas in the disk of this galaxy.

The total integrated flux density for the HI emission in NGC 3184 is 105 Jy km s^{-1}. Assuming a distance of 11.1 Mpc, we can use Equation 8.1 to determine that the total HI mass for this galaxy is 3.1×10^9 $M_\odot$. The stellar mass of NGC 3184 has been estimated by Adam Leroy and colleagues[5] to be 2×10^{10} $M_\odot$. Therefore the HI mass to stellar mass ratio is about 0.15, very similar to that found in the Milky Way.

We discussed in Section 5.1.2 the complications involved with determining the rotation curve for the Milky Way both due to the fact that we are viewing the Milky Way from inside and because we are observing from an orbiting platform. Deriving the rotation curve for external galaxies is much simpler. Figure 8.5 shows a schematic of a rotating, nearly edge-on disk galaxy and the HI spectra that would be observed at three different positions along the galaxy's major axis. These HI spectra provide a measure of the mean rotation speed of the gas at different radial distances from the center. Since most disk galaxies, like the galaxy in the schematic, are not exactly edge-on, we must correct for inclination to determine the true rotation speed of the gas at different radial distances. Because single-dish measurements of most galaxies do not provide adequate resolution to measure the rotation curve, these curves are almost always derived from the HI images obtained with radio interferometers.

Note that the inclination angle, i, is defined as the angle between the plane of the galaxy and the plane of the sky, and therefore the measured velocity must be corrected by a factor of $1/\sin i$. For nearly face-on galaxies, such as NGC 3184 illustrated in Figure 8.4, the inclination correction can be very large and uncertain. Because of the large correction needed for nearly face-on galaxies, observations of galaxies closer to edge-on provide more accurate rotation curves. For this reason we will focus on the galaxy NGC 3198, which is a spiral galaxy at a distance of 13.9 Mpc with an inclination angle of about 70°. HI

[3]2008, Astronomical Journal, vol. 136, p. 2563.
[4]2011, Astronomical Journal, vol. 141, p. 23.
[5]2008, Astronomical Journal, vol. 136, p. 2782.

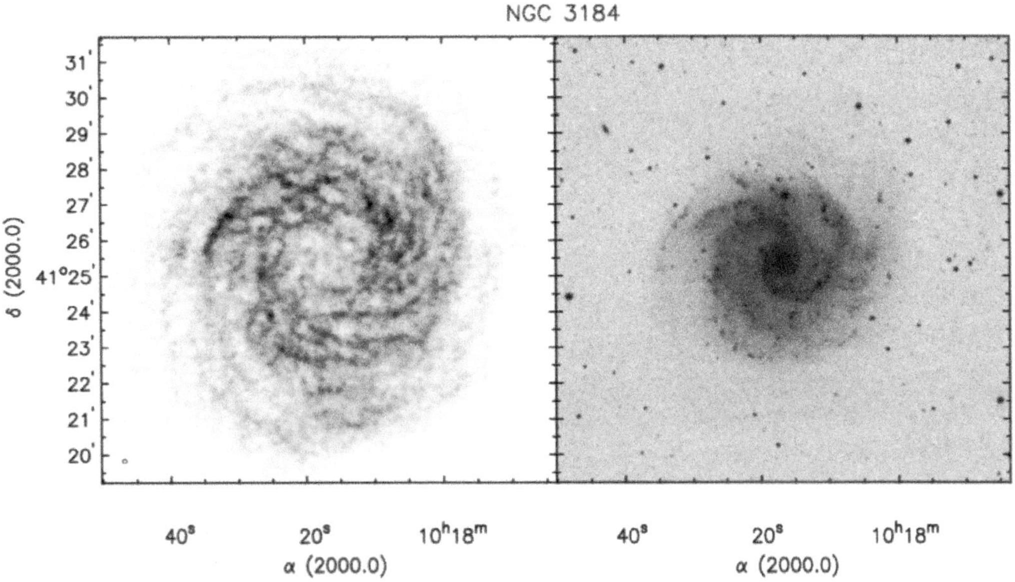

Figure 8.4: On the left is an image of the integrated intensity of the 21 cm HI line for the galaxy NGC 3184, while on the right is an optical image of the galaxy. The HI image was obtained with the JVLA as part of The HI Nearby Galaxy Survey (THINGS) and has an angular resolution of 6″. The HI emission has an angular extent of about 12′, which, at a distance of 11.1 Mpc, corresponds to a spatial extent of about 40 kpc. As is clear from the two images, the galaxy is much larger in HI than in optical light. This figure is from a paper by Fabian Walter and colleagues (2008, Astronomical Journal, vol. 136, p. 2563).

observations with the Westerbork Synthesis Radio Telescope (WSRT) of NGC 3198 were used to compute the rotation curve for this galaxy, which is tabulated in a paper by Tjeerd Sicco van Albada and colleagues[6] and includes a correction for the galaxy's inclination. We plot the rotation curve for this galaxy in Figure 8.6. Since the HI emission often extends well beyond the optical extent of the galaxy, the rotation curves derived from HI observations can be measured to a larger radius than would be possible with just optical spectra.

The gas in disk galaxies moves in nearly circular orbits about the center of the galaxy, bound by gravity. The HI rotation curve provides the orbital speed of gas in its nearly circular orbit about the galaxy as a function of radius, and this information can be used to deduce the distribution of mass as a function of radius. For a gas cloud or a star with mass m, in a circular orbit of radius R, we can equate the inward gravitational force to the centripetal force. For a spherically symmetric mass distribution, the gravitational force experienced by a gas cloud at distance R from the center of the galaxy is determined only by the mass of the galaxy enclosed within radius R, which we denote as $M(R)$. Setting the gravitational force equal to the centripetal force, we have

$$F_{grav} = \frac{GmM(R)}{R^2} = F_{cent} = \frac{mV^2(R)}{R},$$

[6] 1985, Astrophysical Journal, vol. 295, p. 305.

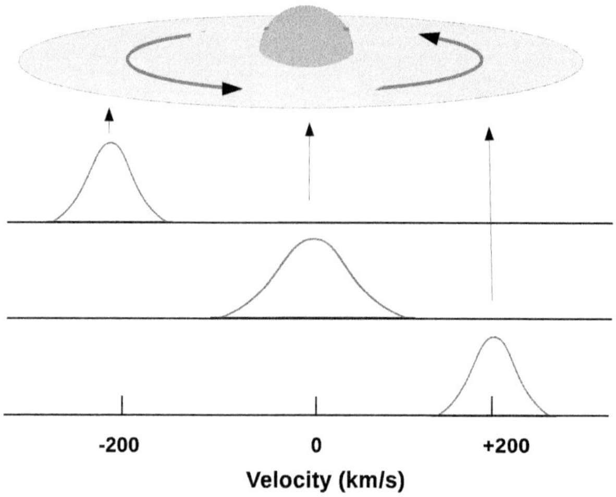

Figure 8.5: A schematic diagram of a rotating disk galaxy viewed nearly edge-on. The HI spectra at three positions along the major axis are shown. The spectra plot intensity of the HI emission versus velocity, where zero velocity is arbitrarily set to be the average (or systemic) velocity of the galaxy. The left spectrum emitted by the side of the galaxy moving toward us is blueshifted relative to the galaxy's average velocity. Similarly the right spectrum emanates from the side of the galaxy moving away from us, and thus is redshifted relative to the galaxy's average velocity. In the center spectrum, most of the orbital velocity of the gas is transverse to the line of sight, and so there is no net Doppler shift and the line is centered on the systemic velocity of the galaxy.

where $V(R)$ is the orbital velocity at radius R. This equation is strictly correct only for spherically symmetric mass distributions; however it is often applied to galaxies regardless of the mass distribution. Even for disk galaxies, such as NGC 3198, this expression is approximately correct. Solving for the enclosed mass we find

$$M(R) \approx \frac{V^2(R)\, R}{G}. \tag{8.2}$$

Since astronomers usually express velocity in km s^{-1}, galactic radius in kpc and mass in $M_\odot$, we can rewrite the above equation with these mixed units as

$$M(R) \approx 2.3 \times 10^5\, M_\odot \left(\frac{V(R)}{\text{km s}^{-1}}\right)^2 \left(\frac{R}{\text{kpc}}\right). \tag{8.3}$$

Therefore measurements of the orbital speed at radius R can be used to determine the mass enclosed within that radius.

Example 8.2:

Based on the rotation curve for NGC 3198 shown in Figure 8.6, estimate the total mass of the galaxy enclosed within a radius of 40 kpc.

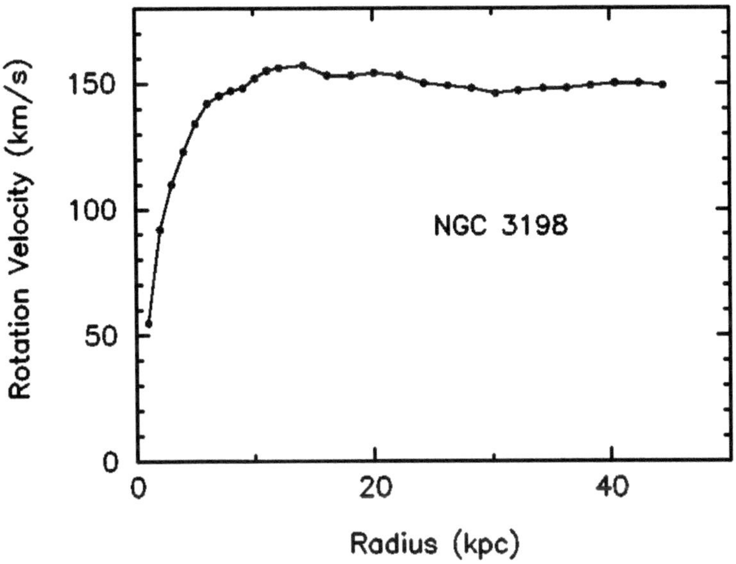

Figure 8.6: The rotation curve for the galaxy NGC 3198 based on HI observations obtained with the Westerbork Synthesis Radio Telescope. The plot shows the rotation speed as a function of galactic radius. The rotation curve was computed and tabulated by Tjeerd Sicco van Albada and colleagues (1985, Astrophysical Journal, vol. 295, p. 305).

Answer:
From the rotation curve we find that at a radius of 40 kpc, the rotation velocity is about 150 km s^{-1}. Using Equation 8.3, we find that

$$M(R) \approx 2.3 \times 10^5 \text{ M}_\odot \left(\frac{150 \text{ km s}^{-1}}{\text{km s}^{-1}}\right)^2 \left(\frac{40 \text{ kpc}}{\text{kpc}}\right) = 2.1 \times 10^{11} \text{ M}_\odot.$$

Note that the rotation curve for NGC 3198 remains relatively flat (constant velocity) at radii beyond about 10 kpc. This is a very significant aspect of the rotation curves of most disk galaxies. *Flat rotation curves* led astronomers to greatly rethink the models of galaxies and the Universe. For a flat rotation curve, Equation 8.3 implies that the enclosed mass increases linearly with radius. The total mass of stars and gas in NGC 3198 has been estimated by Leroy and colleagues[7] to be about 2.6×10^{10} $M_\odot$ — nearly an order of magnitude smaller than the mass estimated in Example 8.2 based on the rotation curve. The mass discrepancy is attributed to the presence of *dark matter*, and this dark matter is the dominant mass component of all galaxies. As in Figure 8.6, the observed velocities are not seen to start decreasing with increasing radii as far as they can be measured. This indicates that the edge of the dark matter distribution has not been reached, since once the radius exceeded the radius of the dark matter then the velocity would decline with $v(R) \propto R^{-1/2}$. The dark matter is therefore believed to reside in large halos that extend far beyond the extent of the stars and gas. The nature of the dark matter is still unknown, but it is unlikely to be composed of ordinary (or baryonic) matter. In NGC 3198, the mass in ordinary matter (stars and gas) makes up at most 10% of the total mass of the galaxy, and thus at least 90% of its mass is in dark matter. Determining the nature of dark matter is one of the top pursuits in astrophysics.

[7] 2008, Astronomical Journal, vol. 136, p. 2783.

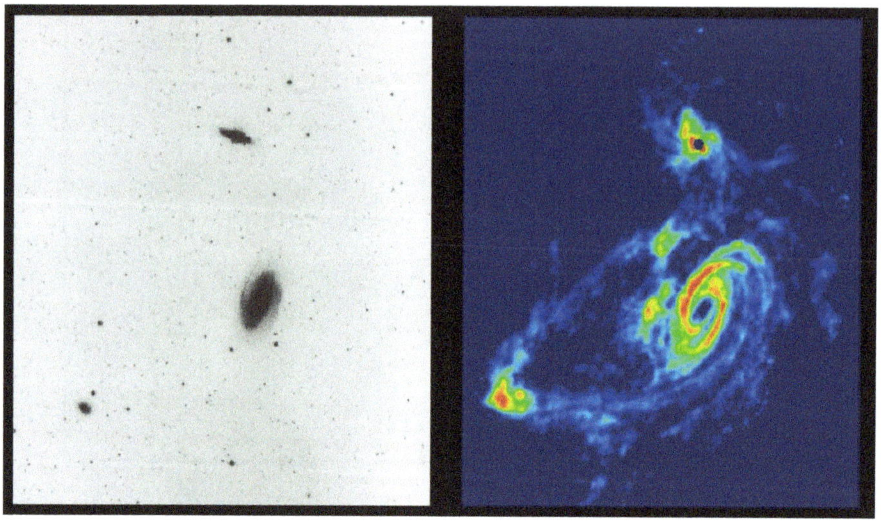

Figure 8.7: On the right is an image of the integrated intensity of the 21 cm HI line obtained with the National Radio Astronomy Observatory (NRAO) Very Large Array of the M81 group and on the left is the same region of the sky at optical wavelengths. The three prominent galaxies, from top to bottom, are M82, M81 and NGC 3077. The HI image shows the galaxies connected by HI filaments revealing gas that has been pulled from the galaxies by tidal interactions during close encounters. The image was provided by Min Yun.

Interferometer images of the HI emission from galaxies also provide valuable information on the distribution of atomic hydrogen. Such images are particularly useful in studying galaxy-galaxy interactions. The extended HI gas disks in galaxies can be readily disturbed by gravitational (or tidal) interactions due to close encounters with other galaxies. These interactions can pull gas out of a galaxy producing gas filaments, called tidal tails, that can extend to hundreds of kiloparsecs beyond the galaxy. A good example of such tidal tails is seen in the nearby M81 galaxy group. An HI image of this group is shown in Figure 8.7 and reveals gas filaments connecting the three prominent galaxy members, M81, M82 and NGC 3077 (for more information see the paper by Min Yun and collaborators[8]). An optical image of this galaxy group is also shown in Figure 8.7 and it shows little sign of any interaction. These interactions between galaxies in clusters and groups are relatively common and can remove gas from galaxies, changing their overall gas content. In more extreme cases, galaxy interactions and collisions can cause major morphological changes in galaxies, such as the transformation of disk galaxies into elliptical galaxies.

8.2 MOLECULAR GAS IN GALAXIES

In addition to atomic hydrogen gas, most galaxies, particularly late-type galaxies, also contain molecular gas. For the reasons discussed in Section 4.2.3, the primary tracers of the molecular gas are millimeter wavelength rotational transitions of carbon monoxide (CO). The lowest frequency rotational transition of CO is at 115 GHz, corresponding to a

[8] 1994, Nature, vol. 372, p. 530.

wavelength of 2.6 mm, a wavelength at which even modest sized radio telescopes achieve angular resolutions better than $1'$. Therefore, for large nearby galaxies, it is possible to resolve the distribution of molecular gas even without the use of radio interferometers. An example of single-dish telescope observations is presented in Figure 8.8, which shows CO J=1-0 spectra obtained along the major axis of the galaxy NGC 3184 (the same galaxy shown in Figure 8.4). The CO emission is strongest at the center of the galaxy and diminishes outward.

8.2.1 Molecular Gas Mass

We can use the empirical relation between the integrated intensity of the CO J=1-0 spectral line and the column density of molecular hydrogen, given by Equation 5.6, to determine the mass of molecular gas in galaxies. Rewriting this equation here

$$N(H_2) = \left(\frac{X_{CO}}{\mathrm{cm^{-2}\ (K\ km\ s^{-1})^{-1}}}\right)\left(\frac{\int T_B\, dv}{\mathrm{K\ km\ s^{-1}}}\right)\ \mathrm{cm^{-2}}.$$

The value for the X-factor derived for the Milky Way (see Section 5.2.2) is $X_{CO} = 2 \times 10^{20}$ $\mathrm{cm^{-2}}$ $\mathrm{(K\ km\ s^{-1})^{-1}}$. If we assume the same X-factor applies to other galaxies, then the integrated intensity of the CO $J = 1$-0 line measured in any position within a galaxy provides an estimate of the column density of molecular hydrogen along the line of sight. For the following discussion it is useful to convert the brightness temperature back to intensity units using Equation 3.13. For the $J = 1$-0 line of CO, we can rewrite the column density as

$$N(\mathrm{H_2}) = 2.45 \times 10^{14}\ \mathrm{cm^{-2}} \left(\frac{X_{CO}}{\mathrm{cm^{-2}\ (K\ km\ s^{-1})^{-1}}}\right)\left(\frac{\int I_\nu\, dv}{\mathrm{erg\ s^{-1}\ cm^{-1}\ Hz^{-1}\ sr^{-1}\ km\ s^{-1}}}\right).$$

Unlike single-dish observations of the 21-cm line of HI, single-dish CO observations often resolve nearby galaxies. We infer the mass of molecular hydrogen within the main beam solid angle of the telescope from the integral of the column density over the area enclosed by the telescope's main beam solid angle. As in the case of HI (see Section 8.1.1), we can convert the integral over area to an integral over solid angle. Therefore, the mass of molecular hydrogen enclosed by the main beam is

$$M(\mathrm{H_2}) = m_{\mathrm{H_2}}\, d^2 \int N(\mathrm{H_2})\, d\Omega,$$

where the integral is over the main beam solid angle, d is the distance of the galaxy and m_{H_2} is the mass of a hydrogen molecule. Inserting the expression for the column density, we find

$$M(\mathrm{H_2}) = 2.45 \times 10^{14}\ \mathrm{g} \left(\frac{X_{CO}}{\mathrm{cm^{-2}\ (K\ km\ s^{-1})^{-1}}}\right)\left(\frac{m_{\mathrm{H_2}}}{\mathrm{g}}\right)\left(\frac{d}{\mathrm{cm}}\right)^2$$
$$\times \left(\frac{\int\int I_\nu\, d\Omega\, dv}{\mathrm{erg\ s^{-1}\ cm^{-1}\ Hz^{-1}\ km\ s^{-1}}}\right).$$

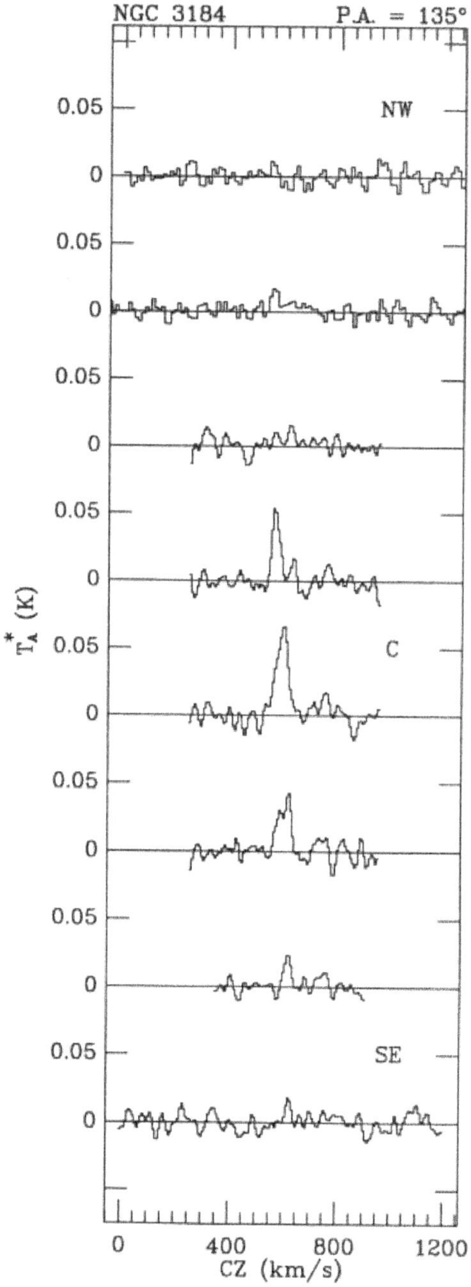

Figure 8.8: A series of CO J=1-0 spectra obtained along the major axis of NGC 3184 with the Five College Radio Astronomy Observatory (FCRAO) 14-m telescope. The spectra show the antenna temperature versus velocity. The spectrum marked with the letter c was obtained toward the center of the galaxy. The spectra above and below the center spectrum are offset in steps of 45″, the beam size of the FCRAO telescope. At the 11.1 Mpc distance, 45″ corresponds to 2.4 kpc. The spectra shown cover the linear extent of the optical image shown in Figure 8.4 along a diagonal from the lower left to the upper right corner. This figure is obtained from the FCRAO Extragalactic CO Survey published by Judith Young and collaborators (1995, Astrophysical Journal Supplement, vol. 98, p. 219).

The integral of intensity over the beam solid angle is flux density, F_ν. Substituting in for the value of the mass of the hydrogen molecule and expressing the value for X_{CO} in terms of that found for the Milky Way and expressing the distance in parsecs and the flux density in jansky, we have

$$M(\mathrm{H_2}) = 7.8 \times 10^3\ \mathrm{M_\odot} \left(\frac{X_{CO}}{2 \times 10^{+20}\ \mathrm{cm^{-2}\ (K\ km\ s^{-1})^{-1}}}\right)\left(\frac{d}{\mathrm{Mpc}}\right)^2\left(\frac{\int F_\nu\, dv}{\mathrm{Jy\ km\ s^{-1}}}\right). \quad (8.4)$$

Thus, if we know the distance to a galaxy and measure the integrated flux density in the CO J=1-0 line, we can estimate the mass of molecular gas.

Mass estimates of molecular gas in galaxies based on Equation 8.4 must assume a value for the X-factor. The X-factor is best measured in the interstellar medium of the Milky Way (see Section 5.2), and for late-type galaxies in the local universe the Milky Way X-factor is often applied. However, in general, one must be careful, as there are many reasons to believe that the Milky Way X-factor may not apply to all galaxies. A good review of this subject is that by Alberto Bolatto, Mark Wolfire and Adam Leroy[9]. Based on estimates of the X-factor in other galaxies, Bolatto and colleagues suggest that the X-factor increases with decreasing metallicity. Recall that astronomers refer to elements heavier than helium as "metals", and the metallicity is the ratio of the abundance of metals to hydrogen. With lower metallicity there are fewer carbon and oxygen atoms to form CO, and thus the emission in CO can be suppressed, even where there are still large amounts of molecular hydrogen. However, since the CO emission is often optically thick, the appropriate value for the X-factor does not simply scale linearly with the abundance of CO, but is more complicated. Therefore, for galaxies with lower metallicities, a larger X-factor may be needed to account for the paucity of CO emission, resulting in a larger molecular gas mass for a given integrated flux density of CO. There is a mass-metallicity relation found for most galaxies, where the metallicity increases with increasing galaxy mass. Therefore, the dependence of the X-factor on the metallicity of the galaxy means that the X-factor may correlate with the mass of the galaxy.

Bolatto and colleagues additionally suggest that the X-factor may be smaller in galaxies with very high rates of star formation, such as starburst galaxies, where the molecular gas properties may be different than in the Milky Way. An example of such a galaxy is the ultra-luminous infrared galaxy Arp 220, which is discussed further in Section 8.2.3. The reason that the X-factor may be smaller in such a galaxy is unclear, but may relate to the large gas motions in excess of what is expected for gas in virial equilibrium. Because of the uncertainty in the X-factor, one must be very cautious in converting the integrated CO flux density using an X-factor into molecular gas mass.

If the entire galaxy observed falls within the telescope beam, then the mass given by Equation 8.4 corresponds to the total molecular hydrogen mass of the galaxy. Alternatively, if the galaxy is resolved, one can obtain the total molecular hydrogen gas mass by integrating over the galaxy. As was the case for the atomic gas, we can calculate the total mass from all molecules by multiplying the hydrogen mass by 1.36. Finally, we note that the product of the distance squared and the velocity integrated flux density of the line is often called the *CO luminosity* and has units of Jy km s^{-1} pc^2.

Example 8.3:

The CO J=1-0 spectrum toward the center of NGC 3184, shown in Figure 8.8, has an integrated flux density of 134 Jy km s^{-1} and the galaxy is at a distance of 11.1

[9] 2013, Annual Reviews of Astronomy and Astrophysics, vol. 51, pg. 267.

Mpc. The diameter of the radio telescope's main beam at the galaxy's distance is about 2.4 kpc. What is the molecular hydrogen mass within the central 2.4 kpc region of this galaxy?

Answer:
We can use Equation 8.4 relating the mass within the telescope main beam to the measured integrated flux density and distance. Therefore, we have

$$M(\mathrm{H_2}) = 7.8 \times 10^3 \ \mathrm{M_\odot} \left(\frac{11.1 \ \mathrm{Mpc}}{\mathrm{Mpc}}\right)^2 \left(\frac{134 \ \mathrm{Jy \ km \ s^{-1}}}{\mathrm{Jy \ km \ s^{-1}}}\right) = 1.3 \times 10^8 \ \mathrm{M_\odot}.$$

8.2.2 Imaging CO in Galaxies

Radio telescopes operating at millimeter wavelengths often have many detectors in their focal plane, called focal plane arrays, so they can rapidly image the sky. One example of a focal plane array is the Heterodyne Receiver Array (HERA) on the IRAM 30-m telescope. The HERA CO-Line Extragalactic Survey (HERACLES) led by Adam Leroy and collaborators[10] used the HERA detector to image 48 nearby galaxies in the CO J=2-1 line, including NGC 3184. An image of the integrated CO intensity of NGC 3184 is shown in Figure 8.9. This image can be compared with images of the HI emission and optical emission from this galaxy shown in Figure 8.4. Note that the CO image and optical image are remarkably similar, with both having nearly the same angular extent and showing very similar spiral features. Thus the starlight and the molecular gas have very similar distributions, unlike the atomic hydrogen which is much more extended. Note that there is very little atomic hydrogen at the center of NGC 3184, while there is a substantial amount of molecular gas. Thus, like the Milky Way, the interstellar medium in the inner region of NGC 3184 is dominated by molecular gas. We see that the distributions of stars, molecular gas and HI gas in NGC 3184 are very similar to what we found for the Milky Way (see Chapter 5). On the other hand, Adam Leroy and colleagues find a total molecular gas mass for NGC 3184 of roughly 1.7×10^9 $\mathrm{M_\odot}$, about one-half of the total atomic gas mass, which is much larger than in the Milky Way where the molecular gas fraction is only about one-tenth.

CO surveys, such as HERACLES, have provided estimates of the molecular hydrogen mass in numerous nearby galaxies. Like the atomic gas mass in galaxies, the molecular gas mass also spans a wide range of values, from less than 10^7 $\mathrm{M_\odot}$ to as large as 10^{10} $\mathrm{M_\odot}$. As we did for the HI mass, we can normalize the molecular hydrogen mass by the stellar mass of the galaxy. In Figure 8.10 we plot the ratio, f_{H_2}, of molecular gas mass to stellar mass as a function of the galaxy's stellar mass. Even the most molecular gas-rich galaxies have molecular gas masses less than their stellar mass. Furthermore, the most gas-poor galaxies have molecular gas masses over a thousand times smaller than their stellar mass. Just as for the atomic gas, the value of f_{H_2} for late-type galaxies is larger than for early-type galaxies. Thus late-type galaxies are more gas-rich in both atoms and molecules than early-type galaxies. We also see that f_{H_2} is smaller for early-type galaxies with larger stellar masses. Thus, the most massive early-type galaxies are more molecular gas-poor than the lower mass early-type galaxies. For the late-type galaxies, though, galaxies with a broad range of stellar masses all have molecular gas masses typically about one-tenth of their stellar mass, as is the case for NGC 3184. The Milky Way has a molecular gas mass of about 1×10^9 $\mathrm{M_\odot}$ (see Section 5.2.2), and thus it has a value of $f_{H_2} = 0.02$, somewhat smaller than most spiral galaxies.

[10] 2009, Astronomical Journal, vol. 137, p. 4670.

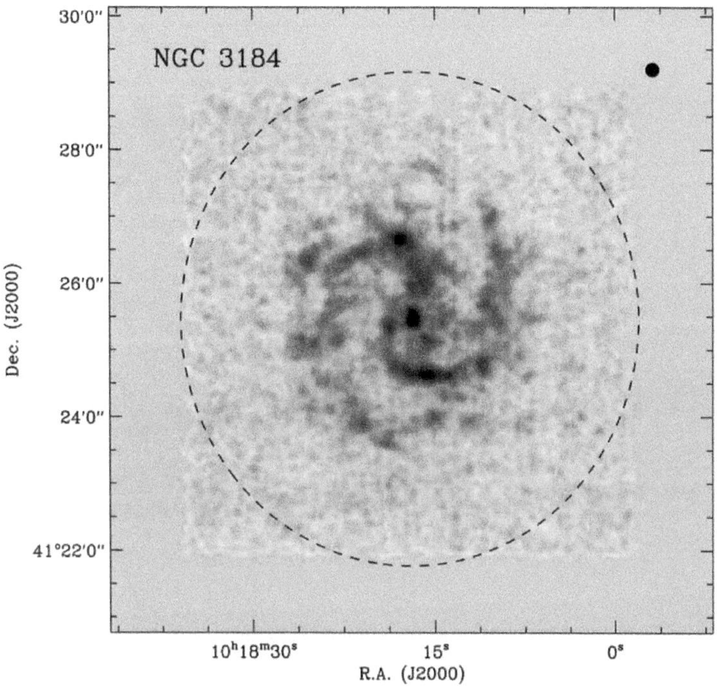

Figure 8.9: An image of the integrated intensity of the CO J=2-1 line for the galaxy NGC 3184. This image of the CO integrated intensity has an angular resolution of $13''$ (shown as a filled black circle in the upper right-hand corner of the image). The distribution of emission is remarkably similar to that seen in optical light as shown in Figure 8.4. The CO image is part of the HERA CO-Line Extragalactic Survey (HERACLES) by Adam Leroy and colleagues (2009, Astronomical Journal, vol. 137, p. 4670) and obtained using the IRAM 30-m radio telescope.

Although data from single-dish telescopes can provide information about the distribution of molecular emission in nearby galaxies, for more distant galaxies or to achieve higher resolution, radio interferometers that can operate at millimeter wavelengths must be used. One such interferometer is the Atacama Large Millimeter/Submillimeter Array (ALMA) located in the Atacama desert in Chile, which can obtain millimeter wavelength images with angular resolutions better than $0.1''$. For nearby galaxies, observations with ALMA can resolve individual molecular clouds.

8.2.3 Other Molecules in Galaxies

In addition to CO, many other molecules have been detected in galaxies. As we described for the Milky Way (see Section 5.2.3), we can use the emission from molecules such as CS, HCN, or HCO^+ to estimate the density and temperature of the molecular gas in these galaxies. Most molecules require much higher density than CO to produce detectable emission (see Section 4.2.3); therefore these other molecules probe the physical conditions in the denser molecular gas. An example of a galaxy with a rich molecular line spectrum is Arp 220, illustrated in Figure 8.11. The study by Thomas Greve and collaborators[11] found that the bulk of the molecular gas in Arp 220 was warm (temperature between 45 and 120 K) and

[11] 2009, Astrophysical Journal, vol. 692, p. 1432.

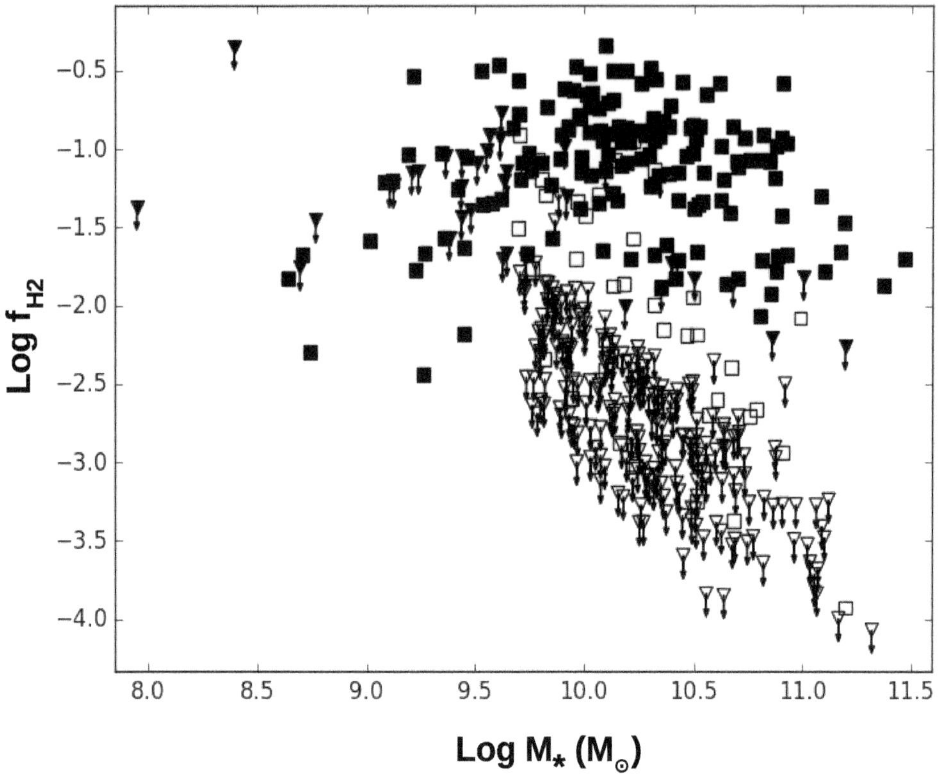

Figure 8.10: Plot of the ratio of molecular hydrogen gas mass to total stellar mass (f_{H_2}) versus the stellar mass for two samples of galaxies. One sample is from the FCRAO Extragalactic CO survey (see paper by Judith Young and collaborators, 1995, Astrophysical Journal Supplement, vol. 98. p. 219), which is a sample of late-type galaxies and indicated in the plot by filled symbols. The other sample is from the ATLAS 3D survey of early-type galaxies (see Michele Cappellari and colleagues, 2011, Monthly Notices of the Royal Astronomical Society, vol. 413, p. 813), indicated by open symbols. For both samples, some of the galaxies were not detected in CO and therefore only upper limits to the molecular gas mass could be estimated; these limits are indicated by triangles with downward pointing arrows. This plot was provided by Michael McCrackan.

dense (densities between 10^5 and 10^6 cm^{-3}). Arp 220 has a peculiar morphology that is likely a result of the collision of two spiral galaxies that are now in the process of merging. Arp 220 is a prototype for ultraluminous infrared galaxies, or ULIRGs. ULIRGS, including Arp 220, are very gas-rich and are forming stars at a prodigious rate. Although Arp 220 is somewhat distant at 77 Mpc, it is one of the brighter galaxies at infrared wavelengths, and numerous molecular species have been detected. At the distance of Arp 220, the observations shown do not resolve individual dense molecular cloud cores. Nevertheless, these observations provide important information about the fraction of dense molecular gas in these galaxies and the average physical and chemical properties of the dense star-forming gas. Interferometers, such as ALMA, can resolve the individual giant molecular clouds in Arp 220.

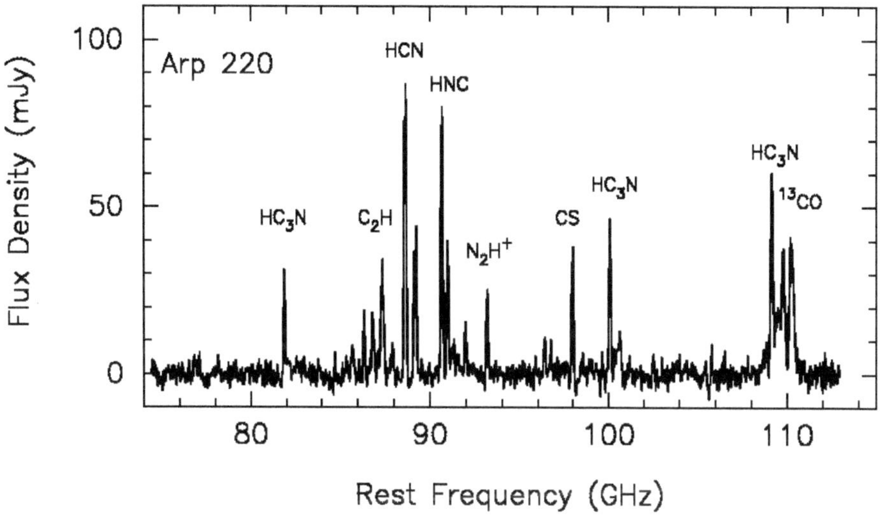

Figure 8.11: A spectrum of the galaxy Arp 220 covering a frequency range from 74 to 102 GHz. The spectrum shows numerous molecular rotational emission lines. Some of the stronger lines have been labelled. These data were obtained using the Redshift Search Receiver (RSR) on the Large Millimeter Telescope (LMT) and were provided by Peter Schloerb.

8.3 RADIO CONTINUUM EMISSION FROM GALAXIES

In addition to spectral line emission, galaxies also produce continuum emission at radio wavelengths. We start by examining the global radio emission from the galaxy NGC 4945. NGC 4945 is a normal spiral galaxy with properties similar to the Milky Way. In Figure 8.12 we plot the integrated flux density of the galaxy as a function of frequency. Each measurement in this figure represents the average flux density over a broad range of frequencies. The poor spectral resolution precludes measuring spectral lines, as the spectral lines emit over a very narrow range of frequencies and so their emission is overwhelmed by the continuum radio emission from the galaxy. The spectrum in Figure 8.12 covers a frequency range from 4×10^8 Hz (0.4 GHz) to 1.2×10^{13} Hz (12,000 GHz) or a wavelength range from 75 cm to 0.0025 cm, and thus covers much of the radio and far-infrared parts of the electromagnetic spectrum.

The radio emission seen in Figure 8.12 arises from several processes. The emission at frequencies below about 100 GHz is produced by a combination of bremsstrahlung (thermal free-free) emission from ionized gas (see Section 3.2.4) and synchrotron emission from relativistic electrons interacting with a magnetic field (see Section 3.3). In most galaxies, at these long radio wavelengths, the synchrotron emission dominates over bremsstrahlung emission. You may recall from Chapter 3 that the synchrotron radiation has a power-law spectrum, decreasing with increasing frequency, similar to that observed for NGC 4945, while bremsstrahlung emission has a relatively flat or increasing spectrum with increasing frequency. At frequencies greater than 100 GHz, the radio emission is dominated by thermal emission from dust. The dust particles are heated by the absorption of ultraviolet and optical light from the stars in the galaxy, and the emission we see is the thermal radiation from these heated dust particles. Since most of the dust is relatively cold, with temperatures of about 20 K (see Section 5.3), the dust has its peak intensity at infrared wavelengths. In star-forming galaxies, such as NGC 4945, ultraviolet light arises from newly-formed stars, but these stars are often still surrounded by dusty cloud material which absorbs much of

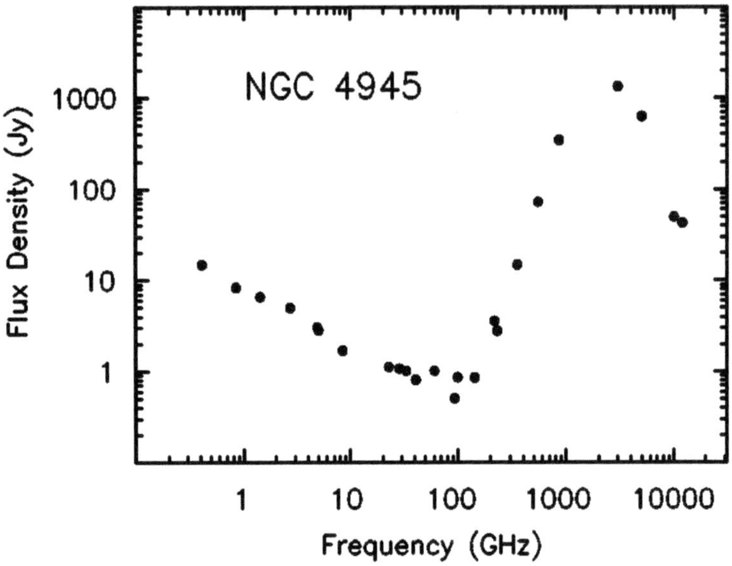

Figure 8.12: The flux density for the galaxy NGC 4945 is plotted as a function of frequency and illustrates the typical radio continuum emission found for normal spiral galaxies. At frequencies below 100 GHz the emission is produced by a combination of bremsstrahlung and synchrotron emission, while at frequencies greater than 100 GHz the emission is produced by thermal radiation from dust. These data were provided by Min Yun.

the light from these newly formed stars. The energy is then re-radiated by the dust at much longer wavelengths. Such galaxies are often most luminous in the infrared, although most of the energy was originally emitted at ultraviolet and optical wavelengths.

8.3.1 Dust Emission

As we discussed in Section 5.3.2 for the Milky Way, measurements of the dust emission at short radio wavelengths can be used to determine both the amount and distribution of dust in galaxies. Furthermore, the dust continuum emission complements the spectral line emission in probing the interstellar medium in galaxies. In Figure 8.13 we show an image of the galaxy NGC 3184 at a wavelength of 500 μm or a frequency of 600 GHz. At this frequency the emission is dominated by thermal dust emission. Earlier we showed images of this galaxy in optical light, HI emission and CO emission (see Figures 8.4 and 8.9). Although the angular resolution of the dust image shown in Figure 8.13 is much poorer than the other images, the dust reveals a spiral pattern and has a similar distribution to the CO emission.

An image of the dust emission can be readily converted to an image representing the mass column density of dust (see Section 5.3.2). The mass column density of dust is a measure of the mass of dust per unit area of the galaxy. By integrating the mass column density over the area of the galaxy, we can determine the total dust mass. The dust emission at radio wavelengths is generally optically thin, so we can start with Equation 5.9 and rearrange it to express the mass column density of dust, σ_d, in terms of the observed dust intensity,

$$\sigma_d = \frac{I_\nu}{\kappa_\nu^m \, B_\nu(T_d)},$$

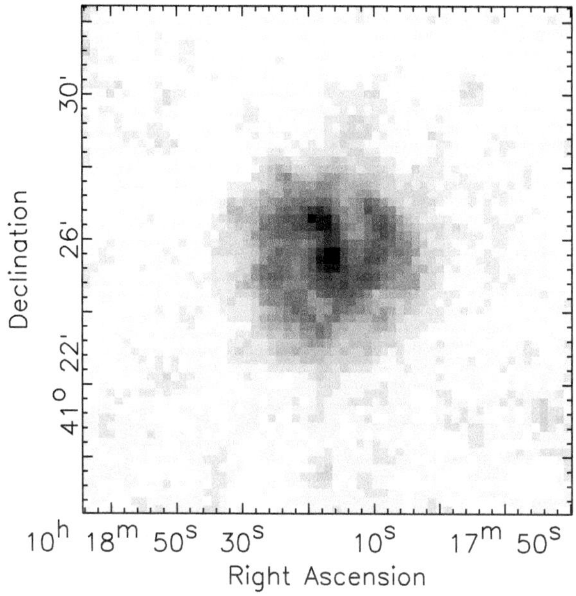

Figure 8.13: An image of the galaxy NGC 3184 at a wavelength of 500 μm obtained with the Herschel Space Telescope. At this wavelength the emission is dominated by the thermal radiation from dust. This image has much poorer angular resolution than the images of this galaxy in optical light, HI emission and CO emission (see Figures 8.4 and 8.9). This image was made using data from the NASA/Extragalactic Database (NED), which is operated by the Jet Propulsion Laboratory, California Institute of Technology, under contract with the National Aeronautics and Space Administration and was obtained as part of the KINGFISH project conducted by Rob Kennicutt and collaborators (2011, Publications of the Astronomical Society of the Pacific, vol. 123, p. 1347).

where κ_ν^m is the dust mass absorption coefficient (see Section 2.1.1). To compute the mass column density of dust, we need to know the dust temperature and the dust mass absorption coefficient or opacity. For dust in the Milky Way, the opacity at a wavelength of 0.1 cm (1000 μm) is approximately 1 cm^{-2} g^{-1}, and scales as $\lambda^{-\beta}$, where β is empirically found to lie between 1 and 2 (see Section 5.3.2). The dust temperature can be determined from data like that shown in Figure 8.12, which provides the flux density as a function of frequency, and Equation 5.10. For example, the dust emission in NGC 4945 peaks at a frequency of about 2000 GHz; if we assume β is 1.5, then the average dust temperature would be about 22 K. Remember that the physical area of a galaxy is related to its distance and the solid angle it subtends (see Equation 1.6). Therefore, the total dust mass, M_d, is given by

$$M_d = \int \sigma_d \, dA = \int \frac{I_\nu}{\kappa_\nu^m \, B_\nu(T_d)} \, d^2 d\Omega,$$

where d is the galaxy's distance and the integral is over the solid angle of the galaxy. Approximating the dust temperature and dust absorption coefficient as roughly constant across the galaxy, we can move them outside the integral and this expression becomes

$$M_d = \frac{d^2}{\kappa_\nu^m \, B_\nu(T_d)} \int I_\nu \, d\Omega.$$

The integral is just the flux density, F_ν, so we can rewrite the dust mass as

$$M_d = \frac{F_\nu \, d^2}{\kappa_\nu^m \, B_\nu(T_d)}. \tag{8.5}$$

Hence if we know the dust temperature and adopt a value of the dust opacity, κ_ν^m, we can use the measured flux density at these high radio frequencies to determine the total dust mass of a galaxy.

Example 8.4:

The total flux density of NGC 3184 at a wavelength of 500 μm is 6.22 Jy (reported by Brent Groves and collaborators[12]). We assume the dust has a temperature of 25 K and scaling the dust opacity given earlier, assuming $\beta = 1.5$, we find that at a wavelength 500 μm the opacity is 2.8 cm^2 g^{-1}. What is the total mass of dust in NGC 3184?

Answer:
Remember that the distance to NGC 3184 is 11.1 Mpc. We can use Equation 8.5 which relates the dust mass to the measured flux density and distance of a galaxy. However, we must first compute the Planck function for the dust at a wavelength of 500 μm or a frequency of 600 GHz (6×10^{11} Hz). At this wavelength and for a temperature of 25 K, we cannot use the Rayleigh-Jeans approximation, but must use the full form of the Planck function (Equation 3.2), which is

$$B_\nu(T_d) = \frac{2h\nu^3}{c^2}\left[\exp\left(\frac{h\nu}{kT_d}\right) - 1\right]^{-1} = \frac{2(6.63 \times 10^{-27} \text{ erg s}^{-1})(6 \times 10^{11} \text{ Hz})^3}{(3 \times 10^{10} \text{ cm s}^{-1})^2}$$
$$\times \left[\exp\left(\frac{(6.63 \times 10^{-27} \text{ erg s}^{-1})(6 \times 10^{11} \text{ Hz})}{(1.38 \times 10^{-16} \text{ erg K}^{-1})(25 \text{ K})}\right) - 1\right]^{-1}$$
$$= 1.47 \times 10^{-12} \text{ erg s}^{-1} \text{ cm}^{-2} \text{ Hz}^{-1}.$$

The dust mass is then given by

$$M_d = \frac{F_\nu \, d^2}{\kappa_\nu^m \, B_\nu(T_d)} = \frac{(6.22 \times 10^{-23} \text{ erg s}^{-1} \text{ cm}^{-2} \text{ Hz}^{-1})\,(3.43 \times 10^{25} \text{ cm})^2)}{(2.8 \text{ cm}^2 \text{ g}^{-1})\,(1.47 \times 10^{-12} \text{ erg s}^{-1} \text{ cm}^{-2} \text{ Hz}^{-1})}$$
$$= 1.78 \times 10^{40} \text{ g or } 8.9 \times 10^6 \text{ M}_\odot.$$

The total mass of atomic and molecular gas in NGC 3184 is estimated to be 4.8×10^9 M$_\odot$; therefore the gas mass is a few hundred times larger than the dust mass. This ratio is somewhat larger than that measured in the Milky Way (see Section 5.3.2).

It should not be surprising that the dust mass, like the gas mass, varies enormously between galaxies, ranging from less than 10^4 M$_\odot$ to greater than 10^9 M$_\odot$. Since the properties of galaxies vary enormously, it makes sense to normalize the dust mass to other galactic properties as we did for the gas mass. For the gas mass we normalized by the galaxy's stellar mass. If we do the same for the dust, we find that the dust mass to stellar mass ratio has many of the same trends we discussed for the gas mass. For instance, the dust mass to stellar mass ratio for early-type galaxies (elliptical and S0 galaxies) is about 10

[12] 2015, Astrophysical Journal, vol. 799, p. 96.

to 100 times smaller than that for late-type galaxies (spiral galaxies). However, it is more common to compare the dust mass to the gas mass and not the stellar mass. The ratio of gas mass to dust mass found for many large nearby galaxies is around 100, similar to the Milky Way. However, this ratio is found to vary with the metallicity of the galaxy (remember that metallicity refers to the abundance of elements heavier than helium relative to hydrogen). The variations of the gas-to-dust ratio with metallicity can be seen in a paper by David Fisher and collaborators[13]. For galaxies with metallicities similar to that of the Milky Way, the gas-to-dust ratio is also similar to the Milky Way. However, galaxies with smaller metallicities are found to have larger gas-to-dust ratios, and *vice versa*. This correlation should not be surprising, since the dust is composed primarily of metals (such as oxygen, silicon, carbon, and iron), and so galaxies containing small amounts of metals relative to hydrogen should have less dust relative to gas and thus a larger gas-to-dust ratio. The gas-to-dust ratio of galaxies varies from less than 10 for the most metal-rich galaxies to over 10,000 for the most metal-poor.

8.3.2 Long Wavelength Radio Continuum Emission

As we have discussed, the radio continuum emission in normal galaxies at longer radio wavelengths is a combination of bremsstrahlung emission from HII regions and synchrotron emission produced by relativistic electrons. Most relativistic electrons in normal galaxies result from Type II or Type Ib supernovae; both types result from core-collapse. Both HII regions and core-collapse supernovae are only produced by massive stars, and since massive stars are very short-lived, the long wavelength radio continuum emission is a tracer of very recent massive star formation (within the last 10^8 yrs). Therefore, the long wavelength radio continuum flux density in a galaxy provides a measure of the recent star formation rate (the mass of stars formed per unit time) in that galaxy; see the review by James Condon[14]. Although the long wavelength radio continuum emission is a consequence of the recent formation of massive stars, if we assume that stars of all masses are formed together according to some initial mass function (the number of stars as a function of mass), we can estimate the total star formation rate based on the measured radio continuum flux density. The star formation rate (SFR) for stars more massive than 0.1 $M_\odot$ is estimated by Condon and collaborators[15] to be

$$SFR \sim 0.15 \, M_\odot \, yr^{-1} \left(\frac{d}{Mpc}\right)^2 \left(\frac{F_\nu(1.4 \text{ GHz})}{Jy}\right), \tag{8.6}$$

where d is the distance and $F_\nu(1.4 \text{ GHz})$ is the flux density of the galaxy at a frequency of 1.4 GHz.

Obtaining images of the radio continuum emission from galaxies at these long radio wavelengths requires radio interferometers. An interferometer image of the 22-cm wavelength continuum emission from NGC 3184 is shown in Figure 8.14. Although the emission at this wavelength is quite weak, one can discern spiral structure similar to that seen in the optical image shown in Figure 8.4. Since the optical light from this galaxy is also dominated by the most massive stars, it may not be surprising that these two images are similar. In Figure 8.14, one can also see at the center of NGC 3184 a bright point of radio continuum emission. This emission is due to weak AGN activity associated with a central supermassive

[13] 2014, Nature, vol. 505, p. 186.
[14] 1992, Annual Reviews of Astronomy and Astrophysics, vol. 30, p. 575.
[15] 2002, Astronomical Journal, vol. 124, p. 675.

blackhole (see Chapter 9). When using the radio continuum emission to estimate the star formation rate it is important to remove any contribution from the AGN.

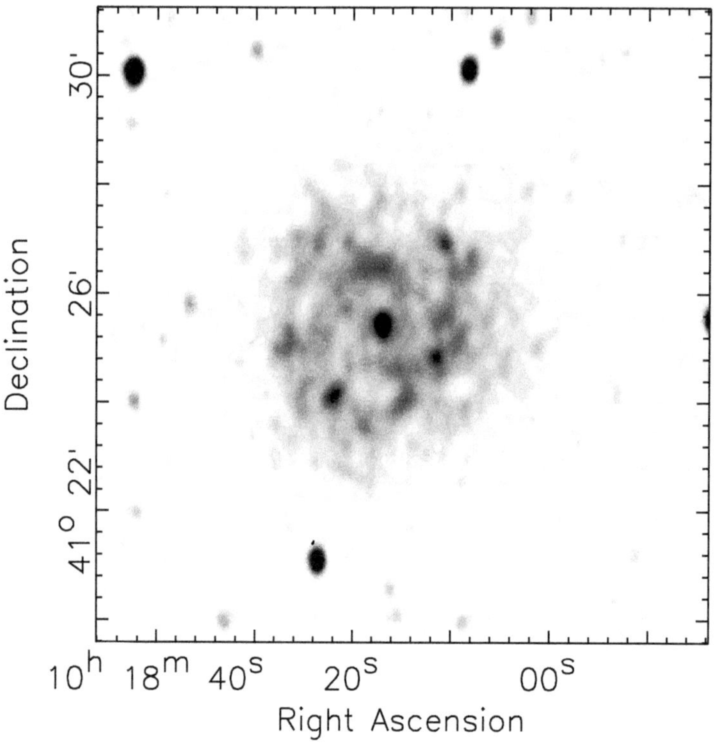

Figure 8.14: An image of the continuum emission from the galaxy NGC 3184 at a wavelength of 22 cm obtained with the Westerbork Synthesis Radio Telescope (WSRT). This image was made using data from the NASA/Extragalactic Database (NED), which is operated by the Jet Propulsion Laboratory, California Institute of Technology, under contract with the National Aeronautics and Space Administration. The data were obtained as part of a study by Robert Braun and collaborators (2007, Astronomy and Astrophysics, vol. 461, p. 455).

Example 8.5:

The integrated flux density of NGC 3184 at a frequency of 1.4 GHz is 0.08 Jy (reported by Braun and collaborators[16]). The AGN adds negligibly to the flux density and can be ignored. NGC 3184 is at a distance of 11.1 Mpc. What is the star formation rate for this galaxy?

Answer:
We can use Equation 8.6 to relate the star formation rate to the integrated flux density at 1.4 GHz and the distance to the galaxy. This gives us

$$SFR \sim 0.15\ \mathrm{M_\odot\ yr^{-1}} \left(\frac{11.1\ \mathrm{Mpc}}{\mathrm{Mpc}}\right)^2 \left(\frac{0.08\ \mathrm{Jy}}{\mathrm{Jy}}\right) = 1.5\ \mathrm{M_\odot\ yr^{-1}}.$$

The star formation rate for NGC 3184 is very similar to that for the Milky Way.

[16] 2007, Astronomy and Astrophysics, vol. 461, p. 455.

The gas and dust in galaxies is the material out of which new stars are born, and since the gas and dust content of galaxies varies enormously, it follows that the star formation rate is also likely to show large variations. However, since galaxies vary in size and total mass, a more meaningful measure is the star formation rate divided by the stellar mass, which is called the *specific star formation rate.* It should not be surprising that early-type galaxies, which contain relatively little gas and dust, have specific star formation rates hundreds of times smaller than late-type galaxies. Within the late-type galaxies there is also a correlation between the specific star formation rate and stellar mass, in that galaxies with larger stellar masses tend to have smaller specific star formation rates. Finally, we note that there are some very gas-rich galaxies in the local universe that appear to be undergoing mergers (such as Arp 220 which was discussed in Section 8.2.3) that have exceptionally high star formation rates of hundreds of $M_\odot$ yr^{-1}; these rapidly star-forming galaxies are called starburst galaxies.

8.4 DISTANT GALAXIES

Our discussion in this chapter has so far been based on radio observations of galaxies in the local Universe. We can also study more distant galaxies. Distant galaxies are of particular interest because, due to the finite speed of light, we are viewing them as they were in the past. Thus, by observing galaxies at very great distances, we can study the Universe at a time when galaxies were much younger and in their early stages of evolution. Furthermore, by observing galaxies at different distances, we can study how galaxies have evolved from the earliest epochs to the present day.

We discussed in Section 1.5 that, due to the expansion of the Universe, galaxies have a cosmological redshift; the observed light from distant galaxies is shifted to longer wavelengths and this redshift increases for galaxies of increasing distance. Observations of the spectral lines emitted by these galaxies can be used to directly measure their cosmological redshifts. The redshift, z, of a galaxy (see Equation 1.20) is defined as

$$z = \frac{\lambda_d - \lambda_e}{\lambda_e},$$

where λ_d is the observed or detected wavelength of the spectral line and λ_e is the wavelength at which the spectral line was emitted.

In Section 1.6, we provided the equations needed to relate the measured redshift to both the age of the galaxy at the time when the light was emitted (Equation 1.27) and to the look-back time. Using the cosmological parameters in Section 1.6, a Universe dominated by dark matter and dark energy (called a ΛCDM cosmology), the relation between cosmological redshift, age of the Universe in units of 10^9 years (Gyr) and look-back time in Gyr is given in Table 8.1. In this cosmological model matter constitutes 30% and dark energy 70% of the mass/energy density of the Universe. This model also assumes a value of the current Hubble constant of $H_o = 70$ km s^{-1} Mpc^{-1} (see Section 1.5). In this model the current age of the Universe is 13.5 Gyr. To date, galaxies have been detected with redshifts larger than 9; thus we are viewing these galaxies only a few hundred million years after the Big Bang.

One class of distant galaxies detected at radio wavelengths is called submillimeter galaxies as they were first detected at a wavelength of about 0.85 mm (see for example the paper by David Hughes and collaborators[17]). Submillimeter galaxies are ultra-luminous in the infrared and are both gas- and dust-rich. Their luminous infrared emission is due to thermal emission from dust heated primarily by newly-formed stars. These galaxies have very large

[17] 1998, Nature, vol. 394, p. 241.

Table 8.1: Cosmological Redshift (Z), Age of the Universe, and Look-Back Time

Z	Age (Gyr)	Look-Back Time (Gyr)
0.0	13.5	0.0
0.1	12.4	1.3
0.2	11.0	2.5
0.5	8.4	5.1
1.0	5.7	7.8
2.0	3.2	10.3
3.0	2.1	11.4
5.0	1.2	12.3
10.0	0.5	13.0

star formation rates, forming stars at rates of hundreds to thousands $M_\odot$ yr^{-1}. Much of the light from these galaxies at shorter wavelengths is obscured by the large amounts of dust present in the galaxy, and for this reason these galaxies are faint or even invisible at optical wavelengths. Submillimeter galaxies typically have large redshifts, between 1 and 6, and so we are viewing them as they were when the Universe was only a few billion years old. There are examples of such ultraluminous infrared galaxies or ULIRGS in the local Universe, such as Arp 220, but such galaxies are very rare. Although current detection thresholds for submillimeter wavelengths permit the detection of only the most luminous submillimeter galaxies, it appears that these extreme galaxies make a significant contribution to the total star formation activity during the early stages of the Universe.

In the rest frame of the galaxy, the dust emission from a submillimeter galaxy peaks in the infrared at a wavelength of about 100 μm. With observations at submillimeter wavelengths, for objects at large cosmological redshifts, because of the shape of the dust emission spectrum and the combination of the shift of the spectrum and the dimming with distance, the observed flux densities of these galaxies are almost independent of distance. Figure 8.15 illustrates the flux density spectrum of a submillimeter galaxy for redshifts of z =1, 2, and 5. The two vertical lines occurring around 270 GHz represent a submillimeter observing frequency range. Note that in general the higher redshift spectrum is shifted to lower flux density and to lower frequency; the redshift causes the observing bandpass to occur closer to the peak of the spectrum. Hence, there is an increase in flux density in the observed frequency range due to the shift of the spectrum that almost exactly compensates for the decrease due to increased distance. Despite the fact that galaxies become dimmer overall at increasing redshifts, at a frequency of 270 GHz, a submillimeter galaxy has an observed flux density independent of redshift. Sensitive imaging surveys at submillimeter wavelengths, therefore, comprise a very effective method for detecting these luminous and dusty galaxies over a broad range of redshifts.

Many surveys have been carried out at submillimeter wavelengths and hundreds of submillimeter galaxies have been detected. An example of a sensitive image of the sky at 1.1 mm wavelength from a study by Kim Scott and collaborators[18] is shown in Figure 8.16. Although this image covers a region only 10′ by 15′ (less that one millionth of the entire sky) it contains about three dozen submillimeter galaxies. The redshifts of these galaxies vary from about 1 to over 5. The brightest source, AzTEC-1, has a flux density of 9 mJy, and is only about 8 times the noise level in the map. With our current sensitivity, we are only able to detect bright submillimeter galaxies. Models suggest that many more fainter galaxies are present in these data, and what appears to be noise in this image may be, in part, emission from these

[18]2008, Monthly Notices of the Royal Astronomical Society, vol. 385, pg. 2225.

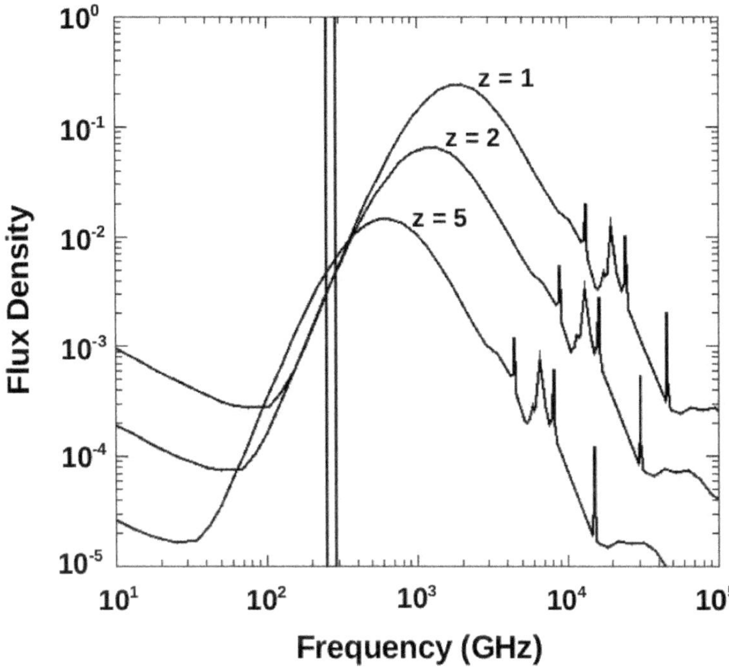

Figure 8.15: A plot of the flux density (in janskys) versus observed frequency for the same ultra-luminous infrared galaxy placed at redshifts of 1, 2, and 5. This plot illustrates how the cosmological expansion of the Universe shifts the peak of the dust emission progressively into an observing band at a frequency of 270 GHz (or a wavelength of 1.1 mm), marked by the double vertical lines. Thus, this ultra-luminous infrared galaxy would have the same flux density at 270 GHz regardless of whether it was at a redshift of 1, 2 or 5. This figure was provided by Min Yun.

fainter galaxies that have been blurred and blended together by the relatively low resolution of this image. The most distant galaxy detected in this survey, AzTEC-3, is at a redshift of 5.3, and thus we are viewing this galaxy only about 1 billion years after the Big Bang. The fact that there are few submillimeter galaxies with redshifts less than 1 in this field indicates that the epoch in the Universe when galaxies commonly had such large star formation rates ended about 8 billion years ago. In fact, the time of maximum star formation in the Universe was about 10 billion years ago, and star formation has been waning ever since.

We can learn more about these galaxies from their spectral line emission. Unfortunately, our current technology does not have the sensitivity to detect the 21-cm HI emission line in these very distant galaxies. The largest redshift for which the 21-cm line of HI has been detected is only 0.3. Future radio telescope arrays, such as the Square Kilometer Array (SKA), will have much larger collecting areas and the sensitivity at longer wavelengths to detect the HI line in more distant galaxies. Fortunately, the rotational lines of CO have much larger fluxes and so are more readily detectable in these submillimeter galaxies. These CO observations have revealed that these galaxies are rich in molecular gas. Detection of the CO spectral lines also is a means of determining the redshift. In fact, for submillimeter galaxies not detected at optical wavelengths, this is the only means of determining a spectroscopic redshift. An example of detection of CO spectral lines in a submillimeter galaxy is shown in Figure 8.17 for the galaxy PJ160917.8. This spectrum shows two spectral lines due to two different rotational transitions of CO and provides a unique determination of the galaxy's redshift of 3.26.

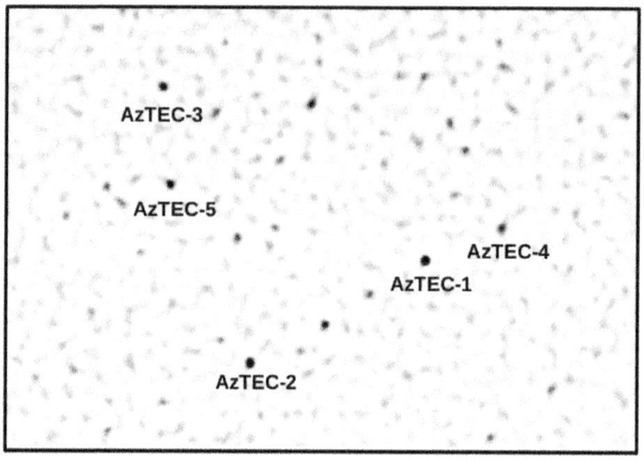

Figure 8.16: An image obtained with the AzTEC camera on the James Clerk Maxwell Telescope (JCMT) at a wavelength of 1.1 mm. The image is about 15 arcminutes by 10 arcminutes in angular size. About three dozen submillimeter sources are detected, with the five brightest sources labeled AzTEC-1 through AzTEC-5. The brightest source, AzTEC-1, has a flux density of about 9 mJy. This study was conducted by Kim Scott and collaborators (2008, Monthly Notices of the Royal Astronomical Society, vol. 385, p. 2225) and the figure was provided by Grant Wilson.

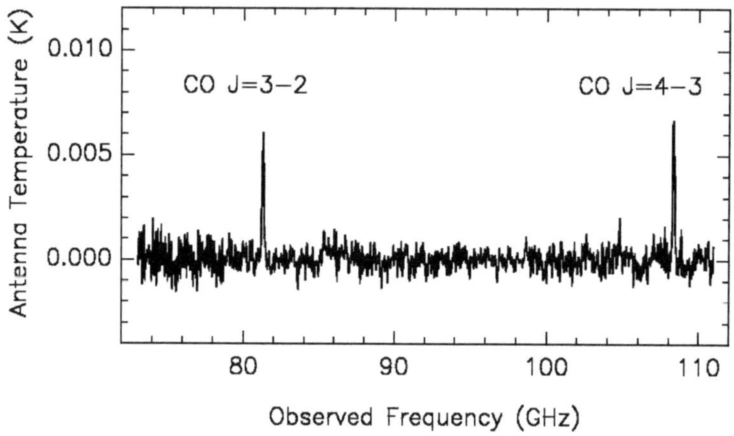

Figure 8.17: The spectrum of the submillimeter galaxy PJ160917.8 obtained with the Redshift Search Receiver on the Large Millimeter Telescope. In the observed frequency range of 75 to 111 GHz two spectral lines are detected, which have been identified as the J=3-2 and J=4-3 rotational transitions of CO. Note that the emitted frequencies of these lines are at 345.8 and 461.0 GHz, respectively, and therefore these lines, due to cosmological expansion, have been shifted to the observed frequencies of 81.3 and 108.3 GHz, respectively, indicating a redshift of 3.26. This study was conducted by Kevin Harrington and collaborators (2016, Monthly Notices of the Royal Astronomical Society, vol. 458, p. 4383) and the data used to make this figure were provided by Min Yun.

In addition to providing important information on the redshifts of submillimeter galaxies, the detection of the CO lines can also provide a measure of the molecular gas content in these galaxies. However the mass of molecular gas is much more uncertain in these distant galaxies, as the conversion of the CO line flux to a molecular gas mass is more complicated than for local galaxies for a number of reasons. First, it is common that observations are made at millimeter wavelengths, so the rotational transition of CO that is detected is not the J=1-0 transition, which is the transition used in the empirical calibration. Therefore, one must estimate the flux density of the J=1-0 transition from the observations of the higher rotational lines of CO. Second, we cannot use the simple relation given in Equation 8.4, because of several cosmological corrections that must be accounted for; these are discussed in a paper by Philip Solomon and collaborators[19]. To add to the complexity, one needs a value for the X-factor (see Section 8.2.1) that is appropriate for these extreme galaxies in the early Universe, and this is poorly known. Finally, galaxies like PJ160917.8 have yet another complication because they can be gravitationally lensed by a foreground galaxy. One result of a gravitational lens is to amplify the observed flux density, and thus the amplification must be accounted for to obtain the correct flux density. Despite all of these corrections, PJ160917.8 is still likely one of the most luminous galaxies ever detected and the total molecular gas mass is estimated to be in excess of 10^{11} $M_{\odot}$ (see the paper by Kevin Harrington and collaborators[20]).

CO emission has been detected in numerous submillimeter galaxies, so it is now possible to ask how the gas content of galaxies has changed over cosmic time. In Section 8.2.2, we discussed that local late-type or disk galaxies typically have a molecular gas mass about one-tenth that of their stellar mass. Surveys, such as that carried out by Roberto Decarli and collaborators[21], have shown that at large redshifts, disk galaxies have molecular gas mass to stellar mass ratios ten times larger than local disk galaxies. Such gas-rich submillimeter galaxies are responsible for the ten-fold increase in the cosmic star formation rate density at redshifts between 2 and 3 as compared with today. Thus, as these submillimeter galaxies evolve with time, their gas is converted into stars, decreasing their gas mass leading to a decline in the star formation rate.

[19] 1997, Astrophysical Journal, vol. 478, p. 144.

[20] 2016, Monthly Notices of the Royal Astronomical Society, vol. 458, p. 4383.

[21] 2016, Astrophysical Journal, vol. 833, p. 70.

QUESTIONS AND PROBLEMS

1. The 21-cm HI spectrum for the galaxy NGC 5936, at a distance of 50 Mpc, is shown in Figure 8.18. If we assume the emission is optically thin, estimate the mass of HI in this galaxy.

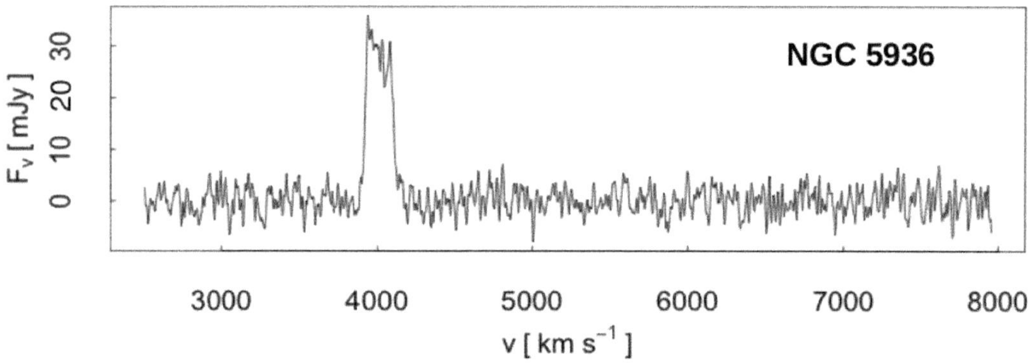

Figure 8.18: HI 21-cm spectrum of NGC 5936. This spectrum was obtained from the NASA/IPAC Extragalactic Database (NED) and is operated by the Jet Propulsion Laboratory, California Institute of Technology, under contract with the National Aeronautics and Space Administration.

2. Using the information from Section 5.1.2, calculate the total mass of the Milky Way interior to the Sun's orbit.

3. The rotation curve of the Milky Way is believed to remain relatively flat (a constant rotation speed of about 230 km s^{-1}) out to a radius of at least 80 kpc. If this is correct, estimate the total mass of the Milky Way enclosed within 80 kpc of its center. The total mass of stars and interstellar medium in the Milky Way is estimated to be about 7×10^{10} $M_\odot$. What fraction of the mass of the Milky Way is dark matter?

4. The integrated flux density for the CO J=1-0 emission in NGC 5936 is measured (from a paper by Young and collaborators)[22] to be 250 Jy km s^{-1}. Using the X-factor for the Milky Way and a distance of 50 Mpc, what is the total mass of H_2? This galaxy is a barred Sb galaxy like the Milky Way. How does its ratio of molecular to atomic gas mass compare with that of the Milky Way?

5. The continuum emission from the galaxy NGC 5936 (at a distance of 50 Mpc) has been measured at a wavelength of 850 μm to have a flux density of 152 mJy (from Dunne and colleagues)[23]. If the mass absorption coefficient at this wavelength is $\kappa_\nu^m = 1.3$ cm^2 g^{-1} and if we assume the dust has a temperature of 30 K, what is the total mass of dust in this galaxy? Using the results from Questions 1 and 4, what is the gas-to-dust ratio in this galaxy?

6. The total stellar mass in NGC 5936 is estimated to be 6×10^{10} $M_\odot$. Considering the answers to Questions 1 and 4, do the atomic and molecular gas contents of this galaxy fit the trend for late-type galaxies seen in Figures 8.3 and 8.10?

[22] 1995, Astrophysical Journal Supplement, vol. 98, p. 219.
[23] 2000, Monthly Notices of the Royal Astronomy Society, vol. 315, p. 115.

7. The flux density measured for NGC 5936 (at a distance of 50 Mpc) for the long wavelength continuum emission at a frequency of 1.4 GHz is 139 mJy (measurement from paper by Condon and collaborators)[24]. Estimate the star formation rate in this galaxy.

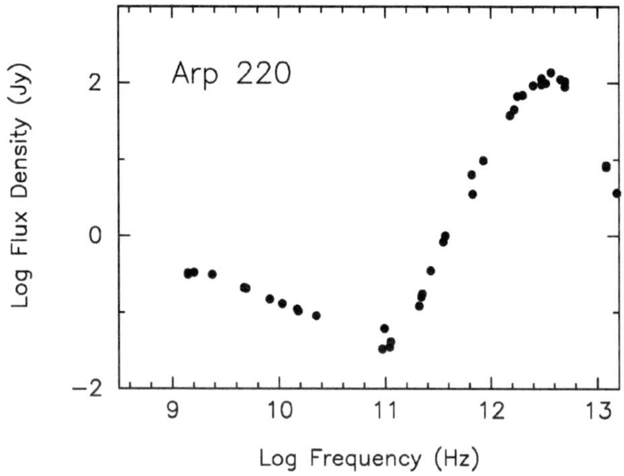

Figure 8.19: A plot of the continuum flux density as a function of frequency for Arp 220. These data were provided by Min Yun.

8. A local ultra-luminous infrared galaxy (ULIRG) is Arp 220 at a distance of approximately 77 Mpc. A plot of the continuum flux density for this galaxy as a function of frequency is shown in Figure 8.19. Estimate the star formation rate for Arp 220. How does the star formation rate compare with that in the galaxy NGC 5936 from Question 7?

9. If the dust temperature in Arp 220 is 50 K, use the data in Figure 8.19 to estimate the dust mass. Remember that the distance to Arp 220 is 77 Mpc.

10. The emitted frequencies of the CO J=4-3 and J=3-2 transitions are separated by 115.2 GHz. If we observe the CO emission from galaxies at z = 0.3, 1.0, 2.0, 5.0, what would be the observed frequency separation of these two CO lines? Explain how this might help in determining the redshift of a galaxy.

[24] 2002, Astronomical Journal, vol. 124, p. 675.

CHAPTER 9

Radio Galaxies and Quasars

Most of the strongest radio signals at centimeter to meter wavelengths that we detect from space, after the Sun and several supernova remnants, emanate from objects beyond the Milky Way. In the centers of some galaxies, highly energetic processes, completely unrelated to stars, occur and produce an assortment of observable phenomena. Studies of these galaxies, known as *active galactic nuclei* (or AGN), have yielded a number of classifications and categorizations. In this chapter we focus on active galaxies that exhibit strong radio-frequency emission. In particular, we give an overview of the AGN classified as *radio galaxies* and *quasars*.

A full and proper discussion of AGN is so extensive and involves so many details that an entire text book (such as *Active Galactic Nuclei: An Introduction* by Ian Robson[1] and *Active Galactic Nuclei: From the Central Black Hole to the Galactic Environment* by Julian Krolik[2]) is warranted. Furthermore, much that is known about AGN also involves studies at other wavelength regimes and so a review of the radio astronomical studies of AGN is really just a subset of the full discussion. Here, we will describe radio galaxies in general, with emphasis on inferring their physical conditions from radio observations. For the sake of brevity, we must exclude discussions of some important topics, such as polarization observations. As a prelude to our discussion we give an overview of the basic model of AGN.

9.1 BRIEF OVERVIEW OF ACTIVE GALACTIC NUCLEI

Active galactic nuclei were originally discovered at visible wavelengths in the 1940s, when Carl Seyfert noticed that some spiral galaxies had especially bright nuclei. These galaxies came to be known as *Seyfert galaxies.* Visible-light spectroscopy of the nuclei of Seyfert galaxies reveal that the radiation includes non-stellar light, indicating that some energetic processes unrelated to normal stars occur in the centers of these galaxies. In addition to a flat continuum, an assortment of emission lines is evident. In some Seyferts, subclassified as Seyfert 1 galaxies, the spectra display two sets of emission lines as indicated by their widths. Very broad emission lines, suggesting velocities of order thousands of km s^{-1}, are seen — but only from permitted transitions, while narrower emission lines, implying velocities of hundreds of km s^{-1}, are seen from both permitted and forbidden transitions. The lack of broad forbidden emission lines suggests that the gas moving at these velocities is dense enough that the upper states of these transitions are de-excited by collisions. The other Seyferts, called Seyfert 2 galaxies, display only the narrower emission lines. In the 1980s, Robert Antonucci and Joe Miller[3] suggested that the difference between Seyfert 1 and

[1] 1996 John Wiley & Sons: Chichester, England.
[2] 1999 Princeton University Press: Princeton, NJ.
[3] 1985, Astrophysical Journal, vol. 297, p. 621.

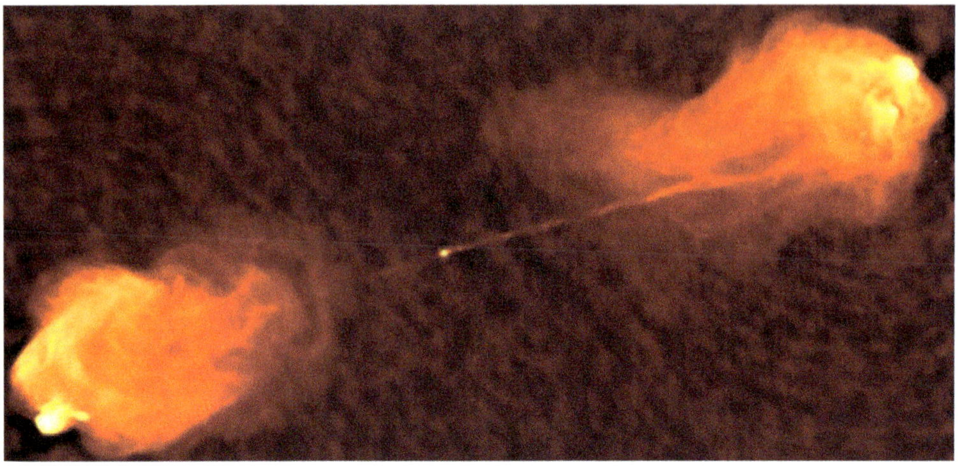

Figure 9.1: Image of the canonical radio galaxy Cygnus A obtained from a 1983 VLA observation at 6 cm by R. Perley, C. Carilli, and J. Dreher. Defining characteristics which include a core component, relativistic jets, large radio lobes, and hot spots are evident in this figure and discussed in the text. Image courtesy of NRAO/AUI.

Seyfert 2 galaxies arises primarily from a difference in viewing angle. The broad emission lines emanate from the very central region of the AGN which is sometimes obscured at visible wavelengths, possibly by a large circumnuclear torus of dust and gas, in which case a Seyfert 2 spectrum results. A Seyfert 1 spectrum occurs when we have a direct line of sight to the broad emission line region. The narrow emission lines are produced farther out in the host galaxy, outside of the circumnuclear torus.

By the early 1950s, with the development of radio interferometry, the positions of some of the brightest radio sources were determined with sufficient accuracy to identify their optical counterpart. Some of these bright radio sources were found to be coincident with known large elliptical galaxies, such as Virgo A and Centaurus A, while the radio source Cygnus A was coincident with a very faint peculiar galaxy. Subsequent surveys of the radio sky revealed many more radio-bright objects associated with known galaxies. These objects became known as *radio galaxies*. Radio interferometric observations of many of these radio galaxies revealed a structure containing two extremely large amorphous volumes of radio emission located well outside the visible limit of the elliptical galaxy. Figure 9.1 displays a radio image of Cygnus A (detected in the Third Cambridge catalog and designated 3C405) from a 5-GHz VLA observation. Figure 9.2 shows an overlay of radio and visible wavelength images of another radio galaxy, Hercules A (3C348). The two large volumes of radio emission, commonly referred to as the *radio lobes*, can be seen to either side and outside the limits of the visible image of the host galaxy. The stellar extent of the galaxy is small in comparison to the distance of the lobes. The lobes are much larger than the visible extent of the host galaxy and can be separated from each other by hundreds of kpc, while the central galaxy is of order many tens of kpc in diameter.

Higher resolution and more sensitive radio observations in the 1970s revealed compact sources of radio emission in the cores of many of these galaxies and that the lobes are connected to the core of the host galaxy by long thin lines of faint radiation; these features are also apparent in Figure 9.1. These lines, now referred to as *jets*, are streams of particles ejected from the nuclei of these galaxies at relativistic speeds. Also apparent in Figure 9.1,

Figure 9.2: Radio-wavelength image of Hercules A overlaid on a visible-wavelength photo. Taken from the NRAO web page. Credit: NASA, ESA, S. Baum and C. O'Dea (RIT), R. Perley and W. Cotton (NRAO/AUI/NSF), and the Hubble Heritage Team (STScI/AURA).

in the lobes, are bright spots of emission; these *hot spots* are believed to be collision sites between the jet particles and the extragalactic medium.

Visible-wavelength spectra of radio galaxy nuclei are similar to those of Seyfert galaxy nuclei; both show a power-law continuum with broad and/or narrow emission lines. Additionally, significant radio emission is associated with most Seyfert galaxies, although they rarely display the large radio lobes. Radio galaxies and Seyfert galaxies, therefore, are two expressions of the same general phenomenon. For this book, we will focus on the radio galaxies.

The huge intensities of the radio emission from radio galaxies lead to absurdly large brightness temperatures, implying that this radiation is produced by non-thermal processes. Additionally, the radio spectra of these galaxies are found to follow decreasing power laws (see Figure 9.3) and so they are inferred to be synchrotron radiation sources (see Section 3.3). Evidently, they must harbor a tremendous number of relativistic charged particles and large-scale magnetic fields in order to produce the amount of radio emission we detect. The synchrotron emission will be dominated by the lowest-mass charged particles, which at least include electrons. The jets may also include positrons, but since the emission from positrons would be identical to that from electrons, we can model the observed radiation considering only electrons. Since these objects must be electrically neutral, we know that positive charges, be they protons or positrons, must be contained in the same volume.

The decreasing power-law spectra can be represented mathematically by

$$F_\nu \propto \nu^\alpha, \tag{9.1}$$

where α is the *spectral index* (also discussed in Section 3.3.3). For decreasing flux density with increasing frequency α is negative. Many authors define spectral index such that the exponent in Equation 9.1 contains a negative sign, in which case α is positive for a decreasing

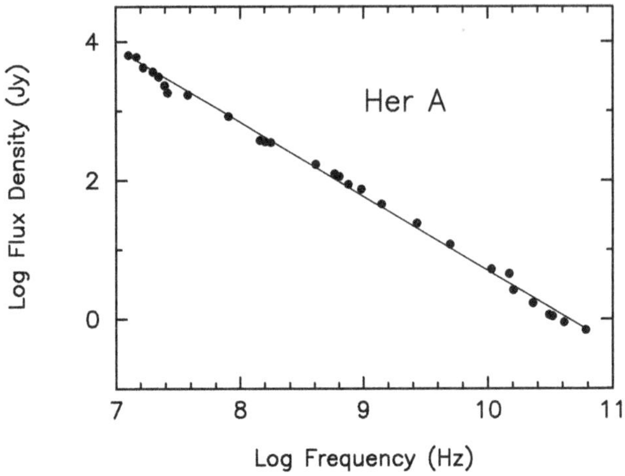

Figure 9.3: Radio-wavelength spectrum of the radio galaxy Hercules A (3C348). The axes are plotted logarithmically so the straight line indicates a power-law relation, $F_\nu \propto \nu^\alpha$. The flux densities decrease with increasing frequency indicating $\alpha < 0$. The data used for the plot were obtained from the NASA Extragalactic Database (NED).

spectrum. There is no established convention for the sign of the spectral index, and so an author should make clear which definition is used.

Example 9.1:

Calculate the spectral index of Hercules A using the plot in Figure 9.3.

Answer:
The best-fit line through the data goes through $\log(F_\nu) = 0$ when $\log(\nu) = 10.7$ and through $\log(F_\nu) = 3$ when $\log(\nu) = 7.8$. We can solve for the spectral index by taking the logarithm of Equation 9.1 and forming the ratio of the two points on the plot, i.e.

$$\alpha = \frac{\log(F_{\nu 2}) - \log(F_{\nu 1})}{\log(\nu 2) - \log(\nu 1)}.$$

This calculation is identical to finding the slope of the line on the log-log plot. Inserting the data from the plot, we have

$$\alpha = \frac{0 - 3}{10.7 - 7.8} = -1.0.$$

A linear least-squares fit to all the data yields a line slope of $\alpha = -0.94$.

Many AGN known today, called quasars, were not initially recognized as galaxies. The term quasar can be understood from the history of their discovery. Some bright radio sources were determined to be coincident with objects that appeared star-like at visible-wavelengths. The radio emission from these objects, though, was clearly non-stellar in origin. Visible-wavelength spectra of these objects contained broad emission lines that could not be identified (at first). Since it was unclear what these objects were, but they looked like stars, they were called *quasi-stellar radio sources*, or *quasars* for short. Later studies revealed many other point-like objects with similar visible-wavelength spectra but with little or no detectable radio emission, and so a more general term, *quasi-stellar objects*, or *QSOs*, was added to the AGN nomenclature.

In the early 1960s, Jesse Greenstein and Maarten Schmidt[4] realized that the visible spectra of the quasi-stellar objects were very highly redshifted – more redshifted than any other objects known. Such high redshifts implied that these objects were at cosmological distances from us (see discussion of cosmological redshifts in Section 1.5). These objects were, therefore, the most distant objects known at that time. They were so far away that not only would ordinary stars be impossible to detect but even whole galaxies would appear quite faint. Therefore, these quasi-stellar objects were inferred to be very luminous active nuclei of distant galaxies, whose stellar component was not detected in the glare of the bright core. Although they appear as faint as ordinary stars, taking into account their distances, the luminosities of these AGN can be as large as hundreds of times that of the entire Milky Way galaxy. Modern observational techniques permit the detection of the underlying host galaxy of a quasar despite the enormous glare of the core. Repeated imaging of quasars with extremely high resolution radio observations using very long baseline interferometry techniques (see Volume I, Chapter 6) revealed significant outward motions (see discussion in Section 9.4.2), confirming that these objects contain radio jets. Hundreds of thousands of QSOs, with a wide variety of characteristics, have now been discovered.

Quasars, as a rule, are found at large redshifts. In fact, one of the first quasars discovered, 3C273, at $z = 0.158$, is considered a relatively low-redshift quasar. The number density of quasars per unit volume is found to increase toward higher redshifts, peaking between redshifts of 1 and 2. Active galactic nuclei, in general, are more common at higher redshifts, typically with higher luminosities at higher redshifts. The redshift distribution depends on the particular defining AGN parameters and on the limitations of the survey, but the tendency for AGN to be more common in the past is definite in all studies. Figure 9.4 shows a histogram of redshifts of optically selected broad-emission line QSOs. The paucity of QSOs at small redshifts and the rapid increase in number towards redshifts between 1 and 2 are clearly seen in this histogram.

The radiation from active galactic nuclei covers all parts of the electromagnetic spectrum. For example, QSOs can be discovered also in surveys with X-ray telescopes. A plot of the product of flux density and frequency (to give units of flux) vs. frequency across the entire EM spectrum for a typical quasar is roughly constant, indicating about equal amounts of energy emitted at all frequencies. The total luminosity of a quasar, therefore, is exceptionally high, several orders of magnitude greater than that of an entire large galaxy.

Since quasars and active galaxies tend to be at large redshifts, determination of physical quantities associated with them, such as luminosity or linear size, requires calculations of cosmological distances, as discussed in Section 1.6.

Example 9.2:

(a) The quasar 3C273 has a redshift $z = 0.158$. What is the look-back time for observations of this object? Assume $H_0 = 2.27 \times 10^{-18}$ s^{-1}, $\Omega_M = 0.3$ and $\Omega_\Lambda = 0.7$.

(b) What is the proper distance to 3C273?

Answer:

(a) We first need the age of the Universe when light with redshift z was emitted,

[4] 1964, Astrophysical Journal, vol. 140, p. 1.

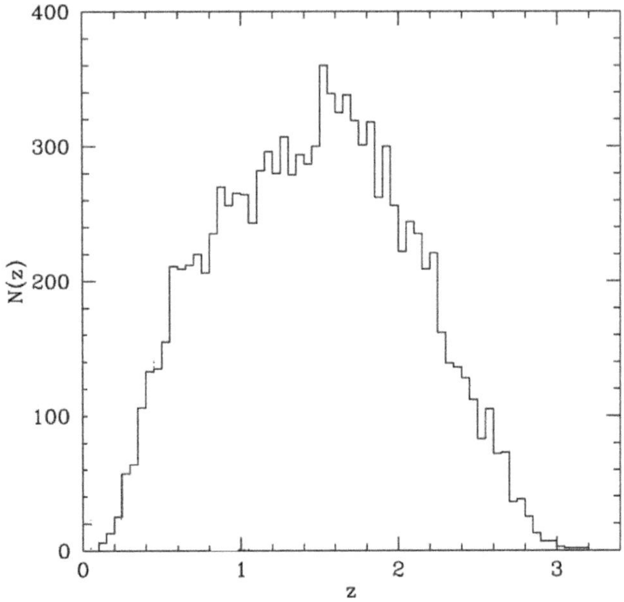

Figure 9.4: The number per unit co-moving volume of optically-selected, broad-emission line QSOs in the 2dF QSO redshift survey by Croom et al. 2001. The number density of quasars in this study is found to peak around $z \approx 1.5$. (Modified image taken from Croom et al. 2001, Monthly Notices of the Royal Astronomical Society, vol. 322, p. L29.)

which is given by Equation 1.27. This gives us

$$t_e = 11.1 \ \sinh^{-1}\left[\frac{1.53}{(1+0.158)^{3/2}}\right] \ \text{Gyr} = 11.5 \ \text{Gyr}.$$

Since the current age of the Universe assuming these cosmological parameter values is about 13.5 Gyr, the look-back time for this object is 2 billion years.

(b) We obtain the proper distance from the numerical integration in Equation 1.29. Using the Mathematica[5] code in Appendix B this yields

$$d_p = 6.52 \times 10^8 \ \text{pc} = 652 \ \text{Mpc}.$$

Calculating cosmological distances can be cumbersome because it requires a numerical integration, and a common temptation is to use the Hubble-Lemaître law approximation to roughly estimate the distance. If we do that with 3C273 we get

$$d = \frac{cz}{H_0} = \frac{0.158 \times 3 \times 10^5 \ \text{km s}^{-1}}{70 \ \text{km s}^{-1} \ \text{Mpc}^{-1}} = 677 \ \text{Mpc},$$

which is 4% larger than the correctly calculated value. The error in estimates based on the Hubble-Lemaître law approximation increases with increasing redshift.

[5]Wolfram Research, Inc., Mathematica, Version 11.3, Champaign, IL (2018).

Example 9.3:

The flux we detect from 3C273 in just the visible window is approximately 2.1×10^{-10} erg s^{-1} cm^{-2}. Calculate the luminosity of the radiation we detect in the visible from 3C273. (Note, because of its redshift, some of the radiation we detect in the visible is emitted in the ultraviolet. This question asks us to focus only the radiation we detect in the visible.) Convert the answer to solar luminosities.

Answer:
The luminosity relates to detected flux by

$$L^V_{3C273} = F^V_{3C273} \times 4\pi d^2_{L,3C273} = (2.1 \times 10^{-10} \text{erg s}^{-1} \text{ cm}^{-2}) \times (4\pi d^2_{L,3C273})$$

where the superscript V indicates flux and luminosity only in the visible window, and $d_{L,3C273}$ is the luminosity distance of 3C273. We use the luminosity distance because we are converting flux to luminosity. The luminosity distance is related to the proper distance by Equation 1.23 and the proper distance we calculated in Example 9.2. The luminosity distance, then, is

$$d_L = 6.52 \times 10^8 \text{ pc } (1 + 0.158) = 7.55 \times 10^8 \text{ pc} = 2.33 \times 10^{27} \text{ cm}.$$

Putting this value for $d_{L,3C273}$ into the equation we get

$$L^V_{3C273} = (2.1 \times 10^{-10} \text{erg s}^{-1} \text{ cm}^{-2})\ 4\pi(2.33 \times 10^{27} \text{ cm})^2$$
$$= 1.43 \times 10^{46} \text{ erg s}^{-1}.$$

Converting to solar luminosities, where 1 L$_\odot$ = 3.83×10^{33} erg s^{-1}, $L^V_{3C273} = 3.7 \times 10^{12}$ L$_\odot$. The visible luminosity of the entire Milky Way galaxy is about 2.5×10^{10} L$_\odot$. We see, then, that the radiation output of 3C273 at visible wavelengths is on the order of that emitted by about a hundred Milky Way galaxies!

Example 9.3 only addressed the radiation we detect at visible wavelengths. This is not really a fair comparison, though, because the emission of 3C273 is significant across the entire electromagnetic spectrum while the emission from the Milky Way is stronger in the visible and infrared than in other bands. The total radiative output of 3C273 relative to the Milky Way, therefore, is significantly larger than what we found in this example.

9.2 AGN MODEL

The model of an active galactic nucleus involves the infall of gas towards a supermassive black hole, generally with $M \approx 10^6$ to 10^9 M$_\odot$. The heating of gas as it spirals in toward a black hole is an efficient mechanism for converting gravitational potential energy into radiation. Numerous studies of the kinematics of the central regions of galaxies provide strong evidence that supermassive black holes commonly exist in galactic centers. As the interstellar gas falls inward, its angular momentum causes it to form a rotating disk of gas around the black hole; this is termed an *accretion disk*. Because of conservation of energy, as the gas falls inward its gravitational potential energy is converted to kinetic energy which is converted by collisions of the gas in the disk to thermal energy. The accretion disk, therefore,

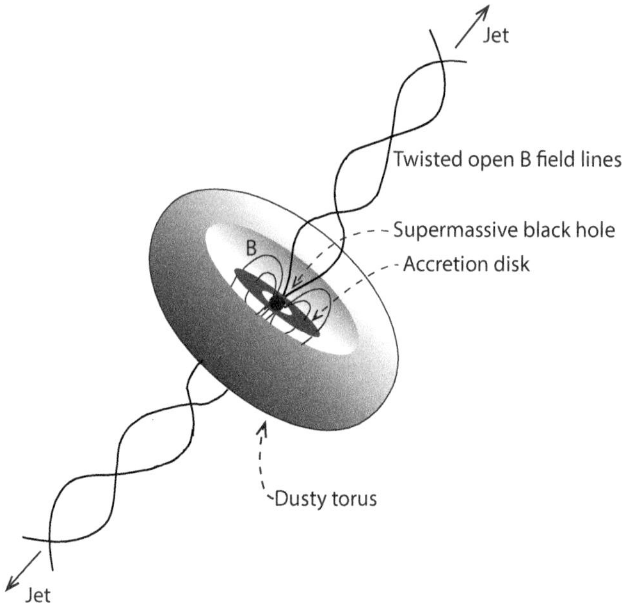

Figure 9.5: Schematic of the standard model of active galactic nuclei. A disk of gas, called an accretion disk, orbits a black hole. A magnetic field dragged in with the infalling matter threads through the disk. The open field lines become twisted by the rotation. A dusty torus sometimes obscures the view of the central region where the broad optical emission lines are produced.

is heated to high temperatures with higher temperatures at smaller radii, closer to the black hole. By the time the gas gets close to the black hole it is sufficiently hot to emit X-rays.

In addition to the small size of the central engine, AGN models must also explain the jets of particles streaming away from the black hole at relativistic speeds. Currently no model for the formation of jets has successfully explained all the details, but the mechanism necessarily involves a combination of magneto-hydrodynamics and general relativity. One of the more successful models was developed by Roger Blandford and Roman Znajek[6], in which rotational energy of the black hole is tapped to power the jets. The material that falls into the black hole carries large amounts of angular momentum with it, so that over millions of years the black hole obtains a large rotation rate. The infalling matter also carries along a magnetic field that intensifies as the gas collapses to smaller volumes. Recall that the intense synchrotron radiation we detect from these sources requires that they contain strong magnetic fields. A rotating supermassive black hole in a magnetic field produces an interesting effect related to formation of the radio jets. By the *Blandford-Znajek effect*, just outside the black hole an electric field is induced, which accelerates charges away from the black hole. These charges get attached to open magnetic field lines, which will be twisted into helical shapes. The points where the magnetic field lines are anchored to the rotating matter are dragged along in the rotation, but because of the finite speed of transfer of information about the rotation, the field lines further out are delayed in their rotation. The bases of the field lines, then, twist ahead of the lines further out and so the open magnetic field lines at the poles end up as helices, as shown in Figure 9.5. We can think of a charged particle following a magnetic field line similar in behavior to a bead on a wire. Imagine a wire bent into a helical shape increasing in diameter with distance from its base, oriented

[6]1977, Monthly Notices of the Royal Astronomical Society, vol. 179, p. 433.

vertically with the small end at the floor, and rotating at high speeds about its axis. The bead will be swung outward and upward. A charged particle attached to an open, helical magnetic field line of a rotating system, similarly, will be pushed outward and accelerated away from the black hole. An aspect of the jets that is more difficult to explain theoretically is how the magnetic field lines and path of ejected particles are focused into and remain in finely-collimated beams.

Extensive studies, which we discuss in Section 9.4.2, show that the particles in the jets are ejected at relativistic speeds. As the jets collide with the ambient medium around the AGN, they slowly dig tunnels into the medium, like a high-power water hose sprayed into the sand at the beach. Provided the ejection of particles into the jets continues, the tunnel grows, and after millions of years these jets reach great distances, ending in the lobes where they crash into the intergalactic medium, hundreds of kpc from the black hole. The jet particles recoil from the collisions but continue to radiate producing the large lobes, causing most of the radio emission that we see.

We have explained that the visible-wavelength luminosities of an AGN can be much greater than those of entire large galaxies (see Example 9.3). We now discuss the radio-wavelength emission. A lower limit of the total radio flux emitted by Hercules A (spectrum shown in Figure 9.3) can be estimated by integrating the measured flux densities over the plotted frequencies:

$$F = \int_{\nu_{\rm min}}^{\nu_{\rm max}} F_\nu \, d\nu.$$

In Figure 9.3 we see that the flux density at 10^9 Hz is 100 Jy, and we calculated in Example 9.1 that the spectral index is $\alpha = -0.94$. The flux density curve, then, is given by

$$F_\nu = 100 \text{ Jy} \left(\frac{\nu}{10^9 \text{ Hz}} \right)^{-0.94}$$

and the total flux contained in the plot is

$$F = \int_{10^7 \text{ Hz}}^{10^{11} \text{ Hz}} 100 \text{ Jy} \left(\frac{\nu}{10^9 \text{ Hz}} \right)^{-0.94} d\nu.$$

Substituting in $x = \nu/(10^9 \text{ Hz})$ and $dx = d\nu/(10^9 \text{ Hz})$ we can rewrite this as

$$F = 10^{11} \text{ Jy Hz} \int_{10^{-2}}^{10^2} x^{-0.94} \, dx = 10^{11} \text{ Jy Hz} \left[\frac{x^{0.06}}{0.06} \right]_{0.01}^{100}$$

$$= 9.33 \times 10^{11} \text{ Jy Hz} = 9.33 \times 10^{-12} \text{ erg s}^{-1} \text{ cm}^{-2}.$$

To convert to a luminosity, we use the redshift of Her A, $z = 0.155$, to calculate its proper distance using the numerical integration in the appendix. We get $d_p = 1.98 \times 10^{27}$ cm. Including the factor of $1 + z$ in the luminosity distance, the luminosity of Her A over the spectral range shown in Figure 9.3 is

$$L = (9.33 \times 10^{-12} \text{ erg s}^{-1} \text{ cm}^{-2}) \, 4\pi \left(1.98 \times 10^{27} \text{ cm } (1 + 0.155)\right)^2 = 6.13 \times 10^{44} \text{ erg s}^{-1},$$

which is greater than 10^{11} $L_\odot$. This is not much smaller than the visible luminosity we calculated for 3C273 in Example 9.3.

The amount of energy radiated each second must be a miniscule fraction of the energy contained in the source. The total energy needed to explain the radio-frequency observations of a large radio galaxy like Hercules A or Cygnus A can be calculated in a fairly straightforward manner and leads to an astonishing result. Since the radiation we detect is synchrotron radiation, presumably from relativistic electrons in a magnetic field, we need to account for energy in both the electrons and the magnetic field. The energy density in the electrons is $\int E\ N(E)\ dE$, where $N(E)$ is the number density of electrons of energy E, while the energy density in a magnetic field is $B^2/8\pi$. Integrating both energy densities over the radio emission volume, the total energy is

$$E_{\rm total} = (\text{Radio emission volume}) \left[\frac{B^2}{8\pi} + \int E\ N(E)\ dE\right].$$

There is, additionally, energy stored in the source in the other particles, such as protons, that do not contribute substantially to the emission.

Let's first consider the energy density in the radiating electrons for an observation at frequency ν_0 in which we detect a flux density $F_\nu(\nu_0)$. As discussed in Sections 3.3.2 and 3.3.3, the synchrotron spectrum emitted by electrons of a specific energy is narrowly peaked (see Figure 3.12), and so the electron energy can be related to the observing frequency by Equation 3.27, which says that

$$E \propto \nu_0^{1/2}\ B^{-1/2}.$$

Meanwhile, the observed flux density is proportional to the number density of electrons of energy E and the power emitted by each electron,

$$F_\nu(\nu_0) \propto N(E)P,$$

and the radiation power emitted by each electron, P, given by Equation 3.28, has dependencies given by

$$P \propto B^2 E^2.$$

Substituting for E into the power emitted by each electron we have

$$P \propto B^2 E^2 \propto B^2\ (\nu_0\ B^{-1}) \propto \nu_0\ B.$$

The dependence of the observed flux density on B, then, is

$$F_\nu(\nu_0) \propto N(E)\ P \propto N(E)\ \nu_0\ B.$$

Thus, for a measured flux density at a given frequency we have a relation between $N(E)$ and B,

$$N(E) \propto B^{-1}.$$

The energy density of the electrons radiating at this frequency, then, depends on B as

$$E\ N(E) \propto B^{-1/2}\ B^{-1} \propto B^{-3/2}.$$

Hence, for a given flux density at a particular frequency, the energy density in the electrons depends on the magnetic field to the $-3/2$ power. Meanwhile, the energy density in the magnetic field depends on the square of the magnetic field. The total amount of energy needed is the sum of the energies contained in both the magnetic field and the electrons, integrated over volume. Remember that the volume of the radio lobes is enormous, leading to very large energies.

A plot of the total energy, the energy in the electrons, and in the magnetic field, as a function of magnetic field is depicted in Figure 9.6. Because of the competing dependencies on B there is a specific magnetic field value for which the total energy required to explain

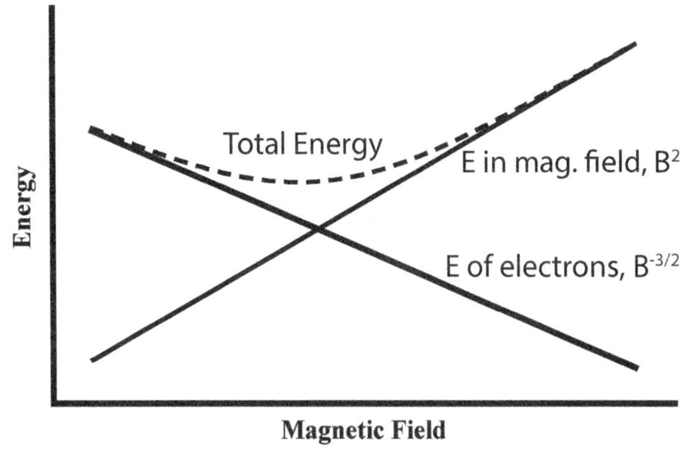

Figure 9.6: The total energy required to explain the observed synchrotron emission as a function of the magnetic field. The energy contained in the electrons ($\propto B^{-3/2}$) decreases with increasing magnetic field strength while the energy contained in the magnetic field ($\propto B^2$) increases. The minimum in the total energy (dashed line) required is near the equipartition energy case.

the observed flux density is a minimum. In some cases, the minimum energy can be as high as 10^{61} erg, the equivalent of 10^{10} supernovae or of order the rest mass energy of 10^7 $M_\odot$.

It so happens that the magnetic field strength where the total energy is minimum also corresponds closely to the case when the total energy in the electrons is equal to the total energy in the magnetic field. To be precise, the minimum total energy occurs when the energy in the electrons is 4/3 times that in the magnetic field. This can be seen in Figure 9.6 by noting that the position of the total energy minimum is at a slightly lower magnetic field than the position where the electron and magnetic field energy lines cross. The situation with an equal amount of energy in the electrons as in the magnetic field is called the *equipartition energy* case. Nature tends to move in the direction of equipartition of energy, so this situation is considered a reasonable assumption. As you'll see in the following sections of this chapter, the assumption of energy equipartition is often used to estimate some physical parameters of radio galaxies.

Example 9.4:

Later in this chapter, in Example 9.9, we estimate the magnetic field in the lobes in Cygnus A to be of order 6×10^{-5} gauss. The redshift of Cygnus A is $z = 0.056$ and the angular diameter of each lobe is approximately 0.4 arcmin. Estimate the minimum total energy in the lobes of Cygnus A.

Answer:
Since the minimum total energy is approximately equal to the equipartition total energy, we can calculate the energy in the magnetic field and set the electron energy equal to this. The minimum total energy, then, is double the magnetic field energy. The magnetic field energy is

$$E_B = \frac{\langle B^2 \rangle}{8\pi} \times \frac{4\pi R^3}{3}.$$

To determine the radius, R, of a lobe, we use the redshift and angular size information. A redshift of 0.056 corresponds to a proper distance (using the numerical integration in the Appendix and assuming $H_0 = 70$ km s^{-1} Mpc^{-1}, $\Omega_M = 0.3$, and $\Omega_\Lambda = 0.7$) of 2.37×10^5 kpc, and so the angular size distance, $d_A = d_p/(1+z)$, is 2.24×10^5 kpc. The angular size of 0.4 arcmin equals 1.16×10^{-4} radians; therefore, the lobe diameter is $D \approx 1.16 \times 10^{-4}$ radians $\times\ 2.24 \times 10^5$ kpc $= 26.1$ kpc. The energy in the magnetic field, then, is

$$E_B = \frac{(6 \times 10^{-5}\ \text{gauss})^2\ (13\ \text{kpc} \times 3.09 \times 10^{21}\ \text{cm kpc}^{-1})^3}{6}$$
$$= 3.9 \times 10^{58}\ \text{erg}.$$

The total minimum energy in a lobe is twice this value, or of order 10^{59} erg. Note that this does not include the energy contained in the core or in the jets, nor in the non-radiating particles in the lobes.

To produce the extended sources, with jet lengths on the order of tens to hundreds of kpc and large lobes of emission well outside the stellar limits of the host galaxies, this jet activity must continue for millions of years. Moreover, there must be a very efficient and stable jet-collimating mechanism. Sources lacking large radio lobes and/or long jets, on the other hand, may be young in terms of the radio emission structure, or have poor or unstable jet collimation, or be sporadic in their activity.

The fact that AGN are preferentially seen at higher redshifts indicates that this activity was more common in the past. The host galaxies must become inactive and then appear as "normal galaxies." The implication is that many, if not all, nearby galaxies once displayed this activity. Moreover, the large power outputs of these sources could not possibly persist throughout the history of the host galaxy, for this would entail a total energy output, in radiation alone, equal to the rest mass energy in 10^{10} $M_\odot$. This activity, therefore, must be of relatively short duration. When the host galaxy does not show AGN activity, the supermassive black hole still exists in its core but is not actively accreting material. In fact, the fraction of galaxies that are seen as active is roughly the same as the fraction of a galaxy's history that the activity would reasonably last. The logical inference, then, is that all large galaxies harbor supermassive black holes, and the active ones are just the ones in which the supermassive black holes are actively accreting material, while in the quiescent galaxies there is little or no accretion. Investigations of nearby quiescent galaxies confirm that their centers harbor supermassive black holes (see review by John Kormendy and Luis Ho[7]). Additionally, many otherwise normal galaxies are found to display weak to moderate levels of AGN-type activity (see review by Luis Ho[8]).

9.3 MORPHOLOGIES, SIZES, AND SPECTRA OF RADIO GALAXIES AND QUASARS

The image of Cygnus A, shown in Figure 9.1, is considered the canonical picture of a double-lobed radio galaxy. However, it does not represent all extragalactic radio sources, or even a majority of them. The majority, including the quasars, are much smaller. Some *compact extragalactic radio sources* have linear sizes of order tens of parsecs. Most extragalactic radio sources also lack the roughly symmetric radio structure, with two lobes on either side of the core, but instead appear more like a core with a single jet, and usually there is no

[7]2013, Annual Reviews of Astronomy & Astrophysics, vol. 51, p. 511.
[8]2008, Annual Reviews of Astronomy & Astrophysics, vol. 46, p. 475.

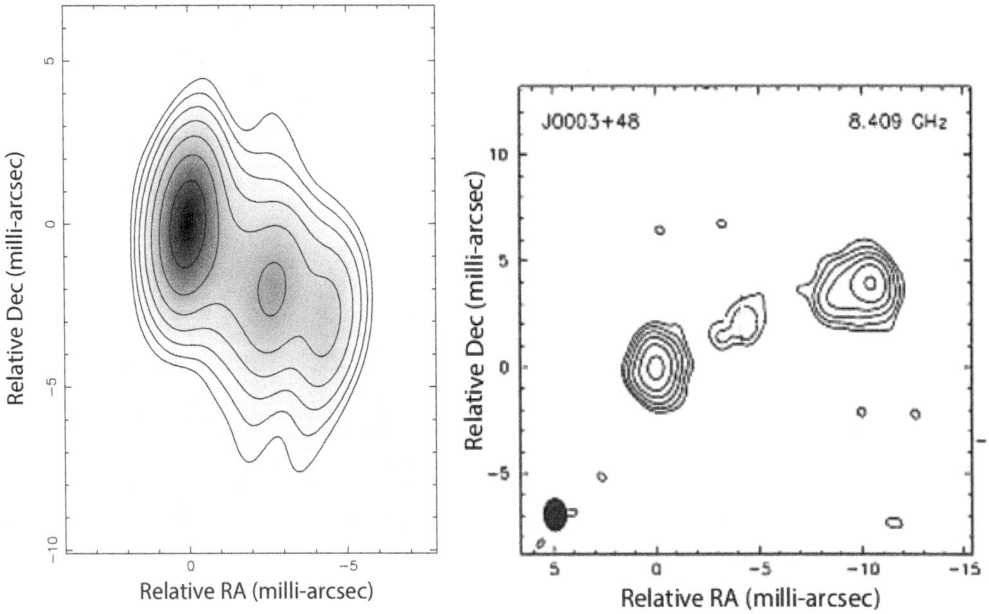

Figure 9.7: VLBI contour maps of (left) a core-jet source – 3C279 at 5.0 GHz (reproduced from data obtained by Marr and collaborators) and (right) a Compact Symmetric Object – J0003+4807 at 8.4 GHz (taken from Peck and Taylor, 2000, Astrophysical Journal, vol. 534, p. 90).

apparent lobe. A radio map of the compact source 3C279 is displayed in Figure 9.7(left) as an example. Sources of this type are referred to as *core-jet* sources. Other compact sources have a symmetric, miniature two-lobe appearance. These sources are referred to as *compact symmetric objects*, or CSOs (they have also been referred to as compact doubles); see Figure 9.7(right) for an example. Note that the unit on the axes in Figure 9.7 is a milli-arcsecond. The discovery of these compact structures required the resolution afforded with the invention of VLBI (see Vol. I, Section 6.14).

The larger, symmetric, *extended* sources, are divided further into two subclasses. In 1974 Bernard Fanaroff and Julia Riley[9] commented that in some large radio galaxies, such as Cygnus A (see Figure 9.1), the jets end in bright spots, or hot spots, believed to be the collision site between the jet particles and the extragalactic medium. In other radio galaxies, such as 3C296, shown in Figure 9.8, the jets fade into the lobes without hot spots to mark the jet ends. The lack of hot spots suggests that less energy is involved in the collision between jet and medium, and hence that these jets, currently, have less power. The radio galaxies with the strong hot spots tend to be significantly more luminous at radio frequencies. Fanaroff and Riley classified the less-luminous radio galaxies with fainter or no hot spots as class I and the others as class II. The AGN nomenclature now includes the terms *FR type I* and *FR type II*. Additionally, FRI sources commonly display two-sided jets, while FRII sources usually exhibit one-sided jets.

Furthermore, there are many galaxies with significant bends in the jets; Figure 9.9 displays three examples. Bends in jets may occur because of the host galaxy's motion through intracluster gas, or due to the orbital motion of the host galaxy, or because of the jet crashing into inhomogeneities in the surrounding medium.

[9]1974, Monthly Notices of the Royal Astronomical Society, vol. 167, p. 31.

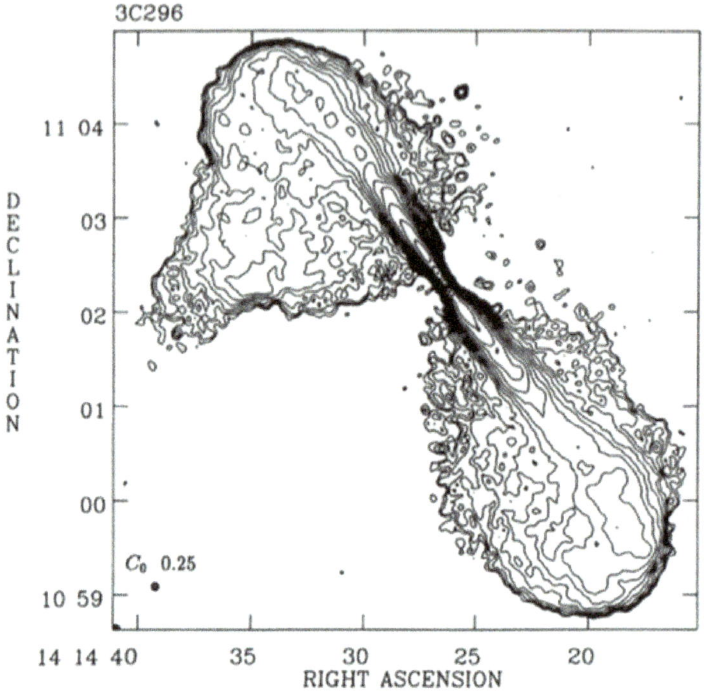

Figure 9.8: VLA contour map of the 1.5-GHz intensity in 3C296, a FR type I radio galaxy. Image taken from Leahy and Perley 1991, Astronomical Journal, vol. 102, p. 537.

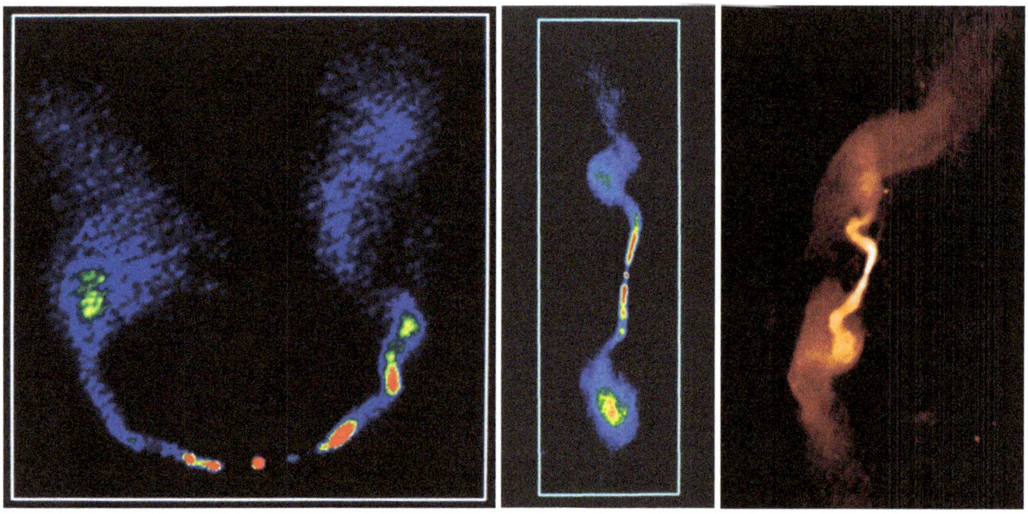

Figure 9.9: Radio galaxies with bent jets: NGC 1265 (image obtained by C. O'Dea and F. Owen), 3C449 (obtained by Perley, Willis and Scott), and 3C31 (obtained by Laing, Bridle, Perley, Feretti, Giovannini, and Parma). All images courtesy of NRAO/AUI.

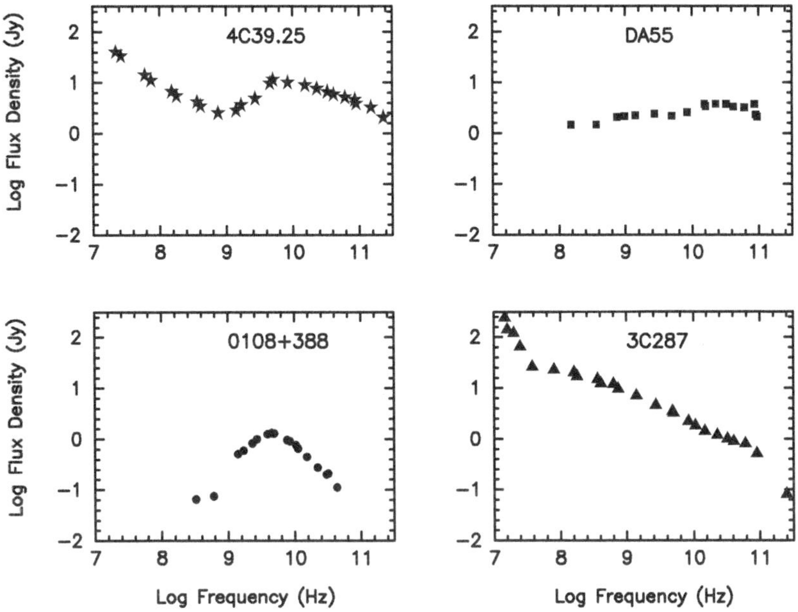

Figure 9.10: Log-log plots of F_ν versus ν of four compact radio sources. The data were obtained from the NASA Extragalactic Database (NED).

The compact radio sources, including the compact cores in extended sources, differ from the large radio lobes in extended sources with regards to their radio spectra. The large, extended sources almost always have the canonical steeply declining power-law spectrum of optically thin synchrotron radiation with spectral indices ranging from -0.5 to -1.0 (as in Figure 9.3). The compact sources usually have flatter spectra, sometimes complex with a number of changes in slope; see the spectra of DA55 and 4C39.25 in Figure 9.10. Some compact sources, however, have steeply declining spectra and are referred to as *compact steep spectrum* (CSS) sources; see the 3C287 spectrum in Figure 9.10. Additionally, some compact sources have spectra that decline steeply at high frequencies but turn over at frequencies of order 1 GHz and have steeply inverted low-frequency spectra with spectral indices as large as $+2$. An example is 0108+388 shown in Figure 9.10; these are known as *gigahertz peaked spectrum* (GPS) sources. An extensive review of these sources was written by Chris O'Dea in 1998[10]. Different spectral slopes are sometimes seen within a single radio galaxy. As shown in Figure 9.11, the core of Cygnus A has a nearly flat spectrum, while the lobes have the steeply declining spectra indicative of optically-thin synchrotron radiation.

The different spectral behavior between compact components and extended components is believed to be an optical depth effect. The extended components are big and diffuse and are transparent, so we see steeply-declining, optically-thin synchrotron spectra. The compact components are dense regions in the cores of the active galaxies and are more likely to be opaque. As discussed in Section 3.3.5, optically thick synchrotron radiation has a positive spectral index. In the simplified case of a synchrotron source with homogeneous structure (uniform density, magnetic field, and electron energy distribution), the synchrotron self-absorption can be important at lower frequencies and produces a power-law spectrum that

[10] 1998, Publications of the Astronomical Society of the Pacific, vol. 110, p. 493.

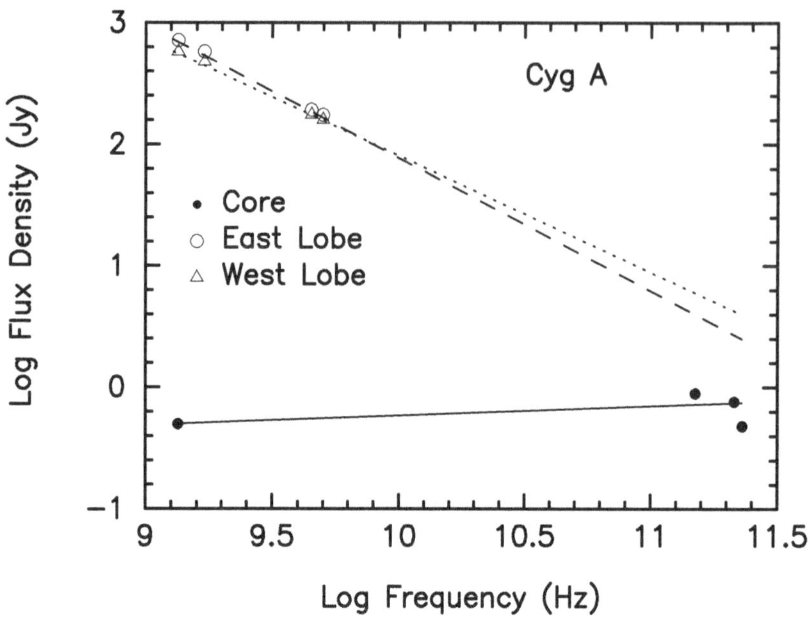

Figure 9.11: Log-log plot of F_ν versus ν of the core (solid line) and the lobes (dashed and dotted lines) in Cygnus A. The data were obtained from the NASA Extragalactic Database (NED).

increases with frequency with a spectral index of +2.5. The spectra observed in compact components never have indices this large because they are inhomogeneous.

9.3.1 Synchrotron Spectrum from an Inhomogeneous Source

We can model an inhomogeneous source as the superposition of many different components, each one homogeneous, but with different physical parameters. For example, imagine a source composed of several concentric spheres each with different electron density and magnetic field that vary systematically with radius r. For simplicity, we will assume that all layers have the same electron energy distribution. Figure 9.12 depicts a model with four layers. We will refer to the core as layer 1 and the surrounding shells as layers 2, 3, and 4, with layer 4 being the outermost shell. The electron density in the source's core, then, is represented by n_1 and its magnetic field by B_1, and let the three shells have densities and magnetic fields of n_2, n_3, and n_4 and B_2, B_3, and B_4. Because astronomical objects tend to have larger densities closer to their centers, and likewise for the magnetic field, let's assume that B_1 and n_1 are the largest values of magnetic field and electron density, while B_4 and n_4 are the smallest, and the values in the middle shells decrease with increasing radius.

Let's now contemplate the expected synchrotron radiation produced by each layer, starting with the outermost shell, layer 4. Its optically-thin intensity will depend on the number of electrons and the power radiated by each electron which, according to Equation 3.28, is proportional to the magnetic field squared. Therefore,

$$I_{\nu,4} \propto n_4 B_4^2,$$

and its spectrum will have a turnover frequency that depends on B as given by Equation 3.32:

$$\nu_{\rm SSA\ turnover} \propto B_4^{1/5} I_\nu^{2/5} \propto n_4^{2/5} B_4.$$

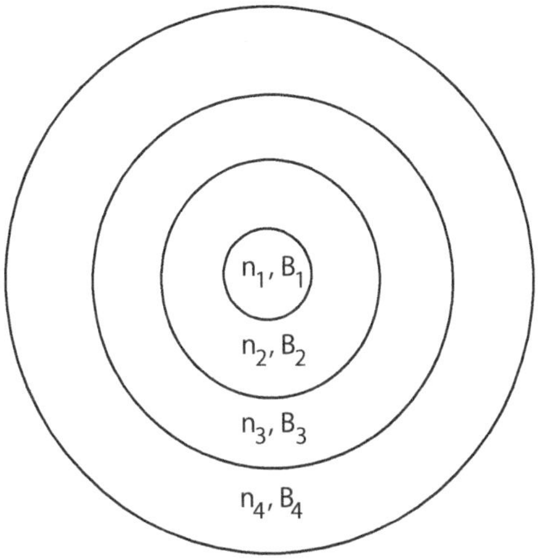

Figure 9.12: A simplified model of an inhomogeneous synchrotron source consisting of a spherical core and three concentric spherical shells, in which each layer has a different electron density, n, and magnetic field, B.

Since n_4 and B_4 are smaller than the other values of n and B, this shell has the lowest turnover frequency, which means at higher frequencies it is transparent and the radiation produced by layer 3 will be seen. Layer 3, with larger n and B, will have a slightly higher turnover frequency, but not as high as layer 2. So, the radiation at the turnover frequency of layer 3 passes through layer 4 and is visible. Likewise, the radiation at the turnover frequency of layer 2 passes through both of the top two layers and so is also visible. Thus, we will detect the radiation at all turnover frequencies from all shells. Furthermore, with larger values of n and B occurring at smaller radii, the intensity of the optically thin part of the spectrum will be greater for the inner shells and weakest for the outermost shell. Putting all this together, the spectrum from this four-layer sphere will look like that in Figure 9.13. This figure shows both the optically-thin spectrum at higher frequencies, and an optically-thick spectrum at lower frequencies, and appears much like the spectrum of 0108+388 shown in Figure 9.10. This model is also applicable when the observation does not include the optically-thin part of the spectrum, as in the case of DA55 or 4C39.25 (also in Figure 9.10).

The composite spectrum in Figure 9.13 has an optically-thin part at the highest frequencies, where all layers in the inhomogeneous source are transparent, and the majority of the observed flux density comes from the inner parts of the source. The spectral index at high frequencies is the same as the optically thin spectral index produced by each layer. The apparent spectral turnover of the composite spectrum in Figure 9.13 occurs at the turnover frequency of the innermost region. At each frequency below the apparent turnover the observed flux density is approximately equal to the turnover flux density from some layer of the source. The radiation at these lower frequencies produced from deeper in the source is blocked by the opaque layers above them and so does not escape, and the radiation from further out in the source will be significantly fainter. The spectral index below the apparent turnover, therefore, is less than +2.5 and can have any value between the optically thin spectral index and +2.5, depending on how the turnover frequency varies with radius, which is governed by how the magnetic field and the electron density vary with radius in the source. Note that observations at different frequencies below the turnover probe different

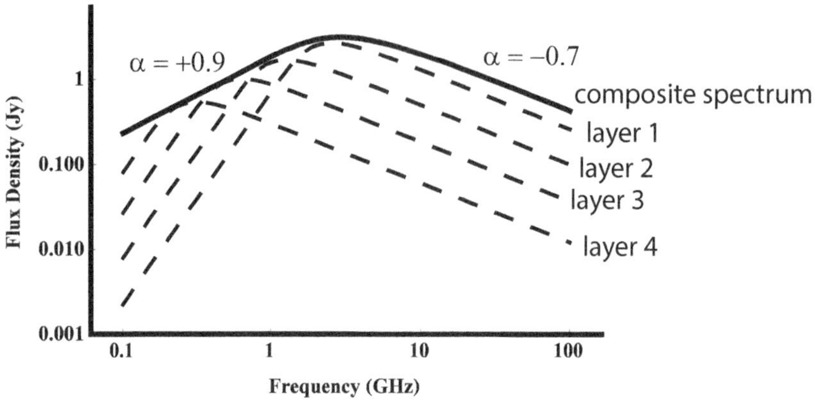

Figure 9.13: The composite synchrotron spectrum from an inhomogeneous spherical source. The dashed lines represent the spectra emitted by each layer in Figure 9.12, while the solid line shows the composite spectrum. At frequencies above the composite spectrum turnover the detected radiation is the sum of that emitted by all layers, while below the turnover we detect the radiation only up to the layer which becomes opaque at that frequency. As a result, the +0.9 slope of the observed spectrum below the turnover is flatter than the +2.5 slope of each of the layers.

layers in the source — the lower-frequency flux density will emanate from further out in the source. Similar to the concept of the Sun's photosphere, the apparent "surface" at any frequency corresponds to the physical depth into the source where the optical depth has reached a large enough value (≈ 1) that photons from deeper in do not escape. This is the depth at which the source becomes opaque, and hence defines the source size at this frequency. We have used a spherical source with radial variations for simplicity in demonstration, but inhomogeneous structures can be used with any geometry and more complex inhomogeneity. An article by A. G. de Bruyn[11] derives in detail the resultant optically thick spectral index for different power-law dependencies of n and B on r and for disks as well as spheres, while flat spectra were modeled with inhomogeneous jets by Alan Marscher[12].

9.3.2 Free-free Absorption of Synchrotron Radiation

It is tempting to assume that inverted spectra in compact sources are always due to synchrotron self-absorption. However, there is another mechanism that can produce an inverted spectrum in compact radio galaxies, which we will now discuss. As explained in Section 3.3.5, the frequency of the turnover due to SSA can be used to estimate the magnetic field in the source, and so if the turnover is not due to SSA, then this estimate will be incorrect. We know that AGN contain large amounts of ionized gas, as revealed by the numerous emission lines of ionized atoms. The synchrotron radiation from compact sources may pass through this plasma. Free electrons, as discussed in Section 3.2.4, are efficient absorbers of radio-frequency radiation. This line-of-sight, ionized gas is at temperatures well below the brightness temperature of the synchrotron radiation, so it is reasonable to expect that the observed synchrotron radiation spectrum will include absorption by these free electrons. This absorption process is referred to as *free-free absorption.*

Recall from Section 3.2.4 that the free-free optical depth depends on frequency as

$$\tau_\nu^{ff} \propto \kappa_\nu^{ff} \propto \nu^{-2.1}$$

[11] 1976, Astronomy & Astrophysics vol. 52, p. 439.
[12] 1980, Nature, vol. 288, p. 12.

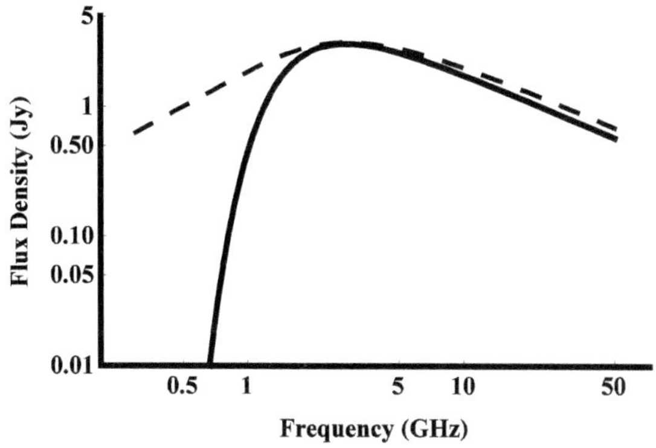

Figure 9.14: A theoretical spectrum of an optically thin synchrotron source with free-free absorption (solid line) due to foreground gas with $T_e = 10^4$ K, $n_e = n_i = 500$ cm^{-3}, and $L = 3$ pc. For comparison, overlaid with a dashed line is the same inhomogeneous SSA spectrum shown in Figure 9.13.

(Equation 3.19). Because the source function of free-free radiation, S_ν^{ff}, will be much smaller than the intensity of the synchrotron radiation, we can ignore the second term in the equation of radiative transfer (Equation 2.16) and so the observed intensity is

$$I_\nu = I_\nu^0(e^{-\tau_\nu^{ff}}),$$

where I_ν^0 is the intensity of the synchrotron radiation. Since τ_ν^{ff} is in the exponent and depends on frequency to the -2.1 power, free-free absorption is characterized by an extreme sensitivity to frequency. It can be negligible at high frequencies and become dominant below some frequency, at which the resultant spectrum falls off exponentially. Figure 9.14 depicts the spectral shape for this situation. Free-free absorption, we see, can mimic SSA in its qualitative dependence on frequency, but the drop off below the turnover frequency is much steeper.

One can calculate the frequency at which free-free absorption is likely to be important. Values of electron densities, temperatures, and sizes of ionized gas regions in AGN are inferred from visible wavelength spectral lines. These values can be used in Equation 3.19 to solve for the frequency where the optical depth equals one. As we demonstrate in Example 9.5, visible-wavelength spectral-line data imply that there may be significant free-free optical depths at radio frequencies.

Example 9.5

Visible-wavelength spectral line studies of AGN suggest that the narrow-line regions have temperatures of order 10^4 K and electron densities of order 10^3 cm^{-3}. How long a path length through this plasma is needed to cause a free-free turnover in the spectrum at 1 GHz?

Answer:
The turnover occurs at the frequency where $\tau_\nu^{ff} \approx 1$. We use Equation 3.19, which we rewrite here:

$$\tau_\nu^{\rm ff} \approx 0.0824 \left(\frac{T}{1\ {\rm K}}\right)^{-1.35} \left(\frac{\nu}{1\ {\rm GHz}}\right)^{-2.1} \left(\frac{EM}{1\ {\rm cm^{-6}\ pc}}\right),$$

where $EM \approx n_e^2 L$ is the emission measure. Inserting the given quantities we have

$$1 \approx \tau_\nu^{\rm ff} \approx 0.0824 \left(\frac{10^4 \text{ K}}{1 \text{ K}}\right)^{-1.35} \left(\frac{1 \text{ GHz}}{1 \text{ GHz}}\right)^{-2.1} \left(\frac{10^6 \text{ cm}^{-6} \text{ L}}{\text{cm}^{-6} \text{ pc}}\right).$$

Solving for L we have

$$L \approx 3 \text{ pc}.$$

Since the narrow-line regions typically extend from about 10 to 100 pc from the center, this is a reasonable depth of free electrons to be along the line of sight to the synchrotron emitting gas.

The cause for the spectral turnovers in GPS sources is probably due to a combination of SSA and free-free absorption. In some parts of the source, such as the very center, the turnover may be due to SSA and in other parts, such as the lobes, the turnover may be free-free absorption. Numerous studies suggest that both processes are important.

Early surveys of radio galaxies and quasars were done at low frequencies; for example, the Third Cambridge radio survey (3C) was performed at 178 MHz[13]. This initial approach to discovering sources meant that the first catalogs of radio sources were biased toward extended sources, such as Cygnus A, which are brighter at lower frequencies. In later years some higher frequency surveys, e.g. at 5 GHz by Pauliny-Toth and Kellermann[14] and by Kühr et al.[15], were undertaken and these surveys indeed revealed a large number of compact sources not detected in the earlier surveys. This is a selection effect to keep in mind: observations at lower frequencies favor the large double-lobed sources while observations at higher frequencies favor compact sources and core regions of AGN. For sources that contain both a core and extended lobes, such as Cygnus A, the lower frequency observations will be more sensitive to the lobes while the higher frequency observations will be more sensitive to the radiation from the core. Additionally, because of the dependence of optical depth on frequency, observations at higher frequencies probe deeper into optically thick compact cores.

9.3.3 Inverse Compton Scattering and the Compton Limit

In synchrotron sources, there must be large numbers of both radio-frequency photons and very high energy electrons, and since the cross-section for a scattering interaction between the photons and free electrons is large, the exchange of energy between the electrons and photons via scattering can be important. Since the electrons must be relativistic to produce synchrotron radiation they have extremely high energies while the radio-frequency photons have very low energies. Therefore, the photons will undergo large gains in energy, bumping them up all the way to the ultraviolet and X-ray regimes. Since these photons were initially emitted at radio wavelengths, this effect causes a reduction in the observed synchrotron intensity at radio wavelengths. This effect is not restricted to synchrotron sources; it is an important consideration when observing photons that pass through any volume of space containing large numbers of high energy charged particles.

This process is called *inverse Compton scattering* because it is the inverse process of that studied by Arthur Compton in the early 1920s. Compton found that when X-rays collide

---

[13] Edge et al. 1959, Memo of the Royal Astronomical Society, vol. 68, p. 37 and Bennett 1962, Monthly Notices of the Royal Astronomical Society, vol. 68, p. 165.

[14] 1968, Astronomical Journal, vol. 73, p. 953; 1972, Astronomical Journal, vol. 77, p. 265; 1972, Astronomical Journal, vol. 77, p. 797.

[15] 1981, Astronomy & Astrophysics Supplement, vol. 45, p. 367.

with low energy electrons, energy is transferred from the photons to the electrons. With regard to observations of radio galaxies, the electrons are relativistic and the photons have low energy. The energy transfer, hence, goes in the opposite direction and so this is termed *inverse* Compton scattering. In the scattering process, the energy of the photon is boosted by a factor ν_f/ν_i, where ν_f and ν_i are the final and initial frequencies of the photons, which depend on the electron's energy, $\gamma m_e c^2$, and the scattering angle. Since we wish to describe the effect for the radiation in bulk, we desire equations with the angle dependence removed. Without derivation, the maximum boosting factor is given by

$$\left(\frac{\nu_f}{\nu_i}\right)_{\max} = 4\gamma^2, \tag{9.2}$$

and the average boosting factor is

$$\left(\frac{\nu_f}{\nu_i}\right)_{\text{avg}} = \frac{4}{3}\gamma^2. \tag{9.3}$$

For example, consider the emission of a synchrotron photon of frequency 43 GHz from a region with a magnetic field of order 0.1 gauss. By Equation 3.26, this requires an electron with $\gamma = 594$. When this photon undergoes inverse Compton scattering with an electron of the same energy it can end up with a frequency as high as 6.1×10^{16} Hz, or a wavelength of 4.9 nm, which is in the X-ray realm.

Inverse Compton scattering provides a means by which the photons and electrons can exchange energy and so, for large photon densities, we also need to consider how this might affect the electrons' energies. Note that the amount of energy an electron transfers to a photon by inverse Compton scattering is many orders of magnitude larger than the amount it loses in emitting a synchrotron photon. Therefore, if the number density of photons is large enough, the electrons can lose energy via inverse Compton scattering at a rate far faster than they can by the synchrotron emission process. The result, then, would be a significant cooling of synchrotron electrons. Thus, there is a limit to how intense the synchrotron radiation can be. Any additional synchrotron photons will be inverse Compton scattered and the scattered electron instantly cooled. This maximum synchrotron intensity is known as the *Compton limit.* In brightness temperature, this limit is (without derivation) $T_{\text{synchr}} \leq 10^{12}$ K.

Of course the loss of energy of the electrons goes hand-in-hand with an up-scattering of photons to UV and X-ray energies. Therefore, sources that are measured to have radio intensities at or near the Compton limit, are also bright at X-ray frequencies. Furthermore, since the frequency of the scattered photon is directly proportional to the frequency of the initial photon, the spectrum of the scattered radiation has the same shape as the initial radiation. For radio galaxies with high intensities, the X-ray spectral indices are found to be strongly correlated with the radio spectral indices.

9.4 INFERRING PHYSICAL CONDITIONS IN AGN

In this section we will outline some of the analytical tools that one can use to infer the physical conditions in radio galaxies and quasars.

Many of the unanswered questions about radio galaxies relate to the initial stages of the jets. What causes the jets? How are they collimated? How are the particles accelerated to such high energies? Because of their great distance, extremely high angular resolution is needed to study the structure of these regions. The highest resolution observations are made using Very Long Baseline Interferometry (VLBI, discussed in Volume I, Chapter 6).

Conventional VLBI observations include wavelengths as short as 1 cm, at which VLBI can obtain a resolution of 8×10^{-10} radians, or a few tenths of a milliarcsecond. This is analogous to being able to see the date on a penny (width about 1 mm) at a distance of 1250 km. A concerted effort to use a global VLBI array at mm and sub-mm wavelengths has recently been undertaken with the goal of resolving the event horizons of two nearby supermassive black holes — SgrA*, the 4×10^6 $M_\odot$ black hole in the center of the Milky Way galaxy, and the 5×10^9 $M_\odot$ black hole in M87 (i.e., Virgo A). The resolution of this instrument, called the *Event Horizon Telescope*, or EHT for short, will be of order tens of micro-arcseconds. At the time of the writing of this volume, the results of the first set of EHT observations have not been reported. At a distance of approximately 16 Mpc, the event horizon of the 5×10^9 $M_\odot$ black hole in M87 has an angular size of 22 micro-arcseconds. The event horizon of SgrA* is expected to be approximately 53 micro-arcseconds. Unfortunately, though, for all other radio galaxies and quasars, at much larger distances, the resolution with VLBI at even the shortest wavelengths will be insufficient to image the AGN engine itself.

9.4.1 Spectral Index Maps

The high resolution maps that have been obtained with VLBI have led to some significant discoveries, as we discuss below. Often, though, the maps are not easy to interpret. Figure 9.15 displays VLBI maps of two compact radio sources. Even though these maps reveal significant detail on scales of a fraction of a parsec, it is far from clear what these structures are. It is not obvious even where in the map the core (the site of the AGN engine) is located. Any feature in the map could be the core or part of a jet. Either or both of the radio galaxies shown in the figure could be a compact symmetric object, with the core located near the center, either in a central feature or obscured by ionized gas, and the features at the extremes could be the ends of the jets. Or, each object might be a core-jet source with a one-sided jet and the core located at one end of the image.

Recall that the core components tend to have flatter or inverted spectra because they are opaque. An effective way of identifying the core component is to create a *Spectral index map*. Recall that the spectral index, α, is the power law index of the spectrum, i.e. $F_\nu \propto \nu^\alpha$. To create a spectral index map, we observe the object at neighboring frequency bands and make maps with identical resolutions. (To be precise, the maps should be sensitive to the same angular scales, see Volume I, Chapter 6.) Then, we combine the maps in such a way that at each position in the map the spectral index is determined. If $F_{\nu 1}$ and $F_{\nu 2}$ are the flux densities at frequencies $\nu 1$ and $\nu 2$, then the spectral index at each position is given by

$$\alpha = \frac{\log(F_{\nu 1}/F_{\nu 2})}{\log(\nu 1/\nu 2)}.$$

The core, then, can be identified as a small region with a positive α. Figure 9.16 shows spectral index maps that were used to identify the locations of the cores in the two radio galaxies shown in Figure 9.15. This method revealed that one of these sources (0108+388) is symmetric with the core near the middle, and so is a compact symmetric object, while the other (VIPS J07334+5605) has its core at one end and so a core-jet source. Even though the morphology of these sources as seen in the maps (Figure 9.15) is similar, the spectral index maps place them in different classes.

Spectral index maps can also reveal dense spots in jets. Note in the image of Cygnus A in Figure 9.1 the spots of intense emission, referred to as hot spots, in the lobes at the ends of the jets. These spots are believed to be the working surfaces where the jet particles collide with surrounding matter. Because of the enhanced density, the spectra of these hot

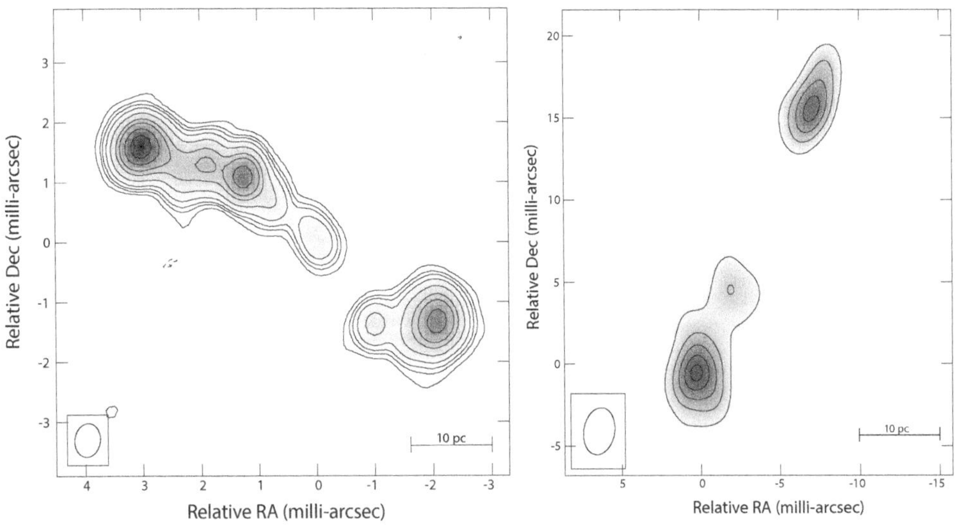

Figure 9.15: VLBI contour maps of two compact sources (left) 0108+388 at 15.4 GHz (taken from Marr et al. 2001, Astrophysical Journal, vol. 550, p. 160) and (right) VIPS J07334+5605 at 5.0 GHz (taken from Tremblay et al. 2016, Monthly Notices of the Royal Astronomical Society, vol. 459, p. 820). Interpretation of the structure of these sources from the maps is not straightforward. Is each a core-jet source or a compact double-lobed source with the core somewhere near the middle?

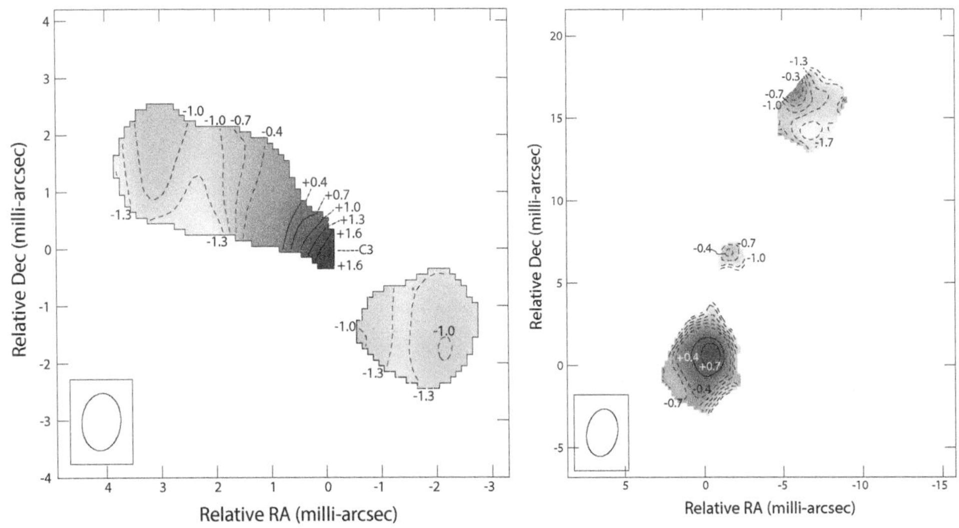

Figure 9.16: Spectral-index maps of the sources in Figure 9.15. Dashed contours indicate negative spectral indices (i.e., a declining spectrum). The locations of the core components are indicated by solid contours, indicating positive spectral indices. In 0108+388 (left) the spectral index between 15.4 and 8.4 GHz in the central component labeled "C3" has a positive spectral index and is inferred to contain the core (taken from Marr et al. 2001, Astrophysical Journal, vol. 550, p. 160). In VIPS J07334+5605 (right) the area of positive spectral index between 8.4 and 5.0 GHz is seen to occur in the feature at the southern end, suggesting that this is a core-jet source (data taken from Tremblay et al. 2016, Monthly Notices of the Royal Astronomical Society, vol. 459, p. 820).

spots are flatter than the lobes and so the spots stand out relative to the lobes in images at higher frequencies more noticeably than at lower frequencies. Similar higher density spots sometimes occur along the jets, and these can be revealed with spectral index maps. The dense spots along the jets are believed to be due to kinks or areas of compression or shocks in the jets. Often, the central engine, which generally does not consume fuel at a constant rate, will undergo an outburst, during which it will eject a new stream of relativistic material, thus producing a bright spot. In a number of cases, a rapid increase in the radio flux density of a radio galaxy has been found to be coincident with the appearance of a hot spot emanating from the core component.

9.4.2 Kinematic Studies

In the 1970's, radio astronomers using VLBI observations were stunned to discover that they could detect significant motions of features in the jets moving outward from the core. In a series of VLBI observations of the same source, the distance between the features in the jet and the core increased with time. This was a big discovery, for starters, because it confirmed the idea that the long thin lines of synchrotron radiation do indeed trace a flow of matter from the core to the lobes.

With the CSOs, in fact, the growth of the entire radio emission structure has been observed, supporting the premise that these are young versions of the Cygnus A-type radio galaxies. Extrapolating the observed growth of a CSO backwards, one generally obtains an age estimate of the radio-emitting structure from hundreds to thousands of years. These objects, therefore, are of special interest for examination of the early stages of this phenomenon.

Repeated mapping of sources, especially the largest flux-density compact sources, revealed an astonishing result. Sometimes the inferred velocities, obtained simply from the change in distance between the core and jet features divided by the time period between observations, were noticeably greater than the speed of light! See Figure 9.17. This is referred to as *apparent superluminal motions* in radio jets. Since motion cannot exceed the speed of light, various alternative explanations were proposed. One model was that the jet features in different observations were not the same bodies, but were random locations in the jet lighting up at random times. This was known as the Christmas tree lights model. However, the apparent motions were always outward and not random. A second model was that these radio galaxies and quasars weren't really as far away as believed, in which case the real motions would be slower because these jets are closer to us, and so the physical distances traveled are smaller.

Eventually a fairly simple and elegant solution was realized, involving a little bit of special relativity. In this model, which is now well-accepted, the features do not really move super-luminally, although they do move relativistically. The reason that the features in the jet appear to move faster than c is because the jet is angled toward the observer, close to the line of sight. The jet material, then, (in the frame of reference of the observer) essentially chases its own light. This causes the time interval between the arrival of light signals at the observer's location to be shorter than the time interval between the emission of light signals from the jet. The apparent transverse speed, which equals the transverse distance traveled divided by the time interval between images, ends up being greater than the true speed because of the smaller time interval in the denominator.

This is easy to show mathematically. Consider a jet, with velocity v pointed at an angle θ relative to the line of sight toward the observer (see Figure 9.18). Consider the light signals emitted toward the observer at two different times: t_1 and t_2. During that time interval, the

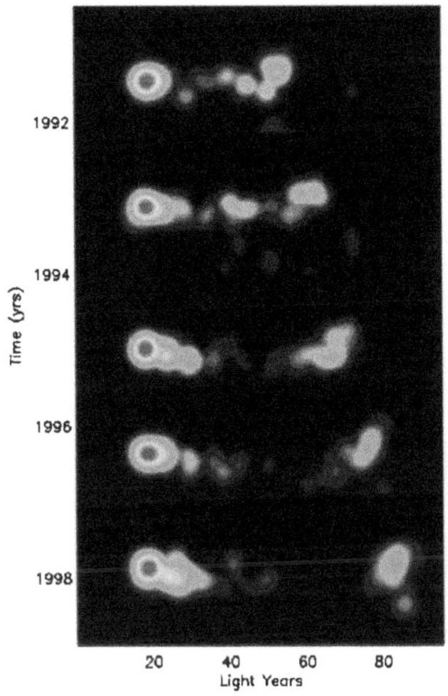

Figure 9.17: VLBI images of the radio galaxy 3C279. Features are seen moving outward, away from the core, at speeds apparently faster than the speed of light. From late 1991 to 1998, with the core placed at the 20 ly mark, a feature shown in these maps to the right of the core component moves from 30 ly to 60 ly from the core and so appears to move 30 ly in a little over 6 years, implying a speed of almost $5c$. Image courtesy of NRAO/AUI.

jet feature moves a distance

$$d = v(t_2 - t_1).$$

By time t_2, the first light signal will have traveled a distance toward the observer equal to $c(t_2 - t_1)$ while the jet feature will have moved along the line of sight a distance equal to $v(t_2 - t_1)\cos\theta$. So the second light signal, emitted at time t_2, will be a relatively short distance behind the first light signal if v is close to c. The distance between light signals is the distance traveled by the first light signal from time t_1 to t_2 minus the distance traveled along line of sight by the jet feature from t_1 to t_2, i.e.

$$\text{distance between light signals} = c(t_2 - t_1) - v(t_2 - t_1)\cos\theta.$$

The observed time interval, Δt_{obs}, between arrival times of the light signals is the distance between the light signals divided by c, so

$$\begin{aligned}\Delta t_{\text{obs}} &= \frac{c(t_2 - t_1) - v(t_2 - t_1)\cos\theta}{c} \\ &= (t_2 - t_1) - \frac{v}{c}(t_2 - t_1)\cos\theta,\end{aligned}$$

or,

$$\Delta t_{\text{obs}} = (t_2 - t_1)(1 - \frac{v}{c}\cos\theta).$$

The apparent velocity of the jet feature, given by the distance traveled perpendicular to the line of sight divided by the time separation between images (i.e. when the two light beams arrived at Earth), is

$$v_{\text{app}} = \frac{\Delta s}{\Delta t_{\text{obs}}},$$

where the distance Δs (see Figure 9.18) is given by

$$\Delta s = v(t_2 - t_1)\sin\theta.$$

So,

$$v_{\text{app}} = \frac{v(t_2 - t_1)\sin\theta}{(t_2 - t_1)(1 - (v/c)\cos\theta)}.$$

The equation for the apparent speed of the jet feature, therefore, is

$$v_{\text{app}} = v\,\frac{\sin\theta}{1 - (v/c)\cos\theta}. \tag{9.4}$$

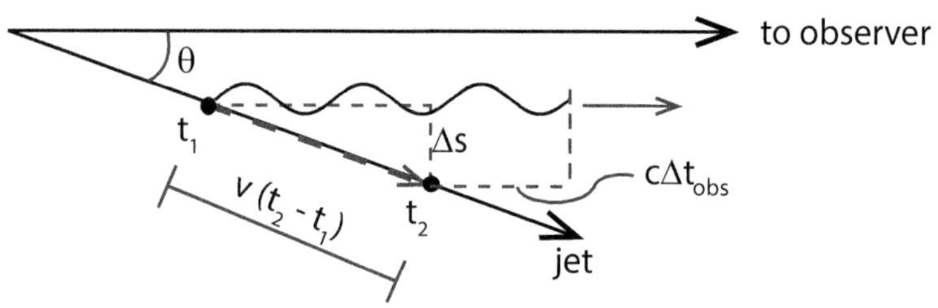

Figure 9.18: Schematic of motion in a jet, angled toward the observer, that leads to an apparent speed greater than c. Photons emitted at times t_1 and t_2 are detected by the observer separated in time by Δt, which is less than $t_2 - t_1$ because of the relativistic motion of the jet angled toward the observer.

It is instructive to examine the dependence of the apparent velocity on the direction angle of the jet in Equation 9.4. As we should expect, when $\theta = \pi/2$, so that the jet is purely in the plane of the sky, then $v_{\text{app}} = v$. Additionally, when $\theta = 0$, so that the jet is pointed directly at us and there is no transverse component, then $v_{\text{app}} = 0$. When the jet is moving away from us, so that $\theta > \pi/2$, then $v_{\text{app}} < v$ and goes to zero when $\theta = \pi$. To see how we can get apparent superluminal motion, we need to explore the dependence of v_{app} on θ between these two extremes. Note that, as the jet is oriented closer to the line of sight, $\theta \to 0$, both the numerator and denominator approach zero. The angle which yields the maximum apparent velocity, given by setting $dv_{\text{app}}/d\theta = 0$ or

$$\frac{d}{d\theta}\left(v\,\frac{\sin\theta}{1 - (v/c)\cos\theta}\right) = 0,$$

is given by

$$v\left(\frac{\cos\theta - (v/c)}{(1 - (v/c)\cos\theta)^2}\right) = 0.$$

This occurs when $\cos\theta = v/c$. Inserting this value for θ in Equation 9.4, we see that the maximum apparent velocity is

$$\max\, v_{\text{app}} = v\frac{\sqrt{1 - (v/c)^2}}{1 - (v/c)^2} = \gamma v,$$

where $\gamma = (1 - (v/c)^2)^{-1/2}$ is the Lorentz factor. For example, if $v = 0.95c$, in which case $\gamma = 3.2$, one will observe an apparent velocity of $3.0c$ if the jet is pointed at an angle $\theta = 18.2° = 0.32$ radians from the line toward the observer.

Example 9.6:

For the case of 3C279, shown in Figure 9.17, where the apparent jet velocity is $5c$, what is the largest angle with respect to the line of sight that is allowable for this jet?

Answer:
Using Equation 9.4, we set $v_{\mathrm{app}} = 5c$ and so we have

$$5 = \frac{(v/c)\sin\theta}{1 - (v/c)\cos\theta}.$$

To solve for an upper limit on θ we set $v/c = 1$.

$$5 = \frac{\sin\theta}{1 - \cos\theta}.$$

Rearranging,

$$5(1 - \cos\theta) = \sin\theta = \sqrt{1 - \cos^2\theta};$$

squaring and rearranging again, we have

$$26\cos^2\theta - 50\cos\theta + 24 = 0.$$

Using the quadratic equation to solve for $\cos\theta$, we find that

$$\theta < 22.6°.$$

In conclusion, the direction of the jet in 3C279 must be oriented within 22.6° of the direction toward Earth.

Example 9.7:

A jet with velocity $v = 0.98c$ is directed nearly toward Earth with $\theta = 11.5°$. What is the apparent speed of features in this jet?

Answer:
Using Equation 9.4, we have

$$v_{\mathrm{app}} = \frac{0.98c\sin(11.5°)}{1 - 0.98\cos(11.5°)} = 4.92c.$$

This is approximately the apparent speed of the feature in Figure 9.17 and so this is a possible configuration for the jet in 3C279.

The explanation given above may seem counter-intuitive. Usually we think of motions close to the line of sight as producing smaller apparent motions. The component of the velocity perpendicular to the line of sight must be small if the angle of the velocity vector relative to the line of sight is small, so you would naturally expect the apparent velocity to be small. So, why does this geometry produce much larger velocities? If you go back through the math above, you can see that the fundamental factor is the small $\Delta t_{\rm obs}$ that results when the jet chases after its own light (in the reference frame of the observer). This is due to the finite and constant speed of light in all reference frames. Even though the component of the velocity perpendicular to the line of sight is indeed small (the numerator in the Equation 9.4), the denominator of the apparent speed is decreased by a larger factor, and so the net result is an increase in apparent velocity. The cause of the small $\Delta t_{\rm obs}$ really is a simple idea; it doesn't even require any prior knowledge of relativity other than the constancy of the speed of light in all reference frames.

This idea, though, led to another conundrum, which may have occurred to you in this discussion. In order for this explanation to work, the jet must be aimed to within a very small angle of the line of sight. In Example 9.6, for example, we found that a jet appearing to move at $5c$ must be aimed within 22.6° of the line of sight. Hence, this jet must be aimed within a cone of 0.15π steradians. Out of the 4π steradians of possible directions that a jet can be beamed, the odds that any individual jet would be beamed close enough to the line of sight to yield apparent superluminal motion are small. Therefore, the fact that many jets studied with VLBI exhibit superluminal motion seems in conflict with expected statistics.

The solution to this issue is also due to a relativistic effect of the beamed jet. In our reference frame the radiation is beamed predominantly along the direction of motion. You can prove this to yourself by using the Lorentz velocity addition equations to determine the direction of the velocity vectors of the light beams in our reference frame. The speeds of the light beams will be c in both frames of reference, so in the transformation of reference frames the directions of the light beam velocities must change. For a simple visualization, imagine a truck moving along the street with pebbles being tossed out of the back in random directions. In the reference frame of the observers on the sidewalk, the total velocity of the pebbles will include the velocity of the truck and so will appear to be moving predominantly in the direction of the truck. With the synchrotron emitting jets, then, even though the total speed of the photons will still be c, their direction in our reference frame will include a component along the jet's motion.

How does this solve the conundrum? This effect makes the jets that are pointed toward us appear much brighter, because a larger number of these jets' photons will be beamed into our telescope. Additionally, these photons will be Doppler shifted toward higher frequencies, which, for optically thin jets, shifts the higher intensity at a lower frequency into the observed passband. The result is that the brighter synchrotron radio sources, which are the ones we study first, will preferentially be the ones whose jets are pointed nearly along the line of sight.

The beaming model provides a solution to one other observational oddity. Almost all of the first jets studied with VLBI appeared one-sided. Models involving intrinsically one-sided jets are difficult to accept, because the net momentum imparted to the central engine by the ejection of relativistic particles for a long period of time would cause the engine to move out of the core of the galaxy (even a 10^9 $M_\odot$ black hole). However, if the jets are beamed toward us, then the jets appear one-sided because the counter jets will be very hard to see as most of their photons are beamed away from us. The observed flux density of a relativistic jet aimed at an angle θ from the direction toward Earth is boosted by a factor Γ given by

$$\Gamma = [\gamma(1-\beta\cos\theta)]^{\alpha-p}, \tag{9.5}$$

where $\gamma = (1-\beta^2)^{-1/2}$, $\beta = v/c$, α is the spectral index (defined as $F_\nu \propto \nu^\alpha$), and p depends on the details of the radiating feature. Models of emission from relativistically beamed jets to determine the value of p involve many complicating details, including geometrical and optical depth factors; these complications are beyond the scope of this book (see papers by Roger Blandford and Arieh Königl[16] and Kevin Lind and Roger Blandford[17]). For most cases $p = 2$ (when looking down a continuous jet emanating from the core) or 3 (when observing an optically-thin component moving along the jet). The ratio, R, of the flux densities of the approaching and receding jets, assuming they are intrinsically identical, then, is

$$R = \left(\frac{1+\beta\cos\theta}{1-\beta\cos\theta}\right)^{p-\alpha}.$$

Example 9.8:

In Example 9.7, we found that the observed motion of the jet feature in 3C279, shown in Figure 9.17, could correspond to a true speed of 0.98c moving in a direction 11.5° from the direction toward Earth. Assume that this is the speed and orientation of this jet component. Assume also this is a single, optically-thin component in the jet ($p = 2$) with a relatively flat spectrum with $\alpha = -0.3$

(a) By what factor is this component's flux density boosted because of its motion toward us?

(b) Assuming that there is a counter jet containing an identical component, by what factor is the approaching component's flux density greater than that in the counter jet?

Answer:
(a) To answer this question we will use Equation 9.5, for which we need to calculate γ. We have $\beta = v/c = 0.98$ and so

$$\gamma = \frac{1}{\sqrt{1-\beta^2}} = \frac{1}{\sqrt{1-0.98^2}} = 5.025.$$

By Equation 9.5 using $p = 3$ and $\alpha = -0.3$ we have, then,

$$\Gamma = [5.025(1 - 0.98\cos(11.5°))]^{-0.3-3} = 205.$$

(b) Likewise, for the component in the jet pointed away from the Earth, we can use $\theta = 180° - 11.5° = 168.5°$. Assuming the same β, then, we get

$$\Gamma = [5.025(1 - 0.98\cos(168.5°))]^{-3.3} = 0.000527.$$

Therefore, a similar component in the counter jet will be $205/0.000527 = 3.89 \times 10^5$ times fainter than the component seen in Figure 9.17 and, hence, undetectable.

With their flux densities enhanced, sources with jets beamed in our general direction compose a significant fraction of the compact radio sources first studied. As a consequence, their apparent properties led to the creation of a subclass of objects known as *blazars*. This term was created as a combination of the names of two groups of sources known as *BL Lac objects* and *optically violent variable quasars*. The properties of these sources include rapid

[16] 1979, Astrophysical Journal, vol. 232, p. 34.
[17] 1979, Astrophysical Journal, vol. 295, p. 358.

variability of the optical and radio emission, steeply declining power-law optical continua and flat radio continua, and strong linear polarization of the radiation. The BL Lac objects, named after the first such object studied, also have visible-wavelength spectra lacking in emission lines, while the optically violent variable quasars do show the broad and narrow emission lines but otherwise are quite similar to the BL Lac objects.

9.4.3 Magnetic Field Estimates

Synchrotron radiation requires a magnetic field and so we know that radio galaxies must contain significant magnetic fields. Inferring the strength of these fields is important to address the energetics as well as to provide information for theoretical models of these sources. There are two methods for inferring magnetic fields, depending on which part of the synchrotron spectrum is observed — the optically thin or optically thick part. Both calculations involve significant assumptions and so only provide order-of-magnitude estimates of the magnetic field strength. The optically thick calculation is preferred and we discuss that method first.

When the spectrum is known to be flat or inverted, such as in a core component, we can infer that absorption is important. Assuming that the absorption is due to synchrotron self-absorption (and not free-free absorption by line-of-sight plasma, discussed in Section 9.3.2) we can use the "photosphere" concept discussed in Section 9.3.1 and consider that the observed emission arises from a layer with optical depth equal to 1. Setting $\tau_{\rm SSA} \approx 1$ and taking into account complicating factors due to beaming, Alan Marscher[18] derived the following expression for estimating the magnetic field from observations at a frequency where SSA is important;

$$B \approx 10^{-5} \text{ gauss } b(\alpha) \left(\frac{\Theta}{1 \text{ mas}}\right)^4 \left(\frac{\nu}{1 \text{ GHz}}\right)^5 \left(\frac{F_\nu}{1 \text{ Jy}}\right)^{-2} \left(\frac{\delta}{1+z}\right), \tag{9.6}$$

where B is the magnetic field strength, $b(\alpha)$ is a parameter containing the dependence on the spectral index (α), Θ is the angular size in milliarcseconds of the region being analyzed, ν is frequency in GHz, F_ν is the flux density of the region in janskys, z is the redshift of the source, and δ is a factor to account for the relativistic motion of the component and is given by

$$\delta = \frac{1}{\gamma(1 - \beta \cos\theta)}, \tag{9.7}$$

where $\gamma = (1 - \beta^2)^{-1/2}$ is the Lorentz factor, $\beta = v/c$ and θ is the direction of the jet relative to the line of sight. Note that θ is defined so that $\theta = 0$ and $\cos\theta = 1$ when the motion is directly toward Earth. The parameter $b(\alpha)$ varies by less than a factor of a few and so has a negligible effect on order-of-magnitude estimates of the magnetic field. We will use the typical value $b(\alpha) \approx 3$. Note that when the object is moving perpendicular to the line of sight ($\cos\theta = 0$) then $\delta = \gamma^{-1} \leq 1$. The Doppler shift associated with motion perpendicular to the line of sight is due to relativistic time dilation. Thus, if we know all the parameter values except β, and hence γ, we can use Equation 9.6 to set an upper or lower limit to B, depending on the geometry of the source. If the region we are analyzing is in a jet in the plane of the sky, then δ is less than one and we get an upper limit to B. If it is in a jet beamed nearly at us, then we get a lower limit to B.

The lobes and jets, however, emit optically thin synchrotron radiation and so Equation 9.6 is not applicable. To infer the magnetic fields in these regions requires an alternative

[18] 1983, Astrophysical Journal, vol. 264, p. 296.

approach. As discussed at the end of Section 9.2, it is reasonable to assume equipartition of energy between the magnetic fields and the radiating electrons, and this assumption allows for a magnetic field estimation. However, this method also requires an assumption about the upper and lower cutoffs of the electron energy distribution, and it yields a magnetic field estimate only for the entire volume of the emitting region being analyzed. This calculation, therefore, is less reliable than that for optically thick emission regions, but it provides the only possible estimate for the optically thin features. With these assumptions in place, George Miley[19] derived the following equation for estimating the magnetic field at a frequency where the synchrotron radiation is optically thin;

$$B \approx 1.4 \times 10^{-4} \text{ gauss } (1+z)^{1.1} \left(\frac{\nu}{1 \text{ GHz}}\right)^{0.22} \left[\left(\frac{F_\nu}{1 \text{ Jy}}\right)\left(\frac{\Theta}{1 \text{ arcsec}}\right)^{-2}\left(\frac{s}{1 \text{ kpc}}\right)^{-1}\right]^{2/7}, \tag{9.8}$$

where s is the line of sight distance through the feature.

Example 9.9:

(a) Use the data displayed in Figure 9.11 to estimate the magnetic field in the lobes of Cygnus A. In Example 9.4, we found that the angular diameter of each lobe is approximately 26.1 kpc.

(b) Data for the core are provided by Ken Kellermann and collaborators[20]. In VLDI observations at 15.4 GHz they measure a peak intensity of 0.6 Jy beam^{-1} with a beam size of 1.0 mas × 0.5 mas. With multi-epoch observations, Kellermann and collaborators[21] measure proper motions of features in the core with an average $\beta \approx 0.5$. Estimate the magnetic field in the core of Cygnus A.

Answer:
(a) As shown in Figure 9.11, each lobe emits 100 Jy at 10 GHz. The lobes are optically thin, so we use Equation 9.8. Assuming the lobes are roughly spherical, we can use the lobe diameter for the value of s. Inserting all the given and inferred values into Equation 9.8 we have

$$B \approx 1.4 \times 10^{-4} \text{ gauss } (1.056)^{1.1} \left(\frac{10 \text{ GHz}}{1 \text{ GHz}}\right)^{0.22}$$
$$\times \left[\left(\frac{100 \text{ Jy}}{1 \text{ Jy}}\right)\left(\frac{(0.4 \times 60) \text{ arcsec}}{1 \text{ arcsec}}\right)^{-2}\left(\frac{26 \text{ kpc}}{1 \text{ kpc}}\right)^{-1}\right]^{2/7},$$
$$\approx 5.9 \times 10^{-5} \text{ gauss}.$$

The intensity in the lobes is not uniform, with the intense hot spots at the ends of the jets surrounded by less intense lobes. So, we can conclude only that an estimate of the average magnetic field in the lobes is of order 6 $\times 10^{-5}$ gauss (or 60 μgauss). This is a much smaller field than that estimated in compact components.

[19] 1980, Annual Reviews of Astronomy & Astrophysics, vol. 18, p. 165.
[20] 1998, Astronomical Journal, vol. 115, p. 1295.
[21] 2004, Astrophysical Journal, vol. 609, p. 539.

(b) The core component, with a flat spectrum, is optically thick, so we use Equation 9.6 with a flux density of 0.6 Jy occurring within a solid angle 1.0 mas × 0.5 mas. Furthermore, the jets in Cygnus A appear to be oriented in the plane of the sky, so we will use $\delta \approx \gamma^{-1}$, and assume that the true speed equals the measured speed. The average measured speed, $\beta \approx 0.5$, yields $\gamma \approx (1 - 0.5^2)^{-1/2} = 1.15$. We have then

$$B \approx 3 \times 10^{-5} \text{ gauss} \left(\frac{1.0 \text{ mas}}{1 \text{ mas}} \times \frac{0.5 \text{ mas}}{\text{mas}} \right)^2$$

$$\times \left(\frac{15.4 \text{ GHz}}{1 \text{ GHz}} \right)^5 \left(\frac{0.6 \text{ Jy}}{1 \text{ Jy}} \right)^{-2} \left(\frac{1}{1.15 \times 1.056} \right)$$

$$\approx 15 \text{ gauss}.$$

Both magnetic field estimation methods require knowledge of the angular size of the feature being analyzed. However, the feature is often unresolved and in this case an upper limit to B may be inferred by using the angular resolution of the observation as an upper limit to the source size. For features that are not Doppler boosted in jets aimed toward us, one can use the Compton limit to infer a lower limit to its angular size. The Compton limit sets an upper limit to the intensity, given in terms of brightness temperature, $T_B \leq 10^{12}$ K, and so when a feature's flux density is known we get a lower limit to its solid angle. The relation between flux density and brightness temperature (using Equations 1.9 and 3.13) is

$$\frac{F_\nu}{\Omega} = \frac{2kT_B}{\lambda^2},$$

where Ω is the solid angle of the feature in steradians. Applying the Compton limit to T_B, then, we have

$$\Omega \geq \frac{\lambda^2 F_\nu}{2k\ 10^{12} \text{ K}}. \tag{9.9}$$

9.4.4 Electron Cooling Timescales and the Nature of Hot Spots in Jets

The synchrotron radiation emitted by the electrons carries energy away and so the electrons gradually lose energy, or cool. An interesting and simple calculation is the time scale over which the electrons lose energy. This provides an estimate of the time that a particular feature in a synchrotron source might radiate, unless a fresh supply of relativistic electrons is injected into it. The cooling time scale is defined by the energy, E, of the electron divided by the rate, dE/dt at which it is losing energy, i.e.,

$$\tau_{\text{cooling}} = \frac{E}{(dE/dt)}.$$

For a synchrotron electron, the power radiated, dE/dt, is given by Equation 3.28. Integrating over all possible directions, the electron's energy loss rate is

$$\frac{dE}{dt} \approx \frac{4e^4}{9m_e^4c^7} E^2 B^2$$

and the cooling time scale is given by

$$\tau_{\text{cooling}} \approx \frac{9m_e^4c^7}{4e^4} \frac{1}{E\ B^2}. \tag{9.10}$$

The cooling time is inversely dependent on E, so higher energy electrons lose their energy faster than the lower energy electrons. Recall from the discussion in Section 3.3.2 that with synchrotron radiation there is a relation between an electron's energy and its frequency of peak emission. In Equation 3.27, the frequency of the peak of the electron's synchrotron emission is given by

$$\nu_{\text{peak}} = 0.0692 \frac{e}{m_e^3 c^5} E^2 B.$$

Since the higher energy electrons cool faster, the synchrotron intensities at the higher frequencies will fade sooner.

Using Equation 3.27 we can solve for the energy that corresponds to a particular frequency of emission. For observations at a particular frequency, we can calculate the cooling time scale of the electrons by rearranging the synchrotron peak frequency equation to solve for E,

$$E = \frac{m_e^{3/2} c^{5/2}}{0.263 \; e^{1/2}} \nu^{1/2} B^{-1/2},$$

and substituting into Equation 9.10. We get

$$\tau_{\text{cooling}} \approx 2.37 \; \frac{m_e^{5/2} c^{9/2}}{4e^{7/2}} \nu^{-1/2} B^{-3/2}$$

$$= 8.58 \times 10^{11} \text{ s} \left(\frac{\nu}{1 \text{ Hz}}\right)^{-1/2} \left(\frac{B}{1 \text{ gauss}}\right)^{-3/2}.$$

In the more convenient units of time in years and frequency in GHz, this is equivalent to

$$\tau_{\text{synch}} \approx 0.86 \text{ yr} \left(\frac{\nu}{1 \text{ GHz}}\right)^{-1/2} \left(\frac{B}{1 \text{ gauss}}\right)^{-3/2}. \tag{9.11}$$

A particularly interesting application of the cooling time is for hot spots seen along the jets, as we demonstrate in Example 9.10. Often the cooling time scale of a hot spot is found to be shorter than the time over which the spot is observed to be radiating. Therefore, these hot spots cannot be individual blobs of electrons ejected from the core. Instead there must be a continuous injection of electrons into the spots, suggesting that the spots are working surfaces of jets. Such working surfaces can occur if there is a bend in a jet, or if the jet becomes more collimated. Note, though, that we have ignored relativistic beaming effects which are important when analyzing a jet feature moving toward us; thus Equation 9.11 is not applicable for beamed jets.

Example 9.10:

Figure 9.19 displays 22.2-GHz VLBI maps of 3C84 (also known as Perseus A) at six different epochs. A small hot spot, labeled "K" for "knot," is seen in each image. As evident in the figure, the knot moved away from the core and its emission was roughly constant over the 5-year observing period. Based on its apparent speed, which was measured to be 0.49 c, the knot was inferred to have been ejected from the core during an outburst around 1970. The magnetic field in the knot was estimated to be $\approx$ 0.4 gauss. Estimate the cooling time scale of the electrons in the knot and compare this to the time since the hot spot exited the core.

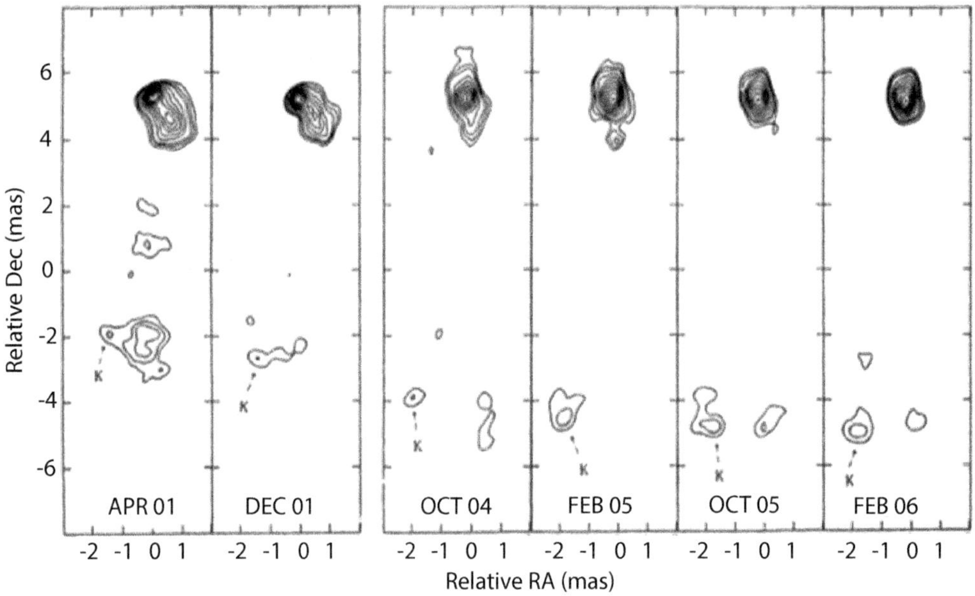

Figure 9.19: A series of VLBI maps, at 22.2 GHz of the radio galaxy 3C84 from 1981 to 1986 showing a strong core component, the very intense feature at the north end, and a one-sided ejection of material to the south. In the ejecta is a small region known to have a small positive spectral index. That small positive α region is easily noticeable as a knot of emission, labeled 'K.' (Figure taken from Marr et al. 1989, Astrophysical Journal, vol. 337, p. 671.)

Answer:
Since the knot's apparent motion was subluminal, with $v_{\mathrm{app}} = 0.49c$, the value for γ cannot be large and so we can apply Equation 9.11. The cooling time scale of the electrons radiating at 22.2 GHz in the knot is therefore

$$\tau_{\mathrm{cooling}} \approx 0.86 \text{ yr} \left(\frac{22.2 \text{ GHz}}{1 \text{ GHz}}\right)^{-1/2} \left(\frac{0.4 \text{ gauss}}{1 \text{ gauss}}\right)^{-3/2},$$

or

$$\tau_{\mathrm{synch\ cooling}} \approx 0.72 \text{ yr}.$$

The knot, though, was seen for a period of 5 years and was ejected from the core 16 years previous to the last image. The authors inferred that an outburst event occurred in 3C84 in 1970 which enhanced the collimation of the jet.

9.5 THE CENTER OF THE MILKY WAY

The nearest large galaxy center that we can study, of course, is that of our own Milky Way. The Milky Way is not an active galaxy, but following the idea that all large galaxies are active sometime in their history, we expect the center of the Milky Way to harbor the skeletal structure of the AGN engine. In particular, the Galactic center should contain a supermassive black hole, and strong evidence for a point mass of over a million solar masses has indeed been found in observations of motions of gas and stars in the central few parsecs. Our view of the Galactic center at visible wavelengths is blocked by dust in the Galactic disk. The Galactic center has, therefore, been extensively studied at radio wavelengths.

The Galactic center is located in the constellation Sagittarius. The extended radio source known as Sagittarius A (or SgrA) is a composite of a several different radio sources; the Galaxy's center is in the western part. The precise dynamical center is located at the position of a point of radio emission called SgrA* (pronounced "saj-A-star").

Encircling the center is a slightly inclined *circumnuclear disk* (or CND) which has been mapped in emission from neutral atoms as well as molecules. High resolution millimeter-wavelength observations, initially by Rolf Güsten and colleagues[22], delineated an almost complete, rotating ring, tilted relative to the plane of the Galaxy by $\approx 70°$. The CND has an inner radius of 1.7 pc and is found to rotate at speeds of 110 km s^{-1}. Inside the CND, first observed in NeII emission lines (at 12.8 microns), is a "minispiral" set of gas streamers with Doppler shift velocities vs. position consistent with gas breaking off the CND and free-falling toward the position of SgrA*[23]. SgrA* is surrounded by a massive cluster of stars. Using numerous high resolution infrared observations with adaptive optics to correct for the atmospheric image distortion, Andrea Ghez and collaborators[24] have revealed the orbital motions of these stars. Some are seen to whip around the position of SgrA*. One star is observed to orbit with a period of only 15 years. The measured orbits require a central mass of between 4 and 5 $\times 10^6$ $M_\odot$, and are consistent with a supermassive black hole located at the position of the radio point source SgrA*. In conclusion, the nucleus of the Milky Way could have in the past been as active as a Seyfert galaxy.

QUESTIONS AND PROBLEMS

1. The plot in Figure 9.20 shows the radio spectrum of a fictitious radio galaxy.
   (a) Use the spectrum to calculate the power-law index of the electron energy distribution (see Section 3.3.3).
   (b) Based on the spectrum, determine whether the source is uniform in structure. Explain.

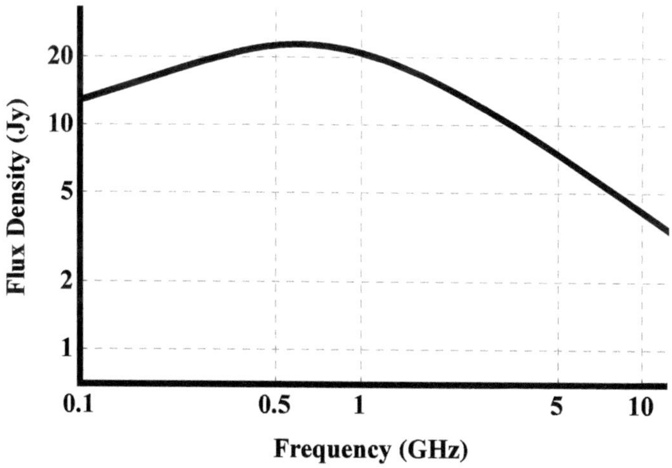

Figure 9.20: For Question 9.1. Log-log plot of a radio galaxy spectrum.

[22]1987, Astrophysical Journal, vol. 318, p. 124.
[23]Serabyn and Lacy 1985, Astrophysical Journal, vol. 293, p. 445.
[24]Ghez et al. 2008, Astrophysical Journal, vol. 689, p. 1044.

2. (a) What power-law index of the energy distribution of relativistic electrons would be needed for optically thin synchrotron radiation in a uniform magnetic field to mimic the low-frequency optically thick thermal emission? In Section 3.3.3, the electron energy distribution was discussed with the power-law index defined and denoted as p. What value of p will result in an optically thin synchrotron spectrum with $\alpha = 2$?
(b) What would the spectral index be for synchrotron radiation emitted by an ensemble of electrons with equal population at all energies, i.e. $N(E)dE \approx$ constant?

3. What is equipartition of energy? Why is it useful to make this assumption in the study of radio galaxies? What implications does a violation of this condition have for the source of energy in radio galaxies?

4. In a large double-lobed radio galaxy, the total flux density of both lobes at 2.30 GHz is 9.00 Jy. Assume that the lobes are spherical and identical. Each lobe has an angular size of 0.0300°, and the galaxy is at an angular-size distance of 16.0 Mpc.
(a) Estimate the average magnetic field in the lobes.
(b) Estimate the total equipartition energy in the lobes.

5. What theoretical model predicts that part of the synchrotron spectrum has a power-law with $F_\nu \propto \nu^{+2.5}$? Why is this never seen and what is seen instead?

6. In Section 9.3.2, we stated that since AGN contain a lot of ionized gas, it is reasonable to expect that a spectral turnover of a compact source may be due to free-free absorption. Why, then, do we not see spectral turnovers of the lobes in the extended sources?

7. (a) In a synchrotron radiation source with a magnetic field of 300 milli-gauss, what is the frequency of the peak of the synchrotron emission by electrons with energy $E = 400\ m_e c^2$?
(b) When these photons are inverse Compton scattered by the same electrons, what is the minimum wavelength of the photons after being scattered?

8. Along a radio galaxy jet, an unresolved knot of enhanced emission has a flux density of 0.200 Jy at 5.00 GHz.
(a) Calculate the lower limit to this feature's solid angle.
(b) Assuming that the feature is circular, calculate the minimum possible angular diameter.
(c) If the source has a redshift of 1.88, which corresponds to a proper distance of 5.00 Gpc (assuming $\Omega_\Lambda = 0.7, \Omega_M = 0.3$, and $H_0 = 70$ km s^{-1} Mpc^{-1}), calculate this feature's minimum possible physical diameter.

9. Imagine a body moving at $0.9c$ in the $+x$ direction in our reference frame and emitting radiation isotropically in its own reference frame. Consider a light beam that is emitted in the $+y$ direction in the reference frame of the body. Using the Lorentz velocity addition equations (refer to a special relativity textbook) determine the direction of the propagation of this light beam in our reference frame. Calculate the angle of the direction of propagation relative to the x-axis. Extrapolating on your result, comment on the beaming of the radiation emitted by relativistically moving bodies.

10. (a) Explain why the measured speeds of superluminal radio jets are not considered a violation of special relativity.
(b) Calculate the apparent speed of a jet feature with true speed $= 0.99c$ in a direction at an angle of 5° relative to the line of sight.

11. Discuss three paradoxes in the study of radio jets that are solved in a model where the radio jets are relativistically beamed in a direction close to being aimed at Earth. Explain how beaming of the jets solves these paradoxes.

12. Imagine a symmetric double-lobed radio-bright quasar at redshift $z = 0.15$ with a circular core component whose spectrum turns over at 8.4 GHz and below the turnover follows a power-law with $\alpha = +0.8$. At 8.4 GHz the core has a flux density of 2.4 Jy and an angular diameter of 0.72 milliarcseconds.
(a) Use these data to calculate the physical diameter and obtain an estimate of the magnetic field in the core.
(b) The measured flux densities and angular sizes of the core component at some frequencies below the turnover are: 1.6 Jy and 0.97 mas at 5.0 GHz; 0.85 Jy and 1.5 mas at 2.3 GHz; and 0.67 Jy and 1.8 mas at 1.7 GHz. Calculate the physical diameter and magnetic field of the core according to the data at each frequency.
(c) Use the answers to (a) and (b) to infer the magnetic field as a function of radius in the core. Fit the data to a power-law dependence for $B \propto r^b$. What is the value of b?

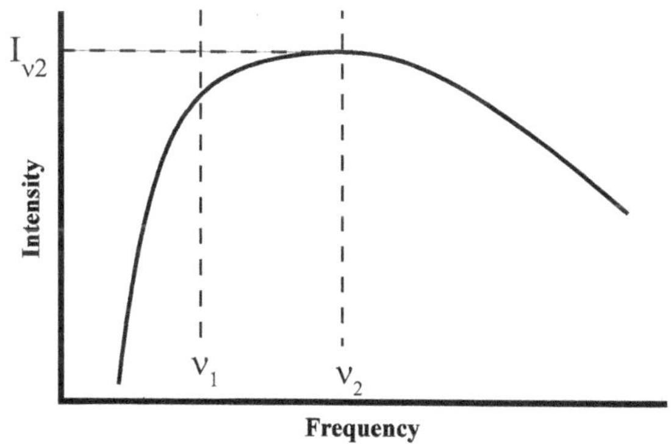

Figure 9.21: For Question 9.13. A radio galaxy spectrum with declining flux densities at high frequencies, a flattening at ν_2, and an exponential drop-off below ν_1.

13. In Figure 9.21, which displays the spectrum of a synchrotron source with a foreground of ionized hydrogen, $\nu_1 = 0.7$ GHz, $\nu_2 = 2.0$ GHz, and $I_{\nu_2} = 1.9 \times 10^{-7}$ erg s^{-1} cm^{-2} Hz^{-1} sr^{-1}. The synchrotron source has an angular size of $\Theta = 1.5$ mas and is at redshift $z = 0.35$. Assume that no relativistic jet motion occurs in this source.
(a) Use the turnover at $\nu 2$ to find the magnetic field in the source.
(b) Use the drop off below $\nu 1$, assuming free-free absorption to find the emission measure of free electrons to the synchrotron source. Assume the temperature of the foreground plasma is 10^4 K.

14. The jets of Cygnus A are about 150,000 light years long and speeds of features in the jets in the core are measured to be about 0.5c.
(a) What is the timescale for the jet particles to travel from the core to the ends of the jets?
(b) Using the fact that the ends of the jets are seen as bright spots at 5.0 GHz, use cooling time scale considerations to obtain an upper limit to the magnetic field in the jets.

CHAPTER 10

Cosmic Microwave Background

ONE of the most important discoveries of the 20th century was the detection of the cosmic microwave background (CMB). Arno Penzias and Robert Wilson serendipitously detected the CMB in 1965 while making measurements at a frequency of 4080 MHz. In their measurements, Penzias and Wilson detected excess power over what they expected regardless of which direction they pointed their telescope in the sky. With very careful work, they concluded that this was not noise power originating in the telescope or detector system, but was isotropic cosmic radiation. The power they measured was equivalent to thermal radiation at a temperature of about 3.5 K. Penzias and Wilson received the 1978 Nobel Prize in Physics for this discovery.

10.1 COSMOLOGICAL MODELS

To understand the significance of this discovery, we need to appreciate the state of cosmology in the 1960's, and that requires starting in the early 20th century. Shortly after Einstein developed his general theory of relativity, in which gravity is attributed to curvature of space-time, he developed a model of the universe. Following the principles of general relativity, the space-time of the universe may be curved and the amount of curvature depends on the density of both mass and energy. Unaware at the time that the Universe was expanding, he found that the Universe could not be static if it contained mass or energy. Believing that the Universe was ever present and unchanging, Einstein inserted another parameter, which he called the cosmological constant and wrote as Λ, which counteracted gravity. By setting the value of the cosmological constant to the right value its effect could balance the curvature caused by the gravity due to mass, thereby yielding a static model of the Universe. Adding a cosmological constant is a simple modification to general relativity and affects physics on only the largest length scales and so did not conflict with measurements on the scales of galaxies and smaller.

The discovery reported by Edwin Hubble (see Section 1.5) in 1929, and now called the Hubble-Lemaître law, showing a relation between a galaxy's recessional velocity and its distance implied that the Universe was expanding. Although the tendency of light from galaxies to be redshifted had been observed previously by Vesto Slipher, and an expanding universe had been proposed independently by Georges Lemaître and Alexander Friedmann after working with Einstein's equations, the Hubble-Lemaître law gave solid observational evidence that the universe was expanding and not static. Einstein later called the cosmological constant his greatest blunder. As you will see, however, the cosmological constant has returned and is a very important parameter today. With the discovery of the Hubble-Lemaître law, new cosmological models were developed to include expansion. One of the two more popular models involved a Universe that was changing with time and had a beginning; this was the Big Bang model. This model was initially proposed by Lemaître and later worked out in more detail by Ralph Alpher, Robert Herman, and George Gamow. The alternative model, first advocated by Hermann Bondi, Thomas Gold, and Fred Hoyle, was the Steady State model. The Steady State model involved an expanding universe as well, but assumed that new matter was being created everywhere at a rate to offset the dilution of matter as the Universe expanded, thereby producing a universe that appeared to be unchanging and had no beginning or end.

In the Big Bang model, the Universe began in an extremely dense and hot state and expanded to the conditions now present in the Universe. The future, shape, and extent of the universe depended on the amount of matter in comparison to the expansion rate. Similar to the calculation of escape velocity, in which the amount of gravitational potential energy is compared to the amount of kinetic energy, the average mass density in the Universe and the expansion rate determined whether the Universe would expand forever or whether it would stop at some point and start contracting. This distinction could also be described in terms of the Universe's curvature and size. If the Universe was dense enough to start contracting in the future, then it would have spherical geometry and be closed. If the Universe's mass density was too small, so that it would expand forever, then it would have hyperbolic curvature, and be open. The average mass density that separates an open universe from a closed universe is called the *critical density*. If the mass density exactly equals the critical density, then the Universe would be flat (i.e., Euclidean) and be open.

In the Big Bang model, Alpher, Herman, and Gamow suggested that radiation associated with the Universe at an early time when the Universe was much hotter and denser could be detectable today. Alpher and Herman[1] made the first prediction of the intensity of this radiation, and suggested that the radiation today would appear as blackbody radiation at a temperature of about 5K, which would be brightest at microwave frequencies. They encouraged astronomers to search for this radiation; however the microwave detection technology at the time was insufficient for such a search to be feasible.

The idea of detecting relic radiation from an early epoch in the Universe was re-derived by Robert Dicke, James Peebles, Peter Roll and David Wilkinson in the early 1960's. They were in the process of developing the microwave equipment to detect this relic radiation when they learned of the discovery by Penzias and Wilson. The two groups compared theory and observation, and the result was a pair of papers published in 1965, one by Robert Dicke and colleagues[2] that gave the background and cosmological implications, and a second by Penzias and Wilson[3] providing details of their detection. The detection of the CMB was the crucial evidence in favor of a Big Bang model over other cosmological models such as the Steady State model, since it showed the Universe did have a hot and dense early phase.

Subsequent studies of the Universe have yielded a number of significant modifications to the Big Bang model. In the 1970's, evidence for dark matter (discussed in Section 8.1.2) became strong enough that dark-matter dominated cosmological models were developed. In the late 1990's, from studies of the redshifts and distances of many very distant Type Ia supernovae, Adam Riess and colleagues[4] and Saul Perlmutter and colleagues[5] found their data required that the expansion rate of the Universe is currently accelerating. Although

[1] 1948, Nature, vol. 162, p. 774.
[2] 1965, Astrophysical Journal, vol. 142, p. 414.
[3] 1965, Astrophysical Journal, vol. 142, p. 419.
[4] 1998, Astronomical Journal, vol. 116, p. 1009.
[5] 1999, Astrophysical Journal, vol. 517, p. 565.

no successful physics model has yet succeeded in explaining this phenomenon, it fits a mathematical model involving the cosmological constant, Λ. The unknown source of Λ is called *dark energy*. The most widely accepted models of the Universe today are those that are dominated by dark energy and secondarily by dark matter. These models collectively are referred to as ΛCDM. The "C" refers to cold, to describe more massive dark matter particles, in contrast to the very low mass candidates, which would have been relativistic, and referred to as "hot", at important epochs of the early universe. The details of the Universe's history and future depend on the amount of dark energy, dark matter, and baryonic matter.

10.2 BLACKBODY NATURE OF THE CMB

With a brightness temperature of about 3 K and filling the sky, the CMB power received by a radio telescope is substantial, and often larger than the power detected from other astronomical sources. As this radiation pervades the Universe, filling every cubic centimeter of space, it is fruitful to estimate how many photons this entails. In particular, let's calculate the ratio of the number of photons to the number of baryons (protons and neutrons). We can obtain this ratio by calculating each of their number densities.

The number density of CMB photons is given by dividing their energy density by the average energy per photon, where the average photon energy is given by Equation 3.5, $\langle E_{\rm ph}\rangle = 2.70\ kT$ and the number density can be obtained from the intensity. The intensity, I_ν (given by the Planck function in Equation 3.2 and with units of erg s^{-1} cm^{-2} Hz^{-1} sr^{-1}), equals the energy density of CMB photons from ν to $\nu + \Delta\nu$ multiplied by c divided by $\Delta\nu$ and by the solid angle of the radiation. We get the total photon energy density by integrating the intensity over all frequencies and multiplying by a solid angle of 4π. The integral of the Planck function over frequency gives

$$\int B_\nu(T)\ d\nu \;=\; \frac{2\pi^4 k^4}{15 h^3 c^2}\ T^4.$$

Hence, we find that the number density of CMB photons in the Universe today is

$$n_{\rm ph} \;=\; 4\pi\ \frac{2\pi^4 k^4\ T_{\rm CMB}^4}{15 h^3 c^2\ \langle E_{\rm ph}\rangle\ c} \;=\; \frac{4\ \sigma T_{\rm CMB}^3}{2.70\ c\ k}, \tag{10.1}$$

where

$$\sigma = \frac{2\pi^5 k^4}{15 h^3 c^2} = 5.67 \times 10^{-5}\ {\rm ergs}^{-1}\ {\rm cm}^{-2}\ {\rm K}^{-4}$$

is the Stefan-Boltzmann constant and the average energy of a photon is $2.70\ kT$.

The number density of baryons can be calculated by dividing the baryonic mass density, $\rho_{\rm baryons}$, by the mass per baryon. Since protons and neutrons have very similar masses, this is given by

$$n_{\rm baryons} \;=\; \frac{\rho_{\rm baryons}}{m_{\rm p}},$$

where $m_{\rm p}$ is the mass of a proton.

The ratio of the number of photons in the CMB to the baryons in the Universe, then, is

$$\frac{n_{\rm ph}}{n_{\rm baryons}} \;=\; \frac{4\ \sigma\ T_{\rm CMB}^3\ m_{\rm p}}{2.70\ c\ k\ \rho_{\rm baryons}}.$$

From several different studies, the mass density of baryons is believed to be only about 5% of the critical density, and the equation for the critical density is

$$\rho_{\rm crit} \;=\; \frac{3H_o^2}{8\pi G}. \tag{10.2}$$

For a value of the current Hubble constant (see Section 1.5), $H_o = 70$ km s^{-1} Mpc^{-1}, the current critical density equals 9.3×10^{-30} g cm^{-3}. Inserting the values for c, m_p, k and setting $\rho_{\rm baryons} = 0.05\ \rho_{\rm critical}$ and $T_{\rm CMB} = 2.725$ K, we find that the ratio of photons to baryons in the universe is approximately

$$\frac{n_{\rm ph}}{n_{\rm baryons}} \approx 10^9.$$

This fairly simple calculation reveals that there are vastly more photons in the Universe than baryons – the particles we often think of as "ordinary" matter. The reason for this very large ratio is easily explained by the nearly complete annihilation of matter and antimatter in the early Universe. This result will be relevant to our later discussion.

In our current cosmological model the CMB radiation we detect today was last influenced by matter about 380,000 years after the Big Bang, at a time when the Universe was transitioning from a fully ionized state to one containing neutral atoms. In the early Universe the temperature was too high for enough neutral atoms to survive, and therefore the Universe was a fully ionized plasma. Because light can readily scatter off charged particles, particularly electrons, the frequent interactions between light and particles resulted in the radiation having a blackbody spectrum whose temperature equaled that of the ionized plasma. With the frequent interactions between light and particles, a photon could not propagate very far before being scattered, and so the Universe was opaque to light at all wavelengths.

As the Universe expanded it cooled. When it reached a temperature of about 3700 K, electrons could remain bound to hydrogen and helium nuclei without being photo-ionized, resulting in neutral atoms. Since radiation does not interact strongly with neutral atoms (only through the spectral line processes discussed in Chapter 4), the Universe became largely transparent to light. This transition from an ionized and opaque Universe to a neutral and transparent Universe is called the *epoch of recombination.* During recombination, photons continued to interact with those electrons that had not yet recombined with nuclei until the Universe cooled to a temperature of about 3000 K. The radiation at this time is termed to have decoupled from matter, and it is these photons we see today.

You might wonder why the Universe needed to cool to temperatures below 3700 K before neutral atoms were common. The ionization energy of hydrogen is 13.6 eV, and you might expect the gas to turn neutral when the photon energies become too low to ionize hydrogen. The average photon energy equals 13.6 eV when the radiation has a temperature of about 58,000 K (see Equation 3.5). At 3700 K, only a very small fraction of the photons in the high frequency tail of the Planck function have energies greater than 13.6 eV. However, because photons outnumber protons by a ratio of about one billion to one, even at temperatures as low as 3700 K, there are sufficient high-energy photons to ionize most of the hydrogen. By the time the Universe cooled to 3000 K the number of hydrogen-ionizing photons decreased to the point where the gas became almost entirely neutral.

Since the time of recombination, the Universe has expanded by a factor of about 1100. Recall from Section 1.5 that we described the expansion of the Universe by the scale factor, R(t), which is defined to be unity today. The CMB radiation suffers a cosmological redshift as the Universe expands, but maintains its blackbody nature, although decreased in temperature as we now demonstrate. As we discussed earlier, a blackbody is defined by its intensity (the Planck function; see Equation 3.2) and the intensity, I_ν, is the energy density of photons of frequencies from ν to $\nu + \Delta\nu$ divided by $\Delta\nu$ times the speed of the photons, c, divided by the solid angle of the radiation. The energy density of photons is the number density of photons times the energy of each photon, $h\nu$. Finally, since the CMB is isotropic the solid angle of the radiation is 4π, and therefore its intensity equals

$$I_\nu = \frac{ch\nu \, n^\nu_{\rm ph}}{4\pi\Delta\nu},$$

where $n^\nu_{\rm ph}$ represents the number density of photons of frequency from ν to $\nu + \Delta\nu$.

Now consider how each of the parameters in the above equation changes as the Universe expands, or as the scale factor R (see Section 1.5) increases. If the radiation was a Planck function of temperature T_1 at time t_1, what will it be like at a later time t_2? First, the number density of CMB photons must vary as the inverse of the volume, therefore

$$n^\nu_{\rm ph}(t_2) = n^\nu_{\rm ph}(t_1)\left(\frac{R_1}{R_2}\right)^3,$$

where R_2/R_1 represents the multiplicative increase in scale factor from time t_1 to time t_2. As discussed in Section 1.5, the wavelength of the CMB photons are stretched proportionally to the increase in scale factor, and so the frequency, ν_2, of a photon at time t_2 is related to the frequency ν_1 by

$$\nu_2 = \nu_1 \frac{R_1}{R_2}.$$

The bandpass, $\Delta\nu$, decreases similarly,

$$\Delta\nu_2 = \Delta\nu_1 \frac{R_1}{R_2}.$$

In the expression for the CMB intensity, only $n^\nu_{\rm ph}$, ν, and $\Delta\nu$ change as the Universe expands. Therefore, since

$$I_\nu \propto \frac{\nu \, n^\nu_{\rm ph}}{\Delta\nu},$$

the intensity changes according to

$$I_\nu(t_2) = \frac{(R_1/R_2)\,(R_1/R_2)^3}{(R_1/R_2)} I_\nu(t_1) = (R_1/R_2)^3 I_\nu(t_1).$$

If the radiation was blackbody at temperature T_1 at time t_1, then the intensity of the radiation at a later time, t_2, is

$$I_\nu(t_2) = \left(\frac{R_1}{R_2}\right)^3 \frac{2h\nu_1^3}{c^2} \frac{1}{{\rm e}^{(h\nu_1/kT_1)} - 1}.$$

Replacing ν_1 with $\nu_2(R_2/R_1)$ we can rewrite this as

$$I_\nu(t_2) = \frac{2h\nu_2^3}{c^2} \frac{1}{{\rm e}^{(h\nu_2 R_2/R_1 kT_1)} - 1}.$$

This is simply the Planck function for a temperature of $T_1(R_1/R_2)$. Therefore, as the Universe expands the CMB emission remains a blackbody but with temperature that decreases as the scale factor increases, $T \propto R^{-1}$. Since the CMB was emitted at temperature $T_e \approx 3000$ K and the scale factor has increased by a factor about 1100, the CMB emission observed today is blackbody radiation at a temperature of about 2.7 K.

The frequency of the peak emission of a blackbody is directly proportional to temperature (Wien's law - see Equation 3.6); therefore when the radiation was emitted, at a temperature of 3000 K, the peak of its emission was at a frequency of 1.8×10^{14} Hz or a wavelength of 1.7×10^{-4} cm (or 1700 nm). However, since the Universe has been expanding during the time the light has traveled to us, the peak is shifted to longer wavelengths. The wavelength of the peak of the CMB radiation observed now, λ_d, is given by Equation 1.19; thus

$$\lambda_d = \frac{\lambda_e}{R(t_e)},$$

where λ_e is the emitted wavelength at the time of recombination and $R(t_e)$ is the scale factor of the Universe at the time of decoupling.

Example 10.1:

At what wavelength is the peak intensity of the CMB emission seen today?

Answer:
The Universe has expanded by a factor of 1100 since the time of decoupling. Since the scale factor, $R(t_e)$, is measured relative to the scale factor today, it has a value of $R(t_e) = 1/1100 = 9.1 \times 10^{-4}$. Therefore the wavelength we observe now for the peak of the CMB emission is

$$\lambda_d = \lambda_e \frac{1}{R(t_e)} = 1.7 \times 10^{-4} \text{ cm} \frac{1}{9.1 \times 10^{-4}} = 0.19 \text{ cm}.$$

Thus, the CMB emission seen today is at microwave wavelengths with its intensity peak at a wavelength of 0.19 cm or 1.9 mm.

One of the first space missions devoted to the study of the CMB was the Cosmic Background Explorer (COBE) which was launched in 1989. COBE had three instruments that studied the CMB radiation, covering wavelengths from about 10^{-4} cm to 1 cm. One instrument on COBE, the Far Infrared Absolute Spectrophotometer (FIRAS), measured the intensity of the CMB covering wavelengths from about 0.05 to 0.4 cm. The observed intensity of the CMB as a function of wavelength measured with the FIRAS instrument on COBE, shown in Figure 10.1, is a perfect blackbody with a temperature of 2.725 K. (Note that the spectrum in Figure 10.1 looks somewhat different from that of a blackbody because I_ν is plotted versus wavelength and not frequency.) The wavelength at which the intensity of the CMB is largest is at about 1.9 mm, as estimated in Example 10.1.

10.3 ANISOTROPIES IN THE CMB

At the time of recombination the Universe was extremely uniform. However, theoretical calculations suggested that even at this early epoch there should be small inhomogeneities in the distribution of matter. Papers published in 1970 by James Peebles and Jer Yu and independently by Rashid Sunyaev and Yakov Zel'dovich suggested that these small inhomogeneities in the distribution of matter would be manifest as small anisotropies in the intensity of the CMB radiation across the sky. Subsequent searches over the next decade failed to detect the anisotropies at the level predicted. In 1982, Peebles realized that the presence of dark matter would alter the magnitude of the predicted anisotropies. With dark matter included, the predicted anisotropies of the CMB were only at the level of about 1 part in 100,000, which was well below the threshold of detection at that time. The discovery had to wait until improvements were made in millimeter-wavelength instrumentation.

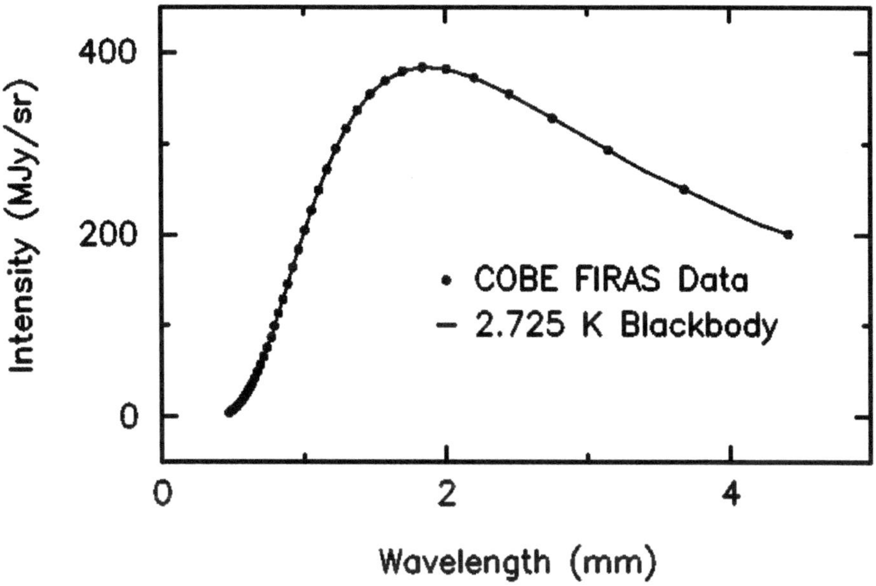

Figure 10.1: The filled circles show the COBE FIRAS result of the intensity (I_ν) of the CMB as a function of wavelength. The error bars on each measurement are much smaller than the size of the filled circle shown in the figure. The intensity is given in units of mega Jy per steradian (1 MJy sr^{-1} is equivalent to 1×10^{-17} ergs s^{-1} cm^{-2} Hz^{-1} sr^{-1}). The solid line is the intensity of a blackbody at a temperature of 2.725 K. The results show that the CMB is a near perfect blackbody. This plot is based on data from the Legacy Archive for Microwave Background Data Analysis (LAMBDA), part of the High Energy Astrophysics Science Archive Center (HEASARC), a service of the Astrophysics Science Division of the NASA Goddard Space Flight Center.

Another instrument on COBE, the Differential Microwave Radiometer (DMR), mapped the intensity of the CMB over the entire sky at several wavelengths in order to search for these anisotropies. The DMR found that the intensity of the CMB is remarkably uniform across the sky, as illustrated in the image in the upper panel of Figure 10.2. This figure shows the entire sky at a frequency of 53 GHz (a wavelength of 5.6 mm); at these wavelengths the emission from the CMB dominates. It is fortunate that the peak of the CMB emission is at a wavelength of a few millimeters — at much longer wavelengths the sky is dominated by synchrotron emission and at much shorter wavelengths by dust emission; however even at a wavelength of 5.6 mm these emissions still must be accounted for.

Although the CMB is nearly isotropic, there are very subtle variations. To see these subtle features we need to remove the isotropic emission. The middle panel in Figure 10.2 shows the sky intensity after the average CMB intensity has been subtracted. We see that in one direction of the sky the CMB intensity or temperature is slightly larger than average (shown in red), while in the opposite direction the intensity or temperature is slightly smaller than average (shown in blue). This is the *CMB dipole anisotropy.* This dipole anisotropy is a result of our motion relative to the CMB. In the direction we are moving the CMB radiation will be slightly more intense and in the opposite direction the intensity will be slightly less intense. Only in the rest frame of the CMB is the intensity isotropic; in all moving frames the CMB will exhibit a dipole anisotropy.

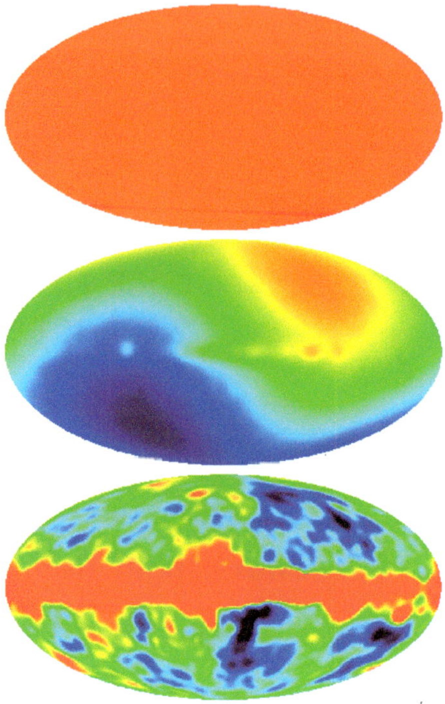

Figure 10.2: Images of the sky obtained with the Differential Microwave Radiometer (DMR) on the Cosmic Background Explorer (COBE) shown in galactic coordinates, with a similar projection to the images shown in Chapter 5. The top panel represents the nearly isotropic emission of the CMB at a frequency of 53 GHz. The middle panel shows the result of subtracting the average emission from the upper panel, revealing the CMB dipole. The lower panel shows the result of modeling and removing the dipole term from the middle panel. What remains is a combination of foreground emission from the Milky Way (the horizontal band of emission, since the plots are in Galactic coordinates) and the first evidence for the predicted anisotropies of the CMB. This figure comes from the Legacy Archive for Microwave Background Data Analysis (LAMBDA) website (https://lambda.gsfc.nasa.gov), part of the High Energy Astrophysics Science Archive Center (HEASARC), a service of the Astrophysics Science Division of the NASA Goddard Space Flight Center.

To relate the radiation intensity in a moving frame relative to a frame fixed with the radiation requires a Lorentz transformation (see textbook by James Pebbles[6]). Skipping over the details of this transformation, we note that the intensity is altered due to two effects. First, in our moving frame the change in the arrival rate of photons produces an intensity change relative to the intensity that would be measured in a frame fixed relative to the CMB. This causes an intensity change for non-relativistic velocities of $(1+\beta\ \cos\theta)$, where $\beta = v/c$ and θ is the angle between the direction of interest and our direction of motion relative to the CMB at velocity v. The second effect is due to a transformation between solid angles in the two reference frames, and this produces an intensity change given approximately by $(1+\beta\ \cos\theta)^2$. Combining these two effects, our motion relative to the CMB produces a net intensity change of $(1+\beta\ \cos\theta)^3$; therefore the intensity we

[6]1993, Principles of Physical Cosmology, Princeton University Press, pp. 151-157.

measure is

$$I_\nu \approx (1+\beta \cos\theta)^3 \, I_\nu^\circ,$$

where I_ν° is the CMB intensity in the rest frame of the CMB.

In addition, our relative motion produces a Doppler effect so that the observed frequency of a CMB photon, ν, is $\nu = \nu_\circ \, (1+\beta \cos\theta)$, where $\nu_\circ$ is the frequency of a photon if we were not moving relative to the CMB. We can write the intensity we observe for the CMB as

$$I_\nu \approx (1+\beta \cos\theta)^3 \, \frac{2h\nu_\circ^3}{c^2} \, \frac{1}{e^{(h\nu_\circ)/(kT_\circ)} - 1},$$

where $T_\circ$ is the temperature of the CMB in the rest frame of the CMB. Substituting in for $\nu_\circ$, we can rewrite the observed intensity as

$$I_\nu \approx (1+\beta \cos\theta)^3 \, \frac{2h\nu^3 \, (1+\beta \cos\theta)^{-3}}{c^2} \, \frac{1}{e^{(h\nu)/[k(1+\beta \cos\theta)T_\circ)]} - 1}.$$

Therefore,

$$I_\nu \approx \frac{2h\nu^3}{c^2} \, \frac{1}{e^{(h\nu)/[k(1+\beta \cos\theta)T_\circ)]} - 1}.$$

As was the case for the expansion of the Universe, the observed CMB emission in our moving frame is still a blackbody; however the temperature of the blackbody is modified such that the apparent blackbody temperature of the CMB in our moving frame is

$$T(\theta) \approx T_o + T_o \frac{v}{c} \cos\theta. \tag{10.3}$$

This equation describes a dipole function. As we already mentioned, in the direction we are moving ($\theta = 0°$) the intensity and the temperature of the CMB appear larger relative to what would be observed if we were at rest relative to the CMB. In the direction opposite of our motion ($\theta = 180°$), the intensity and temperature appear smaller.

Example 10.2:

In the direction of our motion relative to the CMB, the temperature of the CMB is 0.0034 K larger than the average. What is our velocity relative to the CMB?

Answer:
Since this temperature difference is measured in our direction of motion, the angle in Equation 10.3 is $\theta = 0°$ and $\cos 0° = 1$. Using Equation 10.3, we then have

$$\frac{0.0034}{2.725} = \frac{v}{c} = 0.00125.$$

Therefore our velocity relative to the CMB is 370 km s^{-1}.

The CMB dipole reveals that we are moving at a velocity of about 370 km s^{-1} in the direction given by galactic coordinates $l = 264°$ and $b = 48°$ (see paper by Alan Kogut and colleagues[7]). Our motion relative to the CMB is due to a combination of our orbital motion about the center of the Milky Way, the motion of the Milky Way relative to the center of mass of the Local Group and the motion of the Local Group of galaxies through space.

[7]1993, Astrophysical Journal, vol. 419, p. 1.

The dipole anisotropy of the CMB can easily be modeled and subtracted from the middle image of Figure 10.2. What is left after subtraction is shown in the bottom panel in Figure 10.2. Here we see foreground emission from the Milky Way, but more importantly, we see the first evidence for the predicted anisotropies in the CMB. These anisotropies are at the predicted level of 1 part in 100,000. The principle investigators for the FIRAS and DMR instruments on COBE, John Mather and George Smoot, received the 2006 Nobel Prize in Physics for their work on this project.

The maps of the CMB made with the DMR instrument on COBE had an angular resolution of only 7 degrees. Thus, anisotropies in the CMB on smaller angular scales were impossible to measure with COBE. However, there is much interest in and information about anisotropies at smaller angular scales. Following on the success of COBE, two more space missions dedicated to mapping the CMB were launched. The first was the Wilkinson Microwave Anisotropy Probe (WMAP), launched in 2001, and the second was Planck, launched in 2009. Planck produced images of the CMB with an angular resolution about 100 times better than COBE. As was the case with COBE, for both WMAP and Planck one of the first steps necessary to measure the anisotropy was to remove the CMB dipole. In addition, both WMAP and Planck imaged the sky at many different wavelengths so the foreground emission from the Milky Way could be more accurately modeled and removed. An image of the CMB anisotropies from Planck, with the foreground emission removed, is shown in Figure 10.3. Both WMAP and Planck revealed anisotropies in the CMB on much smaller angular scales than COBE could measure. These small intensity variations of the CMB result from density fluctuations at the time of decoupling. These density fluctuations were the seeds for the growth of structure in the Universe; all of the galaxies and galaxy clusters we see today grew from these small fluctuations present in the early Universe.

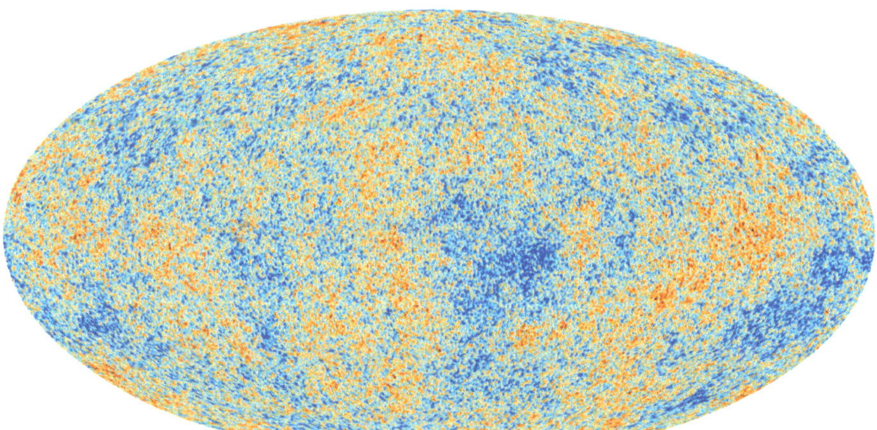

Figure 10.3: An image of the CMB obtained with Planck after the removal of the CMB dipole and foreground emissions. This image clearly shows the CMB anisotropies at an angular resolution nearly 100 times better than the image from COBE shown in Figure 10.2. This image was provided by the ESA/Planck Collaboration. See www.esa.int/Our˙Activities/Space˙Science/Planck.

10.4 COSMOLOGICAL PARAMETERS

The anisotropies produced at or before the time of recombination are called primary anisotropies. Their amplitudes on different angular scales contain important information concerning the nature of our Universe. We next consider how to extract that information from maps of the CMB. From images of the CMB, such as that shown in Figure 10.3, we can compute the CMB power spectrum. A power spectrum is a way of representing the amount of variation present in the intensity of the CMB as a function of angular scale. The CMB power spectrum based on the Planck data is shown in Figure 10.4. Examining the power spectrum we can see that there are certain angular scales at which the CMB has larger intensity variations. The largest peak in the power spectrum, indicating the greatest variation in intensity, is at an angular scale of about 1°.

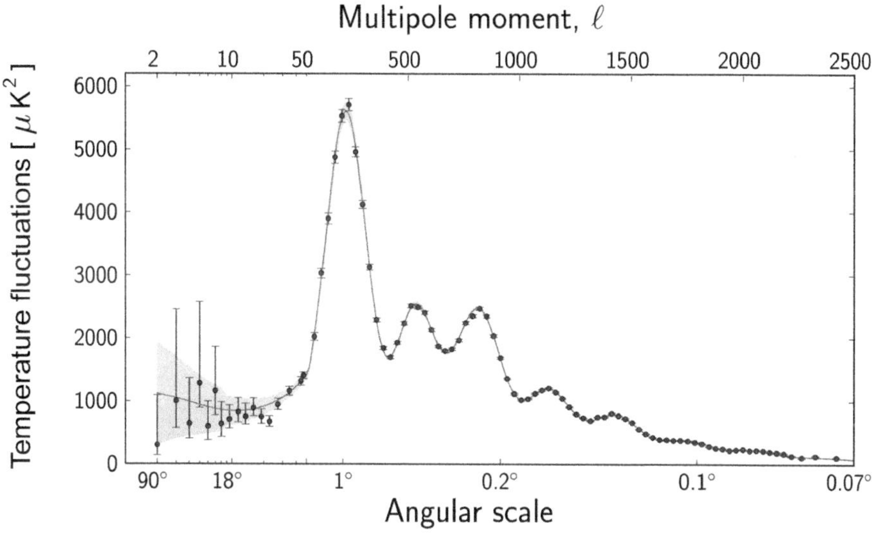

Figure 10.4: A plot showing the amplitude of the anisotropies in the CMB on different angular scales based on data from Planck. We call this the CMB power spectrum and it has important implications about the cosmological model of the universe. This image was provided by the ESA/Planck Collaboration. See www.esa.int/Our˙Activities/Space˙Science/Planck.

The primary anisotropies were produced in the early Universe by the interaction between radiation and matter. Gravity in slightly over-dense regions causes matter to clump and grow in density, while the radiation pressure drives matter apart. This competition results in acoustic oscillations, or sound waves, which cause anisotropies. The first and strongest peak in the CMB power spectrum, at an angular scale of 1°, is referred to as the first acoustic peak. The angular size of this peak provides a measure of the curvature of the Universe. The physical size of the acoustic oscillations at the time of decoupling can be modeled relatively accurately. Since the redshift of the CMB is well determined, the angular size that these anisotropies appear to us today depends on the curvature of the Universe. Light rays traveling in lines parallel in Euclidean space will diverge in a hyperbolic universe and converge in a spherical universe. Similarly, relative to the angular size in a flat universe, the angular size of distant objects will be smaller in a hyperbolic universe and larger in a spherical universe. The measured angular size of 1° of the dominant anisotropies in the CMB suggests that the Universe is spatially flat.

At smaller angular scales there is a second acoustic peak in the CMB power spectrum that is much smaller than the first acoustic peak. The ratio of the height of the first peak to the second peak (and between the odd numbered and even numbered peaks shown in Figure 10.4) provides information about the density of baryons or "ordinary matter". From the Planck observations shown in the figure, we can determine that the baryon density is only about 5% of the critical density. On the other hand, the overall amplitude of the acoustic peaks tells us that the total matter density must be close to 30% of the critical density. Thus, most matter must be dark matter as we found in Chapter 8. Finally, because the geometry of the Universe is close to flat, the total mass/energy density must be closely equal to the critical density. Thus, the remaining 70% of the mass/energy density must be composed of dark energy. At this time, although we can only speculate on the nature of dark energy it is responsible for the accelerating expansion rate.

We can describe the relative amounts of matter and dark energy by the density parameters that we first discussed in Section 1.6. The density parameters give the current effective mass density of each component relative to the critical density defined in Equation 10.2. The matter density parameter, denoted by Ω_M, is equal to the current average density of all matter, baryonic and dark, divided by the current critical density. We characterize the dark energy by its energy density which is given by $\Lambda c^2/(8\pi G)$, where Λ is similar to the cosmological constant first introduced by Einstein. As with the cosmological constant, Λ is often assumed to be constant with time, independent of the scale factor of the Universe. The equivalent mass density of dark energy is then given by $\Lambda/(8\pi G)$. The dark energy parameter, denoted Ω_Λ, is then the equivalent mass density of dark energy divided by the critical density (Equation 10.2), therefore

$$\Omega_\Lambda = \Lambda/(3H_o^2).$$

Although the density of dark energy is often assumed to be constant, the matter density decreases with time, depending on the scale factor as R^{-3}. We used values of H_o, Ω_M and Ω_Λ in Section 1.6 to compute the age of the Universe and the proper distances to objects. The measured anisotropies indicate that Ω_M and Ω_Λ are approximately 0.3 and 0.7, respectively.

As the CMB radiation travels to us over the vast reaches of space, its intensity can be modified; we call these modifications secondary anisotropies. One important example is the thermal Sunyaev-Zel'dovich effect (often just called the thermal S-Z effect), in which the CMB photons are inverse Compton scattered off the hot electrons present in the halos of rich galaxy clusters. This effect produces a distortion of the CMB spectrum that can be readily identified by a decrease in intensity at lower frequencies below the CMB peak and an increase in intensity at frequencies above the CMB peak. The thermal S-Z effect is beneficial in studies of the hot intracluster gas present in rich galaxy clusters.

10.5 CMB POLARIZATION

The CMB spectrum and the CMB anisotropies provide important information about the nature of our Universe; however there is one more measurable quantity beyond the spectrum and spatial distribution and that is polarization. The polarization of the CMB is predicted to vary spatially, and the nature of these spatial variations provides yet more valuable information about our Universe. The polarization of the CMB can be separated into E-mode and B-mode components that completely characterize the polarization of the radiation. The E-mode component is curl-free, while the B-mode component is divergence-free. Measurements of CMB polarization are complicated, both because the polarization signal is weak and because the foreground emission is also polarized.

E-mode polarization arises naturally from scattering of CMB photons off free electrons near the time of recombination, often called the epoch of last scattering. The E-mode polarization was first detected in 2002 by John Kovac and colleagues[8] using the Degree

[8] 2002, Nature, vol. 420, p. 772.

Angular Scale Interferometer (DASI), and has since been confirmed by numerous other experiments including the Planck satellite. The measured E-mode polarization is consistent with that predicted by scattering.

The B-mode polarization (the curl-like pattern) is weaker than E-mode polarization and therefore more difficult to detect. There are two types of B-mode polarization. One type can arise from gravitational lensing where the effect of the lens distorts the E-mode polarization producing B-mode polarization. B-mode polarization due to gravitational lensing was first detected in 2013 by Duncan Hanson and colleagues[9] using the South Pole Telescope (SPT). This B-mode polarization signal was only one part in ten million of the total CMB intensity. Further observations of this type of B-mode polarization will be useful to determine the large-scale structure of the Universe and in principle constrain the sum of the masses of the different neutrino types (electron neutrinos, muon neutrinos and tau neutrinos).

A second type of B-mode polarization is due to the interaction of CMB photons with primordial gravitational waves generated at the time of cosmic inflation. Theories of inflation propose that the Universe expanded by an incredibly large factor at a time before 10^{-30} seconds after the Big Bang. The gravitational waves produced by inflation at this epoch are too faint to be detected by current gravitational wave detectors; however the polarization imprinted on the CMB emission may be detectable. Fortunately, this type of B-mode polarization is distinguishable from B-mode polarization produced by gravitational lensing. Although to date this type of B-mode polarization has eluded detection, if and when it is detected, this polarization signal will be a powerful probe to test various inflation models of our Universe.

QUESTIONS AND PROBLEMS

1. Sometime in the future, when the Universe has expanded by a factor of 10 relative to today, what will be the temperature of the CMB radiation?

2. For the CMB radiation discussed in Question 1, at what wavelength will it have its maximum intensity?

3. What temperature would we measure for the CMB if we observe in a direction 30° ($\theta = 30°$) away from the direction of our motion relative to the CMB? What about in the opposite direction of our motion at $\theta = 180°$?

4. The magnitude of the CMB dipole we measure is 0.0034 K and we interpreted this to mean we are moving at a speed of 370 km s^{-1} relative to the CMB. If an astronomer in another galaxy measured the magnitude of the CMB dipole to be 0.0078 K, what would be their velocity relative to the CMB?

5. When we observe a distant galaxy, we see it as it was in the past when the CMB radiation was hotter. Imagine observing a galaxy at redshift z. Derive the temperature of the CMB at the time when the light we observe was emitted by the galaxy.

[9] 2013, Physical Review Letters, vol. 111, p. 1301.

Appendices

Constants and Conversions

Speed of light = c = 2.998×10^{10} cm s^{-1} = 2.998×10^{8} m s^{-1} = 2.998×10^{5} km s^{-1}

Boltzmann constant = k = 1.381×10^{-16} erg K^{-1} = 1.381×10^{-23} J K^{-1}
= 8.618×10^{-5} eV K^{-1}

Planck constant = h = 6.626×10^{-27} erg s = 6.626×10^{-34} J s = 4.135×10^{-15} eV s

Stefan-Boltzmann constant = σ = 5.671×10^{-5} erg cm^{-2} s^{-1} K^{-4}
= 5.671×10^{-8} J m^{-2} s^{-1} K^{-4}

Universal Gravitational constant = G = 6.673×10^{-8} cm^{3} g^{-1} s^{-2}
= 6.673×10^{-11} m^{3} kg^{-1} s^{-2}

Mass of proton = m_p = 1.673×10^{-24} g = 1.673×10^{-27} kg

Mass of electron = m_e = 9.109×10^{-28} g = m_e = 9.109×10^{-31} kg

Fundamental charge = e = 4.803×10^{-10} esu = 1.602×10^{-19} C

Electron volt (eV) = 1.602×10^{-12} erg = 1.602×10^{-19} J

Rydberg constant = R = 1.0967759×10^{5} cm^{-1} = 1.0967759×10^{7} m^{-1}

Solar Mass ($M_\odot$) = 1.989×10^{33} g = 1.989×10^{30} kg

Solar Luminosity ($L_\odot$) = 3.826×10^{33} erg s^{-1} = 3.826×10^{26} W

Parsec (pc) = 3.086×10^{18} cm = 3.086×10^{16} m

Light-year (ly) = 9.461×10^{17} cm = 9.461×10^{15} m

Astronomical Unit (AU) = 1.496×10^{13} cm = 1.496×10^{11} m

1 year = 3.157×10^{7} s

Jansky (Jy) = 10^{-23} erg s^{-1} cm^{-2} Hz^{-1} = 10^{-26} W m^{-2} Hz^{-1}

Mathematica Code for Calculating Age of Universe and Distances for Given Redshift

THE following is a code to be entered into Mathematica[1]. For a given redshift, z, the code calculates the current age of the Universe, the age of the Universe when the light was emitted, the current proper distance to the source, in both cm and pc, the luminosity distance and angular size distance. The calculation assumes a flat universe with matter and a cosmological constant.

```
inputs:
c = speed of light in cm/s
H = Hubble parameter in km/s/Mpc
WL = Omega of Lambda,
WM = Omega of non-rel matter
R = scale factor at time of emission
dp = proper distance in cm
dppc = proper distance in pc
dL = luminosity distance in pc
dA = angular size distance in pc
tuniv = age of universe in Gyr
tem = age of universe in Gyr when light was emitted

Code:
z = put in z value here
c = 3*10^(10)
WL = 0.7
WM = 0.3
H = 70.0
```

[1]Wolfram Research, Inc., Mathematica, Version 11.3, Champaign, IL (2018).

```
tHubble = (1/H)*3.086*10^(19)/(3.16*10^16)
tuniv = (ArcSinh[(WL/WM)^.5])/(1.5*(1/tHubble)*(WL^.5))
R = (((WM/WL)^.5) Sinh[1.5*((WL)^.5)*(1/tHubble)*t])^(2/3)
tem = (2/3) (14) (WL^(-.5)) ArcSinh[(WL/WM)^(.5)*((1 + z)^(-3/2))]
dp = NIntegrate[(c*3.16*10^16)/R, {t, tem, tuniv}]
dppc = dp/(3.09*10^(18))
dL = dppc*(1 + z)
dA = dppc/(1 + z)
```

Complex-Valued Wave Functions

AN introductory discussion of waves usually involves a cosine or a sine function, or a combination of the two, i.e.,

$$y = A\cos(kx - \omega t + \phi), \tag{C.1}$$

or

$$y = A\sin(kx - \omega t + \phi), \tag{C.2}$$

or

$$y = B_1\cos(kx - \omega t) + B_2\sin(kx - \omega t). \tag{C.3}$$

These are all considered *wave functions* because they meet all the criteria needed for a mathematical description of waves. What are these criteria? The physical conditions that define a wave are provided by what physicists call the *wave equation.* This is a partial differential equation that describes the relation between how the wave changes with time (i.e., its oscillation) and how it changes with distance (i.e., its shape). The wave equation in one dimension is given by

$$\frac{\partial^2 y}{\partial x^2} = \frac{1}{c^2}\frac{\partial^2 y}{\partial t^2},$$

where c is the speed of the wave. This equation describes a function that has a repeating pattern in time as well as in distance and the variations of these patterns are related. The speed, c, is what connects these patterns; as the wave moves in the +x direction, the y-motion at any fixed x (as a function of time) obeys simple harmonic motion, and this results because of the passing of a wave with a simple cosine-shape in the x direction.

The requirement for the mathematical form of the wave function, $y(x,t)$, is that it must satisfy the wave equation. Additionally, because the equation involves second derivatives, the connection between the equation and the function involves two steps of integration, which introduces two constants of integration which are generally determined by the *initial conditions*, such as the initial position and initial velocity. The solution, then, must also contain two free parameters which depend only on the initial conditions. The free parameters in the cosine and sine functions are the amplitude A and the phase ϕ. With Equation C.3, the free parameters are B_1 and B_2. We don't need ϕ in this case, and in fact we are not allowed to have a third free parameter, or the equation is not solved. The k $(= 2\pi/\lambda)$ and $\omega(= 2\pi\nu)$ are not free parameters, because they are constrained by the relation $\omega/k = c$, which is given in the wave equation.

The cosine and sine functions (Equations C.1 and C.2), independently, are both good solutions because they do indeed fit in the wave equation. And, the combination of cosine and sine works also. There is, though, another mathematical function that also works, and that is the complex exponential form, i.e.,

$$y = Ae^{\pm i(kx - \omega t + \phi)}.$$

The exponent can be positive or negative. Let us substitute this function into the wave equation to see that it satisfies the equation. We have

$$\frac{\partial^2 y}{\partial x^2} = \frac{\partial}{\partial x}\left(ikAe^{\pm i(kx-\omega t+\phi}\right) = -k^2 Ae^{\pm i(kx-\omega t+\phi)}$$

and

$$\frac{\partial^2 y}{\partial t^2} = \frac{\partial}{\partial x}\left(i\omega Ae^{\pm i(kx-\omega t+\phi}\right) = -\omega^2 Ae^{\pm i(kx-\omega t+\phi)}.$$

Combining these two equations and since $k^2/\omega^2 = 1/c^2$, we have

$$\frac{\partial^2 y}{\partial x^2} = \frac{1}{c^2}\frac{\partial^2 y}{\partial t^2},$$

and the wave equation is satisfied. This function also has two free parameters, A and ϕ, and, therefore, must be a legitimate mathematical expression of waves.

In fact, this function is very closely related to the cosine and sine functions. A famous equation, known as Euler's formula (which is usually derived in a complex-numbers math class) is

$$Ae^{\pm i\theta} = A\cos\theta \pm iA\sin\theta.$$

So, one can also write the complex exponential form as

$$y = Ae^{\pm i(kx-\omega t+\phi)} = A\cos(kx - \omega t + \phi) \pm iA\sin(kx - \omega t + \phi).$$

The complex exponential form, therefore, is actually the same thing as the wave function in Equation C.3 that includes both the cosine and sine functions, where B_2, the parameter in front of the sine, is set equal to iA, and a phase constant is added. (We need the ϕ, now, because there is only one free parameter in the amplitude.)

As shown in Euler's formula, the complex exponential function can always be broken up into a sum of its real and imaginary parts, where $A\cos\theta$ is the real part and $\pm A\sin\theta$ is the imaginary part. And, if you make a graph in which the x-axis represents real numbers, and the y-axis represents imaginary numbers, as shown in Figure C.1, then Euler's formula shows that the complex exponential function can be viewed as describing a vector of magnitude A that makes an angle θ with respect to the real-numbers axis. If you calculate the components of this vector, what do you get? Answer: exactly what you have in the Euler formula; the component along the horizontal axis is $A\cos\theta$ and the component along the vertical axis is $iA\sin\theta$. With this vector image in mind, the A in front of the exponential is called the *amplitude* and the argument of the exponent, not including the 'i,' is called the *phase angle*.

The two-dimensional system defined by the graph in Figure C.1 is called the *complex plane*. You can always consider the mathematics of real numbers as the same math that occurs in the complex plane but done solely on the *real* axis. Specific to our case here, consider how much easier the complex exponential form is to work with than cosines and sines. The product of two exponentials is simply the exponential of the sum of the exponents, i.e.,

$$e^A e^B = e^{A+B},$$

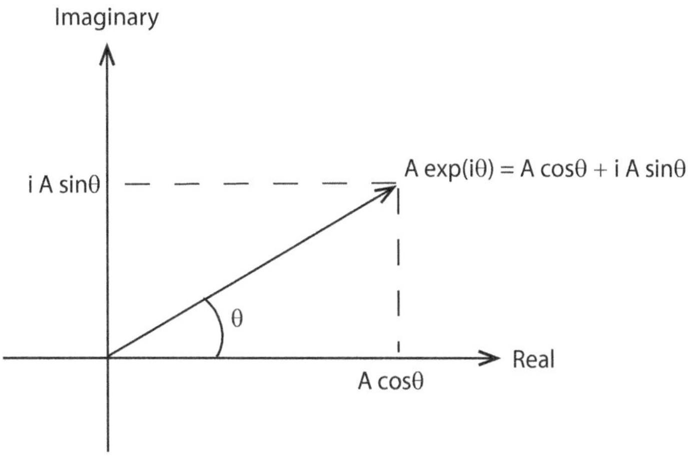

Figure C.1: Graphical representation of a complex number in the *complex plane* in which the Real part is plotted on the x-axis and the Imaginary part on the y-axis.

while the product of two cosines is more complicated,

$$\cos A \cos B = \frac{1}{2}[\cos(A+B) + \cos(A-B)].$$

Also, the derivatives are much easier with exponentials. The derivative of an exponential is still the same exponential, but multiplied by the derivative of the exponent (following the chain rule), whereas the derivative of a cosine turns it into a –sine. Although it may be a less familiar notation, the math is often simpler using complex exponentials.

Example C.1:

(a) Write the number $5e^{-i\pi/3}$ in terms of the real and imaginary parts.

(b) Write the number $2 - i3$ in the complex exponential form.

Answers:
(a) Using Euler's Formula, we have

$$5e^{-i\pi/3} = 5\cos(\pi/3) - i5\sin(\pi/3) = 2.5 - i4.33.$$

Note that we get the same answer if we interpret the negative sign in the exponent as part of the phase angle, that is $5e^{-i\pi/3} = 5e^{i(-\pi/3)}$. Then we have

$$5e^{i(-\pi/3)} = 5\cos(-\pi/3) + i5\sin(-\pi/3) = 2.5 - i4.33.$$

(b) To convert to the complex exponential form, we need to calculate the amplitude and the phase angle, as depicted in Figure C.1. The amplitude is simply the quadrature sum of the components, i.e.

$$A = \sqrt{2^2 + 3^2} = 3.61.$$

The phase angle is given by

$$\phi = \tan^{-1}\left(\frac{-3}{2}\right).$$

There is an ambiguity if we just plug in the numbers and use the calculator. We need to determine which quadrant this angle is in. The real part is positive and the imaginary part is negative so this corresponds to the quadrant below the positive real axis (i.e. the 4th quadrant). Therefore, we can express ϕ either as a negative angle or as a positive angle between $3\pi/2$ and 2π. The former choice is more compact, so we have

$$\phi = -\tan^{-1}(1.5) = -0.983 \text{ radians} = -0.313\pi \text{ radians.}$$

The final answer to the question, then, is

$$2 - i3 = 3.71e^{-i0.313\pi}.$$

APPENDIX D

Derivations of the Effects of Propagation of Radiation in Ionized Media

THE derivations of the interactions of radiation with a plasma require advanced undergraduate electromagnetism. To stay grounded in the fundamentals, we start with a general physics discussion of the response of different media to incident electromagnetic radiation and then apply the general equation to the conditions in the interstellar medium, leading to the equations in Chapter 2.

D.1 BASIC EQUATIONS OF ELECTROMAGNETIC WAVES

In general, electromagnetic radiation can be described by complex wave functions (see Appendix C). The complex wave functions of the $\vec{E}$ and $\vec{B}$ fields propagating in a vacuum are given by

$$\vec{E} = \vec{E}_0 e^{i(\vec{k}\cdot\vec{r}-\omega t)} \text{ and } \vec{B} = \vec{B}_0 e^{i(\vec{k}\cdot\vec{r}-\omega t)}$$

where $\vec{k}$ is the wave number, with a magnitude given by $2\pi/\lambda$ and pointed in the direction of propagation of the wave, and $\omega = 2\pi\nu$ is the angular frequency of the wave. Recall that the directions of the electric and magnetic fields and the direction of propagation are mutually perpendicular. For present purposes, we let the direction of propagation be the x-direction, the direction of the electric field be the y-direction, and the direction of the magnetic field be the z-direction. The electric field wave can, then, be written as

$$\vec{E} = E_0 e^{i(kx-\omega t)}\hat{y}. \tag{D.1}$$

There is, of course, an identical equation for $\vec{B}$, though in the $\hat{z}$ direction.

The focus of the discussion here is the propagation of electromagnetic waves, so we will need to calculate their velocities. Since all electromagnetic signals have a non-zero bandwidth, we must calculate the wave velocity for a range of frequencies. As explained in a physical optics class, the propagation speed of a wave packet is the *group velocity* and is given by

$$v_{\text{group}} = \frac{d\omega}{dk}. \tag{D.2}$$

To get the velocity of a wave, then, we need an expression relating ω and k. We will get this by inserting our expression for the electric field, Equation D.1, into Maxwell's equations, and manipulating to solve for k, which will contain ω.

The goal of this section is to model the propagation of electromagnetic radiation through an ionized medium and so we must take into account the effect of the free charges. The charges respond to the electromagnetic fields of the radiation as well as contribute an electric field. However, the effect of the wave's magnetic field is insignificant in comparison to that of its electric field, and so we can focus on the electric field alone.

For the situations of interest in this book, we need consider only media that have linear responses to $\vec{E}$ and $\vec{B}$, meaning constant values of ϵ, the electric permittivity, μ, the magnetic permeability, and σ, the conductivity. The electric permittivity is a measure of a medium's ability to reduce an applied electric field by charge displacement and the magnetic permeability relates to the amount a medium is magnetized for a given applied magnetic field. In cgs units, in a vacuum $\epsilon = 1$ and $\mu = 1$. (S.I. units include constants, ϵ_0 and μ_0, known as the permittivity and permeability of free space). The conductivity, σ, relates a medium's ability to produce a current in response to an applied electric field. When an electric field $\vec{E}$ is applied to a conductive medium, the amount of current per cross-sectional area $\vec{J}$ (or current density) that results is proportional to $\vec{E}$ and the σ of the medium, as given by Ohm's law,

$$\vec{J} = \sigma\vec{E}. \tag{D.3}$$

As commonly laid out in an upper level electricity and magnetism text, by manipulating two of Maxwell's equations, those of Faraday's Law and Ampere's Law in media, along with Ohm's Law, one can obtain the electromagnetic wave equation in media, which is

$$\vec{\nabla} \times (\vec{\nabla} \times \vec{E}) = -\frac{4\pi\mu\sigma}{c^2}\frac{\partial \vec{E}}{\partial t} - \frac{\epsilon\mu}{c^2}\frac{\partial^2 \vec{E}}{\partial t^2}. \tag{D.4}$$

A similar equation can be derived for $\vec{B}$, but we will not need that equation here.

If we apply this equation to a vacuum, where, in cgs units, $\epsilon = \mu = 1$, and $\sigma = 0$, we recover the more familiar wave equation in vacuum. Following rules of vector differentials, the left-hand side can be rewritten as

$$\vec{\nabla} \times (\vec{\nabla} \times \vec{E}) = \vec{\nabla}(\vec{\nabla} \cdot \vec{E}) - \nabla^2 \vec{E}$$

and $\vec{\nabla} \cdot \vec{E} = 0$ since the net charge of the medium is zero. Substituting in the vacuum values, then, we get

$$\nabla^2 \vec{E} = \frac{1}{c^2}\frac{\partial^2 \vec{E}}{\partial t^2},$$

the well-known vacuum wave equation.

Returning to our derivation of an equation for $k(\omega)$, we insert the complex wave function for $\vec{E}$ into Equation D.4 and apply the derivatives. Let's first contemplate the operation on the left-hand side. By examining Equation D.1 we see that $\vec{E}$ has only a y-component and has a spatial dependence only on x. Applying the first curl, then, the only non-zero term is $\partial E_y/\partial x$ in the z-component, and since the x dependence appears in an exponential multiplied by ik, we have

$$\vec{\nabla} \times \vec{E} = ikE_0 e^{i(kx-\omega t)}\,\hat{z}.$$

Similarly, applying the second curl, we get

$$\vec{\nabla} \times (\vec{\nabla} \times \vec{E}) = +k^2 E_0 e^{i(kx-\omega t)}\,\hat{y}.$$

On the right-hand side we have time derivatives, which are

$$\frac{\partial \vec{E}}{\partial t} = -i\omega E_0 e^{i(kx-\omega t)}\hat{y},$$

and

$$\frac{\partial^2 \vec{E}}{\partial t^2} = -\omega^2 E_0 e^{i(kx-\omega t)}\hat{y}.$$

Putting these expressions into Equation D.4, we have, then,

$$+k^2 E_0 e^{i(kx-\omega t)}\hat{y} = -\frac{4\pi\mu\sigma}{c^2}(-i\omega E_0 e^{i(kx-\omega t)}\hat{y}) - \frac{\epsilon\mu}{c^2}(-\omega^2 E_0 e^{i(kx-\omega t)}\hat{y})$$

and dividing by $E_0 e^{i(kx-\omega t)}\hat{y}$ leads to

$$k^2 = i\frac{4\pi\mu\sigma\omega}{c^2} + \frac{\epsilon\mu\omega^2}{c^2}. \tag{D.5}$$

Equation D.5 is called the dispersion relation and provides the relationship between k and ω and hence the wave propagation speed. Clearly, the propagation speed depends on μ, σ, and ϵ of the medium.

You may be a little confused seeing an expression for k^2 that contains an imaginary term. Just remember that k appears in the exponent of the wave function, which also contains an i and so, as you'll see in the following sections, the real and imaginary parts of k^2 have different physical consequences.

Recall that our goal is to infer the wave's propagation speed, v, which is given by $d\omega/dk$ (Equation D.2). Equation D.5 is a good starting point to get there. However, since this general expression for k^2 is complex, inferring a general expression for k is not straightforward. It is simpler to consider different values of μ, σ, and ϵ, which characterize different types of media. We will then obtain the wave velocity in each case, which leads to some interesting effects.

D.2 APPLICATIONS TO REAL MEDIA

First, to check that Equation D.5 is consistent with what we know to be true, we apply it to a vacuum. In this case, in cgs units, $\sigma = 0, \mu = 1$, and $\epsilon = 1$. Equation D.5 then becomes

$$k^2 = \frac{\omega^2}{c^2},$$

and so

$$\omega = kc.$$

The propagation velocity, given by $d\omega/dk$, equals c, "the speed of light in a vacuum," exactly as we expected.

We can also make sure that applying our equations to a perfect insulator works. A perfect insulator has $\sigma = 0$. The parameters μ and ϵ, in this case, may not equal 1, and so

$$k = \sqrt{\epsilon\mu}\,\frac{\omega}{c}$$

or

$$\omega = \frac{kc}{\sqrt{\epsilon\mu}}.$$

The propagation speed, then, is

$$\frac{d\omega}{dk} = \frac{c}{\sqrt{\mu\epsilon}}. \tag{D.6}$$

In optics, one learns that the index of refraction of a medium is given by

$$n = \frac{c}{v},$$

which, according to Equation D.6, means that the index of refraction of an insulating medium can also be expressed as

$$n = \sqrt{\mu\epsilon}.$$

This is another equation that you may have seen before (in mks units, these would be μ_r and ϵ_r, the relative permeability and permittivity compared to vacuum values). This result, therefore, is also as expected.

D.2.1 Dissipative Media

We now consider a medium that has a non-zero conductivity, σ, meaning that it contains free electrons. Many regions of interest, such as the Sun's atmosphere, the Earth's ionosphere, and the interstellar medium, contain plasmas, and so have non-zero conductivities. The electrons are easily accelerated by the electric field in the wave and hence absorb energy from the wave. The wave, therefore, is dissipated as it passes through the medium. In this case, since $\sigma \neq 0$, Equation D.5 contains both terms, including the imaginary first term, and so k^2 is a complex number. This will lead to a complicated expression, which is not very enlightening by itself, but we will consider specific cases that will enable us to simplify the equation. We start with the general solution.

As with any complex number, we can convert k^2 into the exponential form,

$$k^2 = |k^2| e^{i\phi_{k^2}},$$

where

$$|k^2| = \sqrt{(\mathrm{Re})_{k^2}^2 + (\mathrm{Im})_{k^2}^2}$$

and

$$\phi_{k^2} = \tan^{-1}\left(\frac{(\mathrm{Im})_{k^2}}{(\mathrm{Re})_{k^2}}\right).$$

Applied to Equation D.5, we get a general expression for k in exponential form

$$k = \frac{1}{c}\left[\omega\mu\sqrt{\omega^2\epsilon^2 + (4\pi)^2\sigma^2}\; e^{i\tan^{-1}(4\pi\sigma/\omega\epsilon)}\right]^{1/2}. \tag{D.7}$$

We can obtain an equation for σ by considering Ohm's Law, $\vec{J} = \sigma\vec{E}$, where $\vec{J}$ is related to the number density of charges, n_q, the magnitude of the charges, q, and their drift velocity, $\vec{v}_q$, by

$$\vec{J} = n_q q \vec{v}_q. \tag{D.8}$$

So, we need to derive the resulting velocity of the charges in the medium when the oscillating electric field of the radiation passes through the medium. We do so by considering the acceleration of the charges by the electromagnetic fields in the wave. A charge's acceleration is given by the Lorentz force, in cgs units, as

$$\vec{a} = \frac{q}{m}\left(\vec{E} + \frac{\vec{v}}{c} \times \vec{B}\right).$$

The acceleration due to the radiation's magnetic field is negligible relative to that due to the radiation's electric field, and so we consider only the first term. For a charge at some position, $\vec{r}$, the electric field of the wave oscillates sinusoidally and so we have

$$\vec{a} = \frac{d\vec{v}}{dt} = \frac{q}{m}\vec{E} = \frac{q}{m}\vec{E}_0 e^{-i\omega t}.$$

Integrating, to get an expression for $\vec{v}$, we have

$$\vec{v}(t) = i\frac{q}{\omega m}\vec{E}_0 e^{-i\omega t} = i\frac{q}{\omega m}\vec{E}. \tag{D.9}$$

Equation D.9 appears to suggest that the charges have an imaginary velocity which may seem confusing. However, the right side of Equation D.9 also contains the electric field vector, which is represented by a complex number. Equation D.9 really indicates that there is a phase difference between $\vec{v}$ and $\vec{E}$ (remember that multiplying by i introduces a $\pi/2$ phase shift); thus the velocity of the oscillating charges reaches a maximum a quarter cycle after the wave's electric field is maximum.

We start with Equation D.8 and combining with Equations D.9 and Equation D.3 we get

$$\vec{J} = in_q q\frac{q}{\omega m}\vec{E} = \sigma\vec{E}$$

and, therefore,

$$\sigma = i\frac{n_q q^2}{\omega m},$$

where n_q is the number density of free charged particles, q is the charge of each particle, and m is the mass of a particle.

The conductivity here is purely imaginary. What does this mean? The trick is recalling that the definition of the conductivity is really contained in Ohm's Law (Equation D.3) which contains complex vectors on both sides. As we discussed above with the charge's velocity, the current density is a quarter cycle out of phase with the wave's electric field.

Note the m in the denominator of the above equations. The ions, being far more massive than the electrons, do not respond to the radiation nearly as easily as the electrons. So, the relevant charges to consider are only the electrons and so we will use the following equation:

$$\sigma = i\frac{n_e e^2}{\omega m_e}. \tag{D.10}$$

When we plug this expression for σ into Equation D.5 we get

$$k^2 = i\frac{4\pi\omega\mu}{c^2}i\frac{n_e e^2}{\omega m_e} + \frac{\omega^2\mu\epsilon}{c^2},$$

or

$$k^2 = \frac{\omega^2\mu\epsilon}{c^2} - \frac{4\pi n_e e^2\mu}{m_e c^2}. \tag{D.11}$$

As we will show, this form of the dispersion relation leads to some interesting effects on the propagation of waves. Equation D.11 shows that in a conductive medium, k^2 is the difference of two terms. The frequency of the wave appears in only the positive term and so at frequencies below some critical value, k^2 is negative and hence k is imaginary. We must, therefore, address the high frequency and low frequency realms separately. We address first the low frequency case.

D.2.2 The Plasma Frequency Equation

At frequencies where $\omega^2 < 4\pi n_e e^2/m_e\epsilon$, Equation D.11 shows that k^2 is negative and so k is purely imaginary. We can represent k as $i|k|$, where $|k|$ is the magnitude of k. The expression for the electric field of such a wave in the medium is then

$$\vec{E} = \vec{E}_0 e^{i(i|k|x-\omega t)} = \vec{E}_0 e^{-|k|x} e^{-i\omega t}.$$

This is not a description of a wave propagating in the x-direction, but rather of a quantity that oscillates in time but exponentially decreases with distance into the medium. The wave, therefore, does not penetrate this medium. In general, an imaginary wave number indicates that the wave does not propagate.

Thus, based on Equation D.11, there is a critical frequency below which waves will not propagate. This critical frequency is called the *plasma frequency* and is defined mathematically by

$$\omega_\mathrm{p} = \sqrt{\frac{4\pi n_e e^2}{\epsilon m_e}}. \tag{D.12}$$

D.2.3 Derivation of the Wave Velocity as a Function of Frequency and the Arrival Time of a Pulse

At higher frequencies, where $\omega^2 > 4\pi n_e e^2/m_e\epsilon$ and thus $\omega > \omega_\mathrm{p}$, electromagnetic waves propagate through the plasma, but the plasma causes waves of different frequencies to travel at different speeds. The difference in arrival time of a pulse at two different frequencies yields the *dispersion measure*, discussed in Section 2.2.2. Here we will derive Equation 2.20.

Starting with Equation D.11 we can obtain the group velocity for all frequencies above the plasma frequency. The result is more enlightening, and the math simpler, if we substitute in the plasma frequency, given by Equation D.12, in the second term, so that we have

$$k^2 = \frac{\omega^2\mu\epsilon}{c^2} - \omega_\mathrm{p}^2\frac{\mu\epsilon}{c^2} = \frac{\mu\epsilon}{c^2}(\omega^2 - \omega_\mathrm{p}^2).$$

The group velocity, then, is given by

$$v_\mathrm{group} = \left(\frac{dk}{d\omega}\right)^{-1} = \left[\frac{d}{d\omega}\left(\frac{\sqrt{\mu\epsilon}}{c}\sqrt{\omega^2-\omega_\mathrm{p}^2}\right)\right]^{-1},$$

or

$$v_\mathrm{group} = \left[\frac{\sqrt{\mu\epsilon}}{c}\frac{1}{2}\frac{2\omega}{\sqrt{\omega^2-\omega_\mathrm{p}^2}}\right]^{-1}$$

$$= \frac{c}{\sqrt{\mu\epsilon}}\frac{\sqrt{\omega^2-\omega_\mathrm{p}^2}}{\omega}.$$

Or,

$$v_\mathrm{group} = \frac{c}{\sqrt{\mu\epsilon}}\sqrt{1-\frac{\omega_\mathrm{p}^2}{\omega^2}}. \tag{D.13}$$

Equation D.13 shows that lower frequency waves travel at smaller speeds than those of higher frequencies. Note that as $\omega \to \infty, v_{\text{group}} \to c/\sqrt{\mu\epsilon}$ and as $\omega \to \omega_{\text{p}}, v_{\text{group}} \to 0$, as we expect.

When observing a pulsar in the Galactic plane, then, as the radio frequency pulses travel through the interstellar medium, the higher frequency waves from the pulses will travel faster and reach Earth sooner than the lower frequency waves. The travel time of any wave is given by

$$t_{\text{tr}} = \int_0^L \frac{ds}{v_{\text{group}}} = \int_0^L \frac{\sqrt{\mu\epsilon}}{c\sqrt{1-(\omega_{\text{p}}/\omega)^2}} ds. \tag{D.14}$$

Any observing frequency will be much greater than the plasma frequency of the ISM (since $\omega_{\text{p}}(\text{ISM}) \ll \omega_{\text{p}}(\text{Earth's ionosphere})$), and so $\omega^2 \gg \omega_{\text{p}}^2$. We can, then, do a Taylor series expansion about $\omega_{\text{p}}^2/\omega^2 = 0$, i.e.,

$$\frac{1}{\sqrt{1-(\omega_{\text{p}}/\omega)^2}} \approx 1 + \frac{1}{2}\left(\frac{\omega_{\text{p}}}{\omega}\right)^2.$$

Substituting this into Equation D.14, we have

$$t_{\text{tr}} = \int_0^L \frac{\sqrt{\mu\epsilon}}{c}\left(1 + \frac{1}{2}\left(\frac{\omega_{\text{p}}}{\omega}\right)^2\right) ds,$$

or

$$t_{\text{tr}} = \int_0^L \frac{\sqrt{\mu\epsilon}}{c} ds + \int_0^L \frac{\sqrt{\mu\epsilon}}{c}\frac{1}{2}\left(\frac{\omega_{\text{p}}}{\omega}\right)^2 ds.$$

Substituting in for ω_{p} in the second integral,

$$t_{\text{tr}} = \int_0^L \frac{\sqrt{\mu\epsilon}}{c} ds + \int_0^L \frac{\sqrt{\mu\epsilon}}{c}\frac{1}{2}\left(\frac{4\pi n_e e^2}{\epsilon m_e}\right)\left(\frac{1}{\omega}\right)^2 ds.$$

Assuming μ and ϵ are roughly constant along the line of sight, the only factor in the integrand that may vary along the line of sight is the electron density, n_e, and converting to observing frequency units ($\omega = 2\pi\nu$), this can be rewritten as

$$t_{\text{tr}} = \sqrt{\mu\epsilon}\,\frac{L}{c} + \frac{\sqrt{\mu\epsilon}\,e^2}{2\pi\epsilon m_e c}\frac{1}{\nu^2}\int_0^L n_e ds.$$

As mentioned in Section 2.2.2, this is the travel time without any dispersive effects plus the delay introduced by interactions with electrons in the medium. In general, for applications that are relevant for this book, $\epsilon \approx 1$ and $\mu \approx 1$.

D.3 DERIVATION OF THE ROTATION ANGLE OF POLARIZATION IN MAGNETIZED MEDIA

As discussed in Section 2.2.3, Faraday rotation occurs because free electrons in the presence of a magnetic field respond differently to the RCP waves than the LCP waves. We can model this mathematically by calculating the conductivities of the medium for LCP and RCP waves. The key is to consider the propagation of LCP and RCP waves separately and to calculate the conductivity, defined by

$$\sigma = \frac{\vec{J}}{\vec{E}} = \frac{n_e e \vec{v}}{\vec{E}}.$$

We will do this derivation by solving for the free electron's velocity, $\vec{v}(t)$, induced by the electric field of the waves and the magnetic field in the medium.

For convenience, we let the wave propagation direction be along the z-axis. Then, following the radio astronomy convention for circular polarization, we can express the electric field in the waves, as viewed by charged particles in the plasma (i.e., at a fixed z), by

$$\vec{E} = E_0 e^{i\omega t}(\hat{x} \pm i\hat{y}), \tag{D.15}$$

where the $+$ describes a left circularly polarized (LCP) wave and the $-$ describes RCP. Substituting $e^{i\omega t} = \cos(\omega t) + i\sin(\omega t)$ and manipulating, we find that the real part of Equation D.15 is equal to

$$E_0\left(\cos(\omega t)\hat{x} \mp \sin(\omega t)\hat{y}\right),$$

which may be a more familiar form for describing LCP (using the $-$) and RCP ($+$) waves. In the following, the distinction between LCP and RCP will always be determined by which algebraic sign is used. We will always keep the LCP symbol on top.

The acceleration of the free electrons is given by

$$\frac{d\vec{v}}{dt} = \frac{1}{m_e}\left(-e\vec{E} - \frac{e}{c}\vec{v} \times \vec{B}\right).$$

We can now substitute in for the electric field vector, and for the magnetic field we need only consider the z-component (as explained in Section 2.2.3), which we denote by $B_{||}$ to indicate that this is the component of $\vec{B}$ parallel to the direction of propagation. For the following derivation we will assume that $B_{||}$ is positive. The acceleration of a free electron, then, is

$$\frac{d\vec{v}}{dt} = \frac{-e}{m_e}E_0 e^{i\omega t}(\hat{x} \pm i\hat{y}) - \frac{e}{m_e c}\vec{v} \times (B_{||}\hat{z}).$$

This is a differential equation for velocity with respect to time, and so the next step is to solve for $\vec{v}(t)$. Rearranging to put all the terms with $\vec{v}$ on the left and the other terms on the right yields the first-order, non-homogeneous equation

$$\frac{d\vec{v}}{dt} + \frac{e}{m_e c}\vec{v} \times (B_{||}\hat{z}) = \frac{-e}{m_e}E_0 e^{i\omega t}(\hat{x} \pm i\hat{y}). \tag{D.16}$$

The time dependence of the right-hand side of the equation is all contained in the exponential, $e^{i\omega t}$, and therefore the velocity must have this time dependence. Therefore, we can let

$$\vec{v} = \vec{A}e^{i\omega t} \tag{D.17}$$

and substitute in for $\vec{v}$ in Equation D.16 to solve for the complex vector $\vec{A}$. We have, then,

$$i\omega\vec{A}e^{i\omega t} + \frac{e}{m_e c}\left(\vec{A}e^{i\omega t} \times (B_{||}\hat{z})\right) = \frac{-e}{m_e}E_0 e^{i\omega t}(\hat{x} \pm i\hat{y}).$$

We can cancel each factor of $e^{i\omega t}$ and with a small bit of rearranging we have

$$i\omega\vec{A} + \frac{eB_{||}}{m_e c}\left(\vec{A} \times \hat{z}\right) = \frac{-eE_0}{m_e}(\hat{x} \pm i\hat{y}). \tag{D.18}$$

Every term except the first clearly has no z-component and hence the first term must be independent of z. We know, then, that $\vec{A}$ has only x and y components, i.e. $\vec{A} = A_x\hat{x} + A_y\hat{y}$. The second term, then, is equivalent to

$$\frac{eB_{||}}{m_e c}\left(A_y\hat{x} - A_x\hat{y}\right).$$

Also, conveniently, the fraction in front is the cyclotron frequency, ω_c (Equation 3.22). Equation D.18 then becomes

$$i\omega\left(A_x\hat{x} + A_y\hat{y}\right) + \omega_c\left(A_y\hat{x} - A_x\hat{y}\right) = \frac{-eE_0}{m_e}(\hat{x} \pm i\hat{y}).$$

Gathering all the x-terms and all the y-terms together, this becomes

$$\left(i\omega A_x + \omega_c A_y\right)\hat{x} + \left(i\omega A_y - \omega_c A_x\right)\hat{y} = \frac{-eE_0}{m_e}\hat{x} \pm i\frac{-eE_0}{m_e}\hat{y}.$$

The x-components on the left and right must equal each other as must the y-components. We have therefore two equations and two unknowns, A_x and A_y. Simultaneously solving these two equations, we find that

$$A_x = \frac{ieE_0}{m_e}\frac{1}{(\omega \pm \omega_c)}$$

and

$$A_y = \mp\frac{eE_0}{m_e}\frac{1}{(\omega \pm \omega_c)}.$$

Substituting for $\vec{A}$ back in Equation D.17, our solution for the $\vec{v}(t)$ of the free electron is

$$\vec{v} = \left(\frac{ieE_0}{m_e}\frac{1}{(\omega \pm \omega_c)}\hat{x} \mp \frac{eE_0}{m_e}\frac{1}{(\omega \pm \omega_c)}\hat{y}\right)e^{i\omega t}$$

or

$$\vec{v} = \frac{ie}{m_e}\frac{1}{(\omega \pm \omega_c)}E_0(\hat{x} \pm i\hat{y})e^{i\omega t}.$$

Since the latter part of the right-hand side is the expression for the electric fields of the LCP and RCP waves, we have now that

$$\vec{v} = \frac{ie}{m_e}\frac{1}{(\omega \pm \omega_c)}\vec{E}.$$

Returning to the definitions of current density and conductivity, then,

$$\vec{J} = n_e e\vec{v} = -\frac{in_e e^2}{m_e}\frac{1}{(\omega \pm \omega_c)}\vec{E} = \sigma\vec{E},$$

and therefore,

$$\sigma = -i\frac{n_e e^2}{m_e}\frac{1}{(\omega \pm \omega_c)}$$

(where the + sign is used for LCP waves and the − for RCP). Note that σ differs for the two polarizations and for different frequencies. We now use this expression to solve for the wave number using Equation D.5,

$$k^2 = i\frac{4\pi\omega\mu}{c^2}\left(-i\frac{n_e e^2}{m_e}\frac{1}{(\omega \pm \omega_c)}\right) + \frac{\omega^2\mu\epsilon}{c^2}$$

or

$$k^2 = \frac{4\pi n_e e^2\omega\mu}{m_e c^2}\frac{1}{(\omega \pm \omega_c)} + \frac{\omega^2\mu\epsilon}{c^2}.$$

For convenience, we substitute in the plasma frequency (given in Equation D.12),

$$\omega_{\rm p}^2 = \frac{4\pi n_e e^2}{\epsilon m_e},$$

and so

$$k^2 = \frac{\omega\mu\epsilon\omega_{\rm p}^2}{c^2}\,\frac{1}{(\omega\pm\omega_{\rm c})} + \frac{\omega^2\mu\epsilon}{c^2},$$

or

$$k^2 = \frac{\omega^2\mu\epsilon}{c^2}\left(\frac{\omega_{\rm p}^2}{\omega\,(\omega\pm\omega_{\rm c})} + 1\right).$$

Taking the square root, we have

$$k = \frac{\omega}{c}\,\sqrt{\epsilon\mu}\,\sqrt{1+\frac{\omega_{\rm p}^2}{\omega(\omega\pm\omega_{\rm c})}}. \tag{D.19}$$

The relative phase of a wave a distance L from the wave source is

$$\phi = \int \frac{2\pi}{\lambda} ds = \int k ds \sim kL.$$

But, we want the shift in phase between the LCP and RCP waves, and this all occurs because of the second term inside the radical in Equation D.19. The cycling electrons produce a larger conductivity for the RCP waves than the LCP waves, so the RCP waves will experience a larger shift in phase. (Note, if $B_{||}$ is negative, i.e., pointing opposite the direction of propagation, then the conductivities will be reversed.) The phase difference between the RCP and the LCP waves, then, is

$$\Delta\phi = \frac{\omega}{c}\,\sqrt{\epsilon\mu}\int\left(\sqrt{1+\frac{\omega_{\rm p}^2}{\omega(\omega-\omega_{\rm c})}} - \sqrt{1+\frac{\omega_{\rm p}^2}{\omega(\omega+\omega_{\rm c})}}\right) ds. \tag{D.20}$$

We know that the observing frequency will be much greater than the cyclotron frequency for waves traveling through the interstellar medium; therefore we can use the following Taylor series approximation

$$\frac{1}{\omega\pm\omega_{\rm c}} = \frac{1}{\omega\left(1\pm\frac{\omega_{\rm c}}{\omega}\right)} \approx \frac{1}{\omega}\left(1\mp\frac{\omega_{\rm c}}{\omega}\right).$$

Using this approximation in Equation D.20 gives us

$$\Delta\phi = \frac{\omega}{c}\,\sqrt{\epsilon\mu}\int\left(\sqrt{1+\frac{\omega_{\rm p}^2}{\omega^2}\left(1+\frac{\omega_{\rm c}}{\omega}\right)} - \sqrt{1+\frac{\omega_{\rm p}^2}{\omega^2}\left(1-\frac{\omega_{\rm c}}{\omega}\right)}\right) ds. \tag{D.21}$$

We also know that $\omega \gg \omega_{\rm p}$ and so again using a Taylor series approximation,

$$\sqrt{1+x} \approx 1+\frac{1}{2}x, \text{ for } |x| \ll 1,$$

we can rewrite Equation D.21 as

$$\Delta\phi = \frac{\omega}{c}\,\sqrt{\epsilon\mu}\int\left(1+\frac{1}{2}\frac{\omega_{\rm p}^2}{\omega^2}\left(1+\frac{\omega_{\rm c}}{\omega}\right) - 1 - \frac{1}{2}\frac{\omega_{\rm p}^2}{\omega^2}\left(1-\frac{\omega_{\rm c}}{\omega}\right)\right) ds, \tag{D.22}$$

which reduces to

$$\Delta\phi = \frac{\omega}{c}\sqrt{\epsilon\mu}\int\left(\frac{\omega_\text{c}\omega_\text{p}^2}{\omega^3}\right)ds = \frac{\sqrt{\epsilon\mu}}{c}\omega^2\int\omega_\text{c}\omega_\text{p}^2 ds.$$

Substituting the expressions for ω_c and ω_p back in, and converting ω to ν, we have

$$\Delta\phi = \frac{\sqrt{\epsilon\mu}}{c4\pi^2\nu^2}\int_0^L \frac{eB_{||}}{m_e c}\frac{4\pi n_e e^2}{\epsilon m_e}ds,$$

which simplifies to

$$\Delta\phi = \frac{\sqrt{\mu}\,e^3}{\pi\sqrt{\epsilon}\,m_e^2c^2}\frac{1}{\nu^2}\int_0^L n_e B_{||}ds,$$

where the integral is the rotation measure, as defined by Equation 2.27.

Finally, the polarization angle of a linearly polarized wave is one half the difference in phase of the RCP and LCP waves, and so we find that the rotation of the linearly polarized wave is

$$\Delta\theta = \frac{\sqrt{\mu}\,e^3}{2\pi\sqrt{\epsilon}\,m_e^2c^2}\frac{1}{\nu^2}\int_0^L n_e B_{||}ds.$$

Note that if $B_{||}$ is opposite the direction of propagation then the rotation of the polarization angle is negative.

Fourier Transform

THE Fourier transform is an important and a commonly used operation that has many applications in radio astronomy. The Fourier transform that will be the focus of this appendix is the decomposition of a time-varying signal into its component frequencies. In Chapter 3 we discussed the electromagnetic radiation produced by an accelerated charge. The magnitude of the electric field of the wave varies with time due to the variations in the acceleration of the charge and the function $E(t)$ captures these time variations. We also would like to know the amplitude of the electric field as a function of frequency, $E(\nu)$. $E(t)$ and $E(\nu)$ are related by the Fourier transform.

E.1 MATHEMATICAL DEFINITION

There are three mathematical definitions of the Fourier transform in usage; we adopt, here, one of those three. Consider a signal that is a function of time, $f(t)$. Its Fourier transform is performed by multiplying the function by a complex exponential containing time and its inverse, frequency, and then integrating over all time. Therefore

$$F(\nu) = \int_{-\infty}^{+\infty} f(t)\, e^{-i2\pi\nu t}\, dt. \tag{E.1}$$

The function $F(\nu)$ represents the Fourier transform of $f(t)$. Similarly, $f(t)$ can be obtained from $F(\nu)$ by performing an inverse Fourier transform, which is the same process except that the positive exponent is used, that is

$$f(t) = \int_{-\infty}^{+\infty} F(\nu)\, e^{i2\pi\nu t}\, d\nu. \tag{E.2}$$

The variables t and ν are a Fourier transform pair and are inverses of each other. Thus the product of t and ν is dimensionless (note that time is measured is seconds, while frequency is measured in hertz or inverse seconds).

So what is the Fourier transform actually doing? First we need to recognize that the complex exponential in the Fourier transform above is a complex number whose real and imaginary parts are sinusoids. Euler's formula provides a relation between the exponential and trigonometric representations

$$e^{i\phi} = \cos\phi + i\sin\phi.$$

Therefore the Fourier transform is equivalent to fitting a function, such as $f(t)$, with a series of sinusoids of different amplitudes and frequencies. The Fourier transform, $F(\nu)$, is the function that provides the amplitudes of these sinusoids as a function of frequency. The Fourier transform can be defined in terms of trigonometric functions, or complex exponentials, but the latter are often easier to deal with mathematically.

E.2 EXAMPLE: FOURIER TRANSFORM OF A GAUSSIAN FUNCTION

Consider a function that varies with a Gaussian-shape as a function of t and compute its Fourier transform. We can define the Gaussian function as

$$f(t) = e^{-4\ln 2(t/w)^2}.$$

where w is full width in time of the Gaussian function at half of its maximum value. Note that this function has a peak value at $t = 0$ of 1. Therefore when $t = w/2$, the function has a value of 1/2.

The Fourier transform of $f(t)$ is therefore

$$F(\nu) = \int_{-\infty}^{+\infty} e^{-4\ln 2(t/w)^2} e^{-i2\pi\nu t}\, dt.$$

The solution of this integral is

$$F(\nu) = \sqrt{\pi}\, \frac{w}{2\sqrt{\ln 2}}\, e^{-\frac{(\pi w \nu)^2}{4\ln 2}}.$$

We can define W to be

$$W = \frac{4\ln 2}{\pi} \frac{1}{w}$$

and then rewrite $F(\nu)$ as

$$F(\nu) = \frac{W}{\sqrt{\pi}}\, e^{-4\ln 2(\nu/W)^2}.$$

We see that that the Fourier transform of a Gaussian function is another Gaussian function and this new Gaussian function has a full frequency width at half of its maximum value of W which is inversely proportional to the width w in time.

E.3 APPLICATION TO AN ACCELERATED CHARGE

In Chapter 3 we discussed that accelerated charges radiate electromagnetic waves. For the case of the Coulomb interaction between an electron and a proton, if we know the relative velocity and the impact parameter, we can compute the acceleration of the electron as a function of time. We discussed that the amplitude of the electric field of the radiated wave as a function of time is proportional to the acceleration of the electron as a function of time. Therefore we can compute the time variations of the electric field, $E(t)$. The Fourier transform of $E(t)$ then provides the function $E(\nu)$, which is the amplitude of the electric field of the electromagnetic waves as a function of frequency. The Poynting vector tells us that the square of $E(t)$ or $E(\nu)$ is the power radiated as a function of time or frequency. The power radiated as a function of frequency is the spectrum of the emission.

Similar to the case of the Gaussian function and its Fourier transform, if $E(t)$ is narrowly confined in time, the resultant spectrum will be very broad (see Figure E.1) and if $E(t)$ varies slowly with time, the resultant spectrum will be relatively narrow in frequency. Since

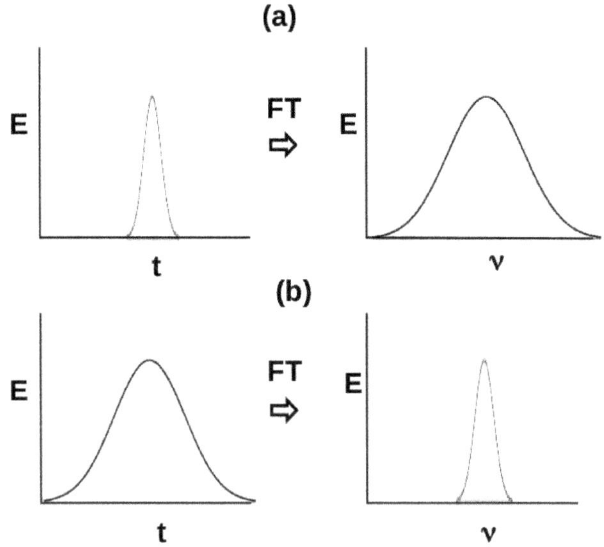

Figure E.1: A schematic showing the Fourier transform relation between Gaussian functions. In (a) the variations of the electric field are sharply peaked in time, and this produces a broad spectrum of frequencies. In (b), the time variation of the electric field is slow, resulting in a narrow range of frequencies.

the accelerations produced by random Coulomb encounters between electrons and protons due to their thermal motions are very brief, the resultant emission will have a broad range of frequencies. Considering that these encounters have varying impact parameters and velocities, it is not surprising that thermal bremsstrahlung emission produces a continuous spectrum.

Index